D0712516

LabVIEW™

Advanced Programming Techniques

SECOND EDITION

LabVIEW™
Advanced Programming Techniques

SECOND EDITION

Rick Bitter
Motorola, Schaumburg, Illinois

Taqi Mohiuddin
Mindspeed Technologies, Lisle, Illinois

Matt Nawrocki
Motorola, Schaumburg, Illinois

CRC Press
Taylor & Francis Group
Boca Raton   London   New York

CRC Press is an imprint of the
Taylor & Francis Group, an nforma business

CRC Press
Taylor & Francis Group
6000 Broken Sound Parkway NW, Suite 300
Boca Raton, FL 33487-2742

© 2007 by Taylor & Francis Group, LLC
CRC Press is an imprint of Taylor & Francis Group, an Informa business

No claim to original U.S. Government works
Printed in the United States of America on acid-free paper
10 9 8 7 6 5 4

International Standard Book Number-10: 0-8493-3325-3 (Hardcover)
International Standard Book Number-13: 978-0-8493-3325-5 (Hardcover)

This book contains information obtained from authentic and highly regarded sources. Reprinted material is quoted with permission, and sources are indicated. A wide variety of references are listed. Reasonable efforts have been made to publish reliable data and information, but the author and the publisher cannot assume responsibility for the validity of all materials or for the consequences of their use.

No part of this book may be reprinted, reproduced, transmitted, or utilized in any form by any electronic, mechanical, or other means, now known or hereafter invented, including photocopying, microfilming, and recording, or in any information storage or retrieval system, without written permission from the publishers.

For permission to photocopy or use material electronically from this work, please access www.copyright.com (http://www.copyright.com/) or contact the Copyright Clearance Center, Inc. (CCC) 222 Rosewood Drive, Danvers, MA 01923, 978-750-8400. CCC is a not-for-profit organization that provides licenses and registration for a variety of users. For organizations that have been granted a photocopy license by the CCC, a separate system of payment has been arranged.

Trademark Notice: Product or corporate names may be trademarks or registered trademarks, and are used only for identification and explanation without intent to infringe.

Library of Congress Cataloging-in-Publication Data

Bitter, Rick.
    LabVIEW : advanced programming techniques / Richard Bitter, Taqi Mohiuddin, Matthew R. Nawrocki. -- 2nd ed.
        p. cm.
    ISBN 0-8493-3325-3 (alk. paper)
    1. Computer programming. 2. LabVIEW. 3. Computer graphics. I. Mohiuddin, Taqi. II. Nawrocki, Matt. III. Title.

QA76.6.B5735 2006
005.1--dc22   2006044686

Visit the Taylor & Francis Web site at
http://www.taylorandfrancis.com

and the CRC Press Web site at
http://www.crcpress.com

Preface and Acknowledgments

As the power of the standard personal computer has steadily evolved, so have the capabilities of LabVIEW. LabVIEW has simplified the working lives of thousands of scientists, engineers, and technicians, and has increased their productivity. Automation has reduced the costs and increased the manufacturing outputs of factories around the world. Cycle times for product development have been shortened and quality of many products has steadily improved. LabVIEW does not get credit for all of these improvements, but has without question played a valuable role in many organizations for accomplishing these goals.

In our earlier experiences with LabVIEW, we found that adequate coverage of key topics was lacking. Subjects that are useful to users without a formal background in computer science such as approaches to software development, exception handling, and state machines were very difficult to find. In addition, newer areas such as multi-threading and ActiveX are even harder to locate and sometimes documentation is non-existent. Part of our intent in this book is to cover these topics that are difficult to find in other books on LabVIEW.

The chapters in this book are written in a manner that will allow readers to study the topic of interest without having to read the contents in sequential order. Users of LabVIEW with varying levels of expertise will find this book beneficial.

Proficiency with a programming language requires an understanding of the language constructs and the tools needed to produce and debug code. The first two chapters provide an overview of LabVIEW's Integrated Development Environment, programming constructs, and main features. These chapters are meant to supplement LabVIEW's documentation, and provide some good background information for programmers new to the language.

Effective programmers have an understanding of programming techniques that are applicable to a large number of programming problems. Programming tools such as state machines that simplify logic of handling various occurrences and the use of instrument drivers are two such programming tools. Exception handling is left out of more applications than we want to discuss (including some of our own), but we have included a chapter specifically on exception handling in LabVIEW.

Advanced programmers understand the operation of the language they are working with and how it interacts with the system. We present a chapter on multi-threading's impact on LabVIEW. Version 5.0 was LabVIEW's debut into the world of multi-threaded capable programming languages. A number of the issues that occur with multi-threading programming were abstracted from the programmer, but a working knowledge of muti-threaded interactions is needed.

Object Oriented Programming (OOP) is commonly employed in languages such as C++ and Java. LabVIEW programmers can realize some of the benefits to such an approach as well. We define key terms often used in OOP, give an explanation of object analysis and introduce you to applying these concepts within a LabVIEW environment.

We also present two chapters on ActiveX and .NET. An explanation of related technologies such as Component Object Model (COM) and Object Linking and Embedding (OLE) is provided along with the significance of ActiveX. A description on the use of ActiveX in LabVIEW applications is then provided. We follow this up with several useful examples of ActiveX/.NET such as embedding a browser on the front panel, use of the tree view control, and automating tasks with Microsoft Word, Excel, and Access.

This book would not have been possible without the efforts of many individuals. First, we want to thank our friends at National Instruments. Ravi Marawar was invaluable in his support for the completion of this book. We would also like to thank Norma Dorst and Steve Rogers for their assistance.

Our publishers at CRC Press, Nora and Helena have provided us with guidance from the first day we began working on this edition until its completion. We haven't forgotten about the first edition publishing support of Dawn and Felicia. If not for their efforts, this book may not have been successful enough to warrant a second edition.

A special thanks to Tim Sussman, our colleague and friend. He came through for us at times when we needed him. Also thanks to Greg Stehling, John Gervasio, Jeff Hunt, Ron Wegner, Joe Luptak, Mike Crowley, the Tellabs Automation team (Paul Mueller, Kevin Ross, Bruce Miller, Mark Yedinak, and Purvi Shah), Ted Lietz, and Waj Hussain (if it weren't for Waj, we would have never written the papers which got us to writing this book).

Finally, we owe many thanks for the love and support of our families. They had to put up with us during the many hours spent on this book. We would like to begin by apologizing to our wives for the time spent working on the second edition that could not be spent on the households! A special appreciation goes out to the loving wives who dealt positively with our absences — Thanks to Claire, Sheila, and Jahanara! Thank you moms and dads: Auradker and Mariam Mohiuddin, Rich and Madalyn Bitter, Barney and Veronica Nawrocki. For moral support we thank Jahanara, Mazhar, Tanweer, Faheem, Firdaus, Aliyah and Asiya, Matt Bitter, Andrea and Jerry Lehmacher; Sheila, Reilly, Andy, Corinne, Mark, and Colleen Nawrocki, Sue and Steve Fechtner.

The Authors

Rick Bitter graduated from the University of Illinois at Chicago in 1994. He has presented papers at Motorola and National Instruments-sponsored symposia. Rick currently develops performance testing applications as a Senior Software Engineer.

Taqi Mohiuddin graduated in electrical engineering from the University of Illinois at Chicago in 1995. He obtained his MBA from DePaul University. He has worked with LabVIEW since 1995, beginning with version 3.1, ranging in various telecommunications applications. He has presented papers on LabVIEW at Motorola and National Instruments conferences.

Matt Nawrocki graduated from Northern Illinois University in 1995. He has written papers and has done presentations on LabVIEW topics at Motorola, National Instruments, and Tellabs.

Contents

1 Introduction to LabVIEW

Programmers develop software applications every day in order to increase efficiency and productivity in various situations. LabVIEW, as a programming language, is a powerful tool that can be used to help achieve these goals. LabVIEW (Laboratory Virtual Instrument Engineering Workbench) is a graphically-based programming language developed by National Instruments. Its graphical nature makes it ideal for test and measurement (T&M), automation, instrument control, data acquisition, and data analysis applications. This results in significant productivity improvements over conventional programming languages. National Instruments focuses on products for T&M, giving them a good insight into developing LabVIEW.

This chapter will provide a brief introduction to LabVIEW. Some basic topics will be covered to give you a better understanding of how LabVIEW works and how to begin using it. This chapter is not intended to teach beginners LabVIEW programming thoroughly. Those wishing to learn LabVIEW should consider attending a National Instruments LabVIEW Basics course. Relevant information on the courses offered, schedules, and locations can be found at www.ni.com/training. If you have prior experience with LabVIEW, you can skip this chapter and proceed to the advanced chapters.

First, VIs and their components will be discussed, followed by LabVIEW's dataflow programming paradigm. Then, several topics related to creating VIs will be covered by explaining the front panel and block diagram. The chapter will conclude with descriptions of icons and setting preferences.

1.1 VIRTUAL INSTRUMENTS

Simply put, a Virtual Instrument (VI) is a LabVIEW programming element. A VI consists of a front panel, block diagram, and an icon that represents the program. The front panel is used to display controls and indicators for the user, and the block diagram contains the code for the VI. The icon, which is a visual representation of the VI, has connectors for program inputs and outputs.

Programming languages such as C and BASIC use functions and subroutines as programming elements. LabVIEW uses the VI. The front panel of a VI handles the function inputs and outputs, and the code diagram performs the work of the VI. Multiple VIs can be used to create large scale applications, in fact, large scale applications may have several hundred VIs. A VI may be used as the user interface or as a subroutine in an application. User interface elements such as graphs are easily accessed, as drag-and-drop units in LabVIEW.

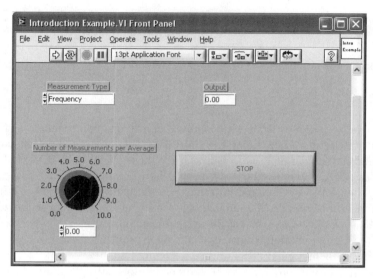

FIGURE 1.1

1.1.1 THE FRONT PANEL

Figure 1.1 illustrates the front panel of a LabVIEW VI. It contains a knob for selecting the number of measurements per average, a control for selecting the measurement type, a digital indicator to display the output value, and a stop button. An elaborate front panel can be created without much effort to serve as the user interface for an application. Front panels and LabVIEW's built-in tools are discussed in more detail in Section 1.5.

1.1.2 BLOCK DIAGRAM

Figure 1.2 depicts the block diagram, or source code, that accompanies the front panel in Figure 1.1. The outer rectangular structure represents a While loop, and the inner one is a case structure. The icon in the center is a VI, or subroutine, that takes the number of measurements per average as input and returns the frequency value as the output. The orange line, or wire, represents the data being passed from the control into the VI. The selection for the measurement type is connected, or wired to the case statement to determine which case is executed. When the stop button is pressed, the While loop stops execution. This example demonstrates the graphical nature of LabVIEW and gives you the first look at the front panel, block diagram, and icon that make up a Virtual Instrument. Objects and structures related to the code diagram will be covered further in Section 1.6.

LabVIEW is not an interpreted language; it is compiled behind the scenes by LabVIEW's execution engine. Similar to Java, the VIs are compiled into an executable code that LabVIEW's execution engine processes during runtime. Every time a change is made to a VI, LabVIEW constructs a wire table for the VI. This wire table identifies elements in the block diagram that have inputs needed for that element

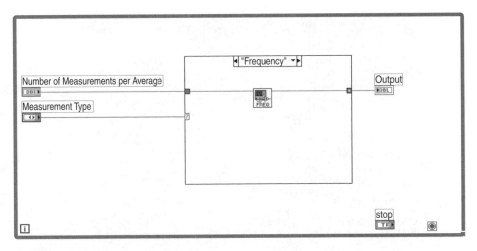

FIGURE 1.2

to run. Elements can be primitive operators such as addition, or more complex such as a subVI. If LabVIEW successfully constructs all the wire tables, you are presented a solid arrow indicating that the VIs can be executed. If the wire table cannot be created, then a broken arrow is presented for the VIs with a problem, and also for each VI loaded in memory that requires that VI for execution. LabVIEW runs in several subsystems, which will be described throughout this book. All that we need to understand now is that the main execution subsystem compiles diagrams while you write them. This allows programmers to write code and test it without needing to wait for a compiling process, and programmers do not need to worry about execution speed because the language is not interpreted.

The wire diagrams that are constructed do not define an order in which elements are executed. This is an important concept for advanced programmers to understand. LabVIEW is a dataflow-based language, which means that elements will be executed in a somewhat arbitrary order. LabVIEW does not guarantee which order a series of elements is executed in if they are not dependent on each other. A process called arbitrary interleaving is used to determine the order elements are executed in. You may force an order of execution by requiring that elements require output from another element before execution. This is a fairly common practice, and most programmers do not recognize that they are forcing the order of execution. When programming, it will become obvious that some operations must take place before others can. It is the programmer's responsibility to provide a mechanism to force the order of execution in the code design.

1.1.3 EXECUTING VIs

A LabVIEW program is executed by pressing the arrow or the Run button located in the palette along the top of the window. While the VI is executing, the Run button changes to a black color as depicted in Figure 1.3. Note that not all of the items in the palette are displayed during execution of a VI. As you proceed to the right along

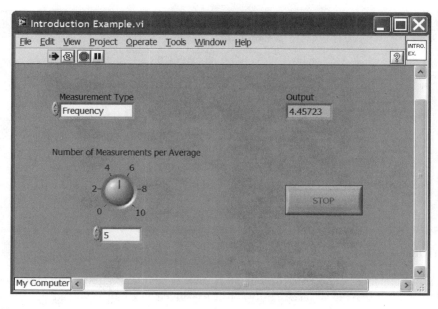

FIGURE 1.3

the palette, you will find the Continuous Run, Stop, and Pause buttons. If you compare Figures 1.1 and 1.3, the last three buttons in Figure 1.1 disappear in Figure 1.3. These buttons are used for alignment of objects on the panel or diagram, and are not available while a program is running. VIs are normally run from the front panel; however, they can also be executed from the block diagram. This allows the programmer to run the program and utilize some of the other tools that are available for debugging purposes.

If the Run button appears as a broken arrow, this indicates that the LabVIEW program or VI cannot compile because of programming errors. When all of the errors are fixed, the broken Run button will be substituted by the regular Run button. LabVIEW has successfully compiled the diagram. While editing or creating a VI, you may notice that the palette displays the broken Run button. If you continue to see this after editing is completed, press the button to determine the cause of the errors. An Error List window will appear displaying all of the errors that must be fixed before the VI can compile. Debugging techniques are discussed further in Chapter 6, which covers exception handling.

The palette contains four additional buttons on the block diagram that are not available from the front panel. These are typically used for debugging an application. The button with the lightbulb is for Execution Highlighting and the three following it are used for stepping through the code. Figure 1.4 shows the code diagram with Execution Highlighting activated. You can see bubbles that represent the data flowing along the wire, from one block to the next. You can step through the code as needed when the Pause button is used in conjunction with Execution Highlighting. As stated earlier, debugging techniques will be covered in Chapter 6.

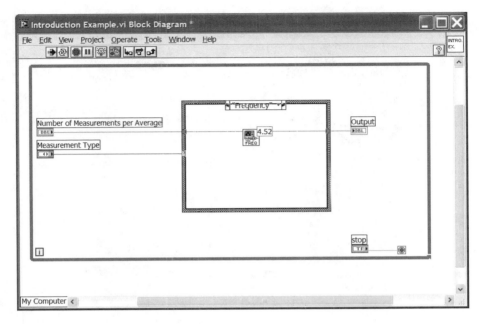

FIGURE 1.4

1.1.4 LabVIEW File Extensions

LabVIEW programs utilize the .vi extension. However, multiple VIs can be saved into library format with the .llb extension. Libraries are useful for grouping related VIs for file management. When loading a particular VI that makes calls to other VIs, the system is able to find them quickly. Using a library has benefits over simply using a directory to group VIs. It saves disk space by compressing VIs, and facilitates the movement of VIs between directories or computers. When saving single VIs, remember to add the .vi extension. If you need to create a library for a VI and its subVIs, you will need to create a source distribution using the LabVIEW Project. If you want to create a new library starting with one VI, you can use Save or Save As. Then select New VI Library from the dialog box. The File Manager can then be used to add or remove VIs from a library.

1.2 LABVIEW PROJECTS

Among other features in LabVIEW 8, the one you should be interacting with daily is the project view. LabVIEW's new project view provides a convenient interface to access everything in a LabVIEW project. Historically, locating all the Vis in an application has required the use of the hierarchy window, but that does not locate some things like LabVIEW libraries and configuration of the application builder. The project explorer provides a tree-driven list of all of these. The set of VI sources and libraries are shown in the first major breakdown: the Source tree. Information related to compilation and installation of an application are kept in

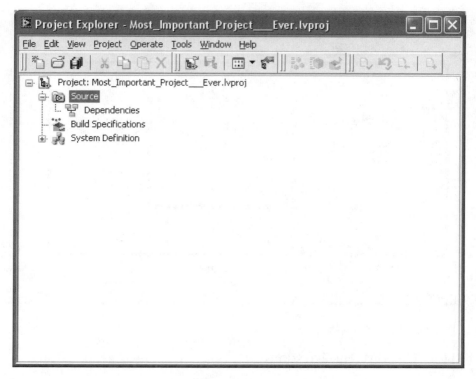

FIGURE 1.5

the second branch of the tree: Build Specifications. Information relating to the target machine environment you are building an application to is located in the last branch: System Definition. Applications that use the same operating system as the development platform will not find the System Definition folder to be of value. If a compile target is something like a Palm Pilot, then this folder is where definitions specific to a Palm based target would be configured. The project window is shown in Figure 1.5.

Among other things worth noting on the project explorer window is the toolbar, which contains buttons to create, save, and save all VIs in the application; compile; the standard cut, copy, and paste buttons; buttons to support compilation of VIs; and buttons to support source code control tools. All of these features will be elaborated on in Chapters 2 and 4.

In general, most work will be done in the Sources branch which provides a listing of all VIs and variables in the project. The Dependencies section is for VIs, DLLs, and project libraries that are called statically by a VI.

1.3 HELP

For beginning users of LabVIEW, there are various sources for assistance to aid in learning the language. Because this book is not a comprehensive guide for beginners, this section was prepared to reveal some of these sources. LabVIEW's built-in help tools will be shown first, followed by outside references and Websites. LabVIEW's online reference is an excellent source of information on the operation of various LabVIEW elements, error code definitions, and programming examples. Few languages can boast of having an online help system that is put together as well as LabVIEW's.

1.3.1 BUILT-IN HELP

The first tool that is available to the user is the Simple Help. This is enabled by selecting this item from the Help pull-down menu. When selected, it activates a balloon type of help. If the cursor is placed over the particular button, for example, a small box pops up with its description. This description contains information such as the inputs and outputs the VI accepts in addition to a short text description of what the VI does. Balloon help is available for all wire diagram elements, including primitive elements, National Instruments-written VIs, and user-developed VIs. This tool is beneficial when first working with LabVIEW. It is also helpful when running VIs in single-stepping mode to find out what each of the step buttons will execute.

The Help window will probably be the most utilized help tool available. It is also activated from the Help pull-down menu by selecting Show Help (Ctrl+H). The Help window displays information on most controls, indicators, functions, constants, and subVIs. The type of information displayed varies depending on the object over which the cursor is located. For many of LabVIEW's functions, descriptions are provided along with inputs, outputs, and default values. When the cursor is placed over an icon of a VI that a user has created, that user must input the relevant description to be displayed by the Help window. The same is true for specific controls and indicators used in an application. This is an element of good documentation practices, which is explained further in Chapter 6.

Figure 1.6 shows the Help window as it appears when the cursor is placed over the "In Range?" function. A brief description of the function is provided in the

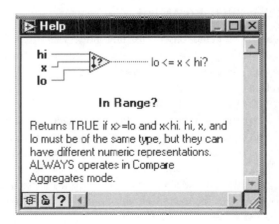

FIGURE 1.6

TABLE 1.1
Websites

http://www.ni.com/support	Technical support and contact information
http://www.ni.com/devzone/idnet/default.htm	Instrument drivers for more than 5000 instruments
http://www.ni.com/support/techdocs.htm	Technical documents, application notes, knowledge base (searchable database), product manuals
mailto://info-labview-on@labview.nhmfl.gov	Submit request for subscription to LabVIEW email user group

window along with the inputs and outputs. The three buttons located in the bottom left corner of the window are used for displaying the simple/detailed diagram, locking help on a specific object, and launching the Online Help for that topic.

The Online Help or Reference can be accessed from the Help menu also. The help files are normally installed with LabVIEW if you choose the typical installation. If you perform a custom installation of LabVIEW, you must ensure that the appropriate box is checked for help files. The Online Reference covers introduction material, overview, information on functions, and advanced topics. It also has a searchable index and word list for specific instances of key words.

1.3.2 WEBSITES

Several other sources are also available for help on LabVIEW-related topics. National Instruments' website offers help through online technical support, documents, free downloads, product demonstrations, the instrument driver network, and the Developer Zone. National Instruments has continuously expanded its online resources, and the result is a full fledged support center. Table 1.1 lists the major websites that will be of value.

1.4 DATA FLOW PROGRAMMING

LabVIEW applications execute based on data flow. LabVIEW applications are broken up into nodes and wires; each element in a diagram that has input or output is considered a node. The connection points between nodes are wires. A node can be a simple operation such as addition, or it can be a very complicated operation like a subVI that contains internal nodes and wires. The collection of nodes and wires comprise the wire diagram. Wire diagrams are derived from the block diagrams and are used by LabVIEW's compiler to execute the diagrams. The wire diagrams are hidden from the programmer; they are an intermediate form used by the compiler to execute code. While you program, the compiler is behind the scenes verifying that diagrams are available to execute. LabVIEW applications that are built using the Application Builder use the execution engine as if LabVIEW were still being used to run the VIs.

A node can be executed when all inputs that are necessary have been applied. For example, it is impossible for an addition operation to happen unless both numbers to be added are available. One of these numbers may be an input from a control and would be available immediately, where the second number is the output of a VI. When this is the case, the addition operation is suspended until the second number becomes available. It is entirely possible to have multiple nodes receive all inputs at approximately the same time. Data flow programming allows for the tasks to be processed more or less concurrently. This makes multitasking code diagrams extremely easy to design. Parallel loops that do not require inputs will be executed in parallel as each node becomes available to execute. Multitasking has been an ability of LabVIEW since Version 1.0. Multitasking is a fundamental ability to LabVIEW that is not directly available in languages like C, Visual Basic, and C++. When multiple nodes are available to execute, LabVIEW uses a process called arbitrary interleaving to determine which node should be executed first. If you watch a VI in execution highlighting mode and see that nodes execute in the desired order, you may be in for a rude shock if the order of execution is not always the same. For example, if three addition operations were set up in parallel using inputs from user controls, it is possible for eight different orders of execution. Similar to many operating systems' multithreading models, LabVIEW does not make any guarantees about which order parallel operations can occur.

Often it is undesirable for operations to occur in parallel. The technique used to ensure that nodes execute in a programmer-defined order is forcing the order of execution. There are a number of mechanisms available to a LabVIEW programmer to force the order of execution. Using error clusters is the easiest and recommended method to guarantee that nodes operate in a desired order. Error Out from one subVI will be chained to the Error In of the next VI. This is a very sensible way of controlling the order of execution, and it is essentially a given considering that most programmers should be using error clusters to track the status of executing code. Another method of forcing the order of execution is to use sequence diagrams; however, this method is not recommended. Sequence diagrams are basically LabVIEW's equivalent of the GOTO statement. Use sequences only when absolutely necessary, and document what each of the frames is intended to do.

Most VIs have a wire diagram; the exceptions are global variables and VIs with subroutine priority. Global variables are memory storage VIs only and do not execute. Subroutine VIs are special cases of a VI that does not support dataflow. We will discuss both of these types of VIs later. LabVIEW is responsible for tracking wire diagrams for every VI loaded into memory.

Unless options are set, there will be exactly one copy of the wire diagram in memory, regardless of the number of instances you have placed in code diagrams. When two VIs need to use a common subVI, the VIs cannot execute concurrently. The data and wire diagram of a VI can only be used in a serial fashion unless the VI is made reentrant. Reentrant VIs will duplicate their wire diagrams and internal data every time they are called.

1.5 MENUS AND PALETTES

LabVIEW has two different types of menus that are used during programming. The first set is visible in the window of the front panel and diagram. On the Macintosh,

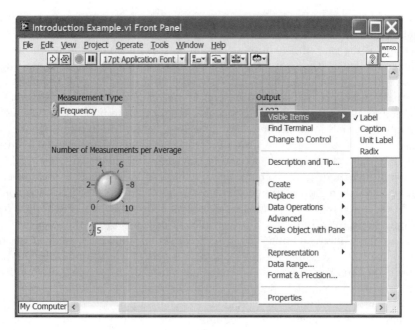

FIGURE 1.7

they are visible along the menu bar when the application is active. These are typical pull-down menus similar to other applications.

The second set of menus are called *pop-up menus* (also referred to as *popping up*). Pop-up menus are made to appear by right clicking and holding down. Macintosh users must hold down the apple key while pressing the mouse button down. The pop-up menu that appears when the cursor is on a blank part of the front panel or block diagram is the Controls palette. Similarly, the Functions palette appears on the block diagram. You can select specific objects on the front panel or block diagram and pop up on them. The menus that appear allow you to customize, modify, or perform other actions on the object. These menus can vary depending on the object that you pop up on. Figure 1.7 shows the pop menu that appears for a digital indicator.

FIGURE 1.8

The Tools palette is made visible by selecting Show Tools Palette from the Windows pull-down menu from either the front panel or block diagram. Figure 1.8 displays the movable Tools palette. The first tool is known as the Operating tool. This is used for editing numbers and text as well as changing values on controls. The arrow represents the Position-

TABLE 1.2
Shortcuts

Shortcut/Key Combination	Description	Menu Item
Tab	Allows you to switch to most common tools without accessing palette.	None
Ctrl, Option, O (Windows, Macintosh, Sun)	Allows duplication of objects. Hold down key, click on object, and drag to new location.	None
Ctrl + E	Lets you toggle between front panel and block diagram.	Show Panel/Show Diagram
Ctrl + H	Displays Help window and closes it.	Show Help
Ctrl + B	Deletes bad wires from code.	Remove Bad Wires
Ctrl + Z	Undo last action	Undo
Ctrl + R	Begins execution of VI.	Run

ing tool for selecting, positioning, and resizing objects on the front panel or block diagram. Next is the Labeling tool for editing text and creating labels. The Wiring tool is depicted by the spool and is used for wiring data terminals. The Object Pop-up tool is located under the arrow. This is exercised for displaying the pop-up menu as an alternative to clicking the right mouse button. Next to this is the tool for scrolling through the window. The tool for setting and clearing breakpoints is located under the wiring tool. The probe tool is used with this when debugging applications. Debugging tools and techniques are explained further in Chapter 6. Finally, at the bottom is the paintbrush for setting colors, and the tool for getting colors is right above it.

LabVIEW incorporates shortcut key combinations that are equivalent to some of the pull-down menu selections. The shortcuts are displayed next to the items in the menu. The key combinations that are most helpful while you are programming with LabVIEW are listed in Table 1.2. There are also some shortcuts that are not found in the menus. For example, you can use the Tab key to move through the Tools palette. This is a quick way to change to the tool you need. The spacebar lets you toggle between the Positioning tool and the Operating tool. The normal key combinations used in Windows and Macintosh for save, cut, copy, and paste are also valid.

1.6 FRONT PANEL CONTROLS

Numerous front panel controls are available in LabVIEW for developing your applications. The Controls palette (shown in Figure 1.9) appears when you make the appropriate selection in the Windows menu. The controls are grouped into categories in a tree. Categories now include things like the modern control palette, the classic control palette, and specific use selections such as express VIs, and application control references. The subpalettes have a lock in the top left corner to keep the window visible while you are working with the controls. When creating a

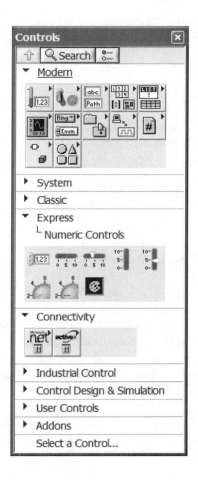

FIGURE 1.9

VI, controls can be simply dragged from the palettes and dropped on the front panel. A terminal, representing the control on the block diagram, then appears for use according to the program. Controls are basically variables that can be manipulated in the code. The following subsections will briefly describe the various control palettes. The Connectivity palette includes .NET and Active X references and will be described in Chapter 8.

1.6.1 USER CONTROL SETS

The first three branches in the control tree are modern, system, and classic. These three sections contain all the controls that an application user would interact with such as data entry controls and file path controls. The controls will behave and present data according to the operating system running the application. There is no need for us as programmers to worry about the epic battle between Windows and Linux and the use of a forward or back slash for directory listings.

The system palette contains fewer controls than the classic and modern palettes; in fact, all system controls are strictly user interface controls. Classic and modern palettes contain additional controls that have appeared in previous versions of Lab-VIEW. The following sections describe the control palettes as they appear in the classic and modern tree sections. The primary difference between them is appearance on the display. A classic numerical control will store data internally as a modern palette control.

1.6.1.1 Numeric

Internally, LabVIEW supports a number of numeric data types. Main types are floating point, integer, and complex numbers. Each type supports three levels of precision. Floating-point numbers are available as single, double, and extended precision. LabVIEW defines the number of digits in the mantissa for single and double precision numbers. Extended precision numbers are defined by the hardware platform LabVIEW is executing on.

Integers are available as byte, word, long word, and quad word precision. Bytes are 8-bit numbers, words are 16-bit numbers, long words are 32-bit numbers, and quad words are 64-bit numbers. Integers may be used as signed or unsigned quantities. LabVIEW supports 64-bit integers on all platforms; 32-bit machines will use code to emulate 64-bit integers. The controls in the Numeric palettes for the classic and modern sets are displayed in Figure 1.10. A full set of controls for allowing a user to enter and view data exists. Simple text representations exist for displaying data in addition to a variety of styled, graphical controls for presentations. User interfaces benefit from a set of controls that are relevant to the application. For example, an application supporting automation of operations at a brewery would benefit from using the tank controls to show the fill level of a primary fermenter. Choice of controls should be used with prudence. An application that contains a

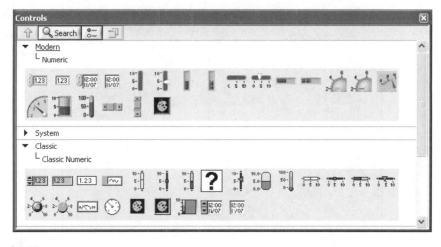

FIGURE 1.10

dizzying array of colors, styles, sizes quickly becomes an eyesore for a user. Use controls to design an appearance that a user will relate to. Palette selection is not mutually exclusive — for example, using a tank in the classic set does not eliminate the ability to use a simple data display from the System set.

Once you have dragged a control or indicator onto the front panel, the pop-up menu can be used to modify its attributes. The type (floating point, integer, unsigned, or complex), data range, format, and representation are typical attributes for a digital control. Representation types that can be displayed for users are decimal, hexadecimal, date/time, and engineering notation. Representation types do not alter the numbers stored in memory; for example, displaying two digits beyond the decimal point does not cause LabVIEW to truncate numbers internally.

Figure 1.11 displays the window that appears when Format & Precision is selected from the pop-up menu. The Numeric Properties pop-up window contains several tabs. The appearance tab contains control configuration properties such as the label and caption visible on the display. The Data Range tab is of importance; it configures the default control value and allows the control to have its valid range of inputs configured. Data validation is critical in any application that is geared towards quality and we strongly encourage all programmers to use this functionality. Data entered outside the minimum and maximum range values can be either coerced to the range or ignored. This functionality does not work if the VI is called as a subVI. In the coercion case, the input data is set to the minimum or maximum range. The control can also be configured to ignore the entry. If the data range functionality is not used, the application should validate ranges in the application itself.

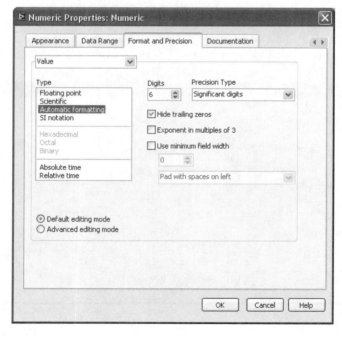

FIGURE 1.11

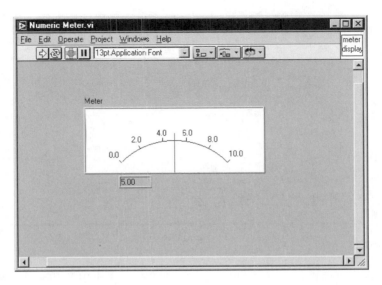

FIGURE 1.12

The format and precision tab affects the display of the data, but not the internal value. The type ranges will determine how data is presented in the control. Internally numerical values are not truncated or rounded when settings in this tab are selected. Floating point data can be shown in various formats, such as truncating the number of digits displayed and integer data can be displayed in decimal, hexadecimal, octal, or binary formats. Nondecimal displays are commonly used and convenient when it comes to data such as fields in communications protocols.

The nondigital objects in the numeric palette have an option to display a digital value with them through the pop-up menu. Just select the Visible Items in the pop up menu and then select Digital Display from the submenu. Figure 1.12 shows the meter with its associated digital indicator for precise readings. The meter, as most controls, can be resized by dragging one of the corners. The scale, markers, and mapping can also be modified on the meter.

1.6.1.2 Boolean

The Boolean palettes for the modern and classic palettes are illustrated in Figure 1.13. These palettes contain various true or false controls and indicators. Buttons to replicate switches, LED indicators, and operating system OK and Cancel buttons are provided. It is unlikely programmers will come up with Boolean indicator requirements that are not captured somewhere in this palette. Some of the controls in this palette are also available in the Dialog palette.

An interesting feature that LabVIEW programmers can use with Boolean controls is the mechanical action of the controls themselves. Configuration options available are switch when pressed, switch when released, switch until released, latch when pressed, latch when released, and latch until released. The major decision is whether the switch should switch or latch. Switching involves a somewhat permanent

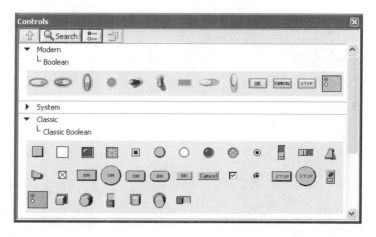

FIGURE 1.13

change. Latching changes the value of the control for a short period of time. The release time is when the user presses the button, and finally lets go. Switch when pressed makes the new value of the Boolean available as soon as the user touches it, and the change stays in place regardless of how long the user holds the button down. Switching when released does not trigger the new value until the user lets go of the control. Switching until released will change the control's value until the user releases the button. When the button is released, it toggles back to its original value.

Latching controls will toggle their value for a short period of time. Unlike switching, latching controls will return to their original value at some point in time. Latch-when-pressed Booleans will make the toggled value available as soon as the user clicks the control. Latch-when-released Booleans are toggled for a short while after the user releases the control. Latch-until-released controls will retain a toggled value while the control is activated by the user, and for a short period of time after the user releases the control.

Boolean controls have a default action of switch when pressed. Latching controls are very helpful in applications that allow users to change the behavior of an application for a short period of time. For example, a test application could have a button titled "e-mail status NOW." This button is not one that should be mechanically switched, where hundreds of e-mails can be sent to your boss when one would have done well. Buttons that switch when released are helpful when users try to time when a VI may have to stop. Also note that the mechanical action of subVIs is completely ignored; LabVIEW itself is not considered a user.

In general, there may not be a lot of material that can be presented on a topic such as programming buttons, but LabVIEW does provide a fair amount of flexibility for programmers as to how users and their programs can interact.

1.6.1.3 String & Path

The String & Path palette for the Modern and Classic control sets is displayed in Figure 1.14. It holds the string control, indicator, and file path control/indicators.

FIGURE 1.14

LabVIEW strings will automatically adjust their size to hold whatever data you place into them. String controls and indicators have a number of options that make them very flexible when programming a user interface. LabVIEW's string display functionality is one of the best in the industry.

Display options are very useful for programmers performing communications work. Many strings that are processed as an implementation of a communications protocol contain nonprintable characters. String displays can be set to show the ASCII or hexadecimal value of the contents. We have used this display option many times when writing drivers and code that use nonprintable arrays of characters. The "slash codes" display option is useful for showing white space used in the string. Spaces would appear as /s in slash code display. Again, this is very useful when writing code that needs to be clearly understood by a user. As an example, when writing code to validate protocol handling and the application needs to generate an Internet Protocol 4 header, it is easier to understand the header information presented in hexadecimal format than it is as a printable string. The first bye is normally 0x45 followed by 0x00. In telecommunications, protocol encapsulation is quite common, such as in IP tunneling. It will not always be practical or necessary to break a message apart field by field. LabVIEW string handling provides tools that make this display trivial where a C# programmer has some work to do.

Information that is sensitive can be protected with the password display option. Similar to standard login screens, password display replaces the characters with asterisks. Few programmers write their own login screens, but there are times when this display is necessary. Later in this book we will demonstrate using an ActiveX control to send e-mail. Before the control can be used to process e-mail, a valid user login must be presented to the mail server. The password would need to be obscured to casual observation.

It is possible to enable scrollbars for lengthy text messages, and also possible to limit values to a single line. If LabVIEW is used to display text files, scrollbars may become a necessary option. Form processing may want to limit the length of data users can insert, and single-line-only mode would accomplish this.

1.6.1.4 Ring & Enum, List & Table

The Ring & Enum and List & Table palettes are displayed in Figure 1.15. You will find the text, dialog, and picture rings along with the enumerated type and selection listbox in the palette. These items allow menu type controls or indicators for the user interface of an application. The text or picture represents a numeric value, which can be used programmatically. The enumerated type has an unsigned number representation and is especially useful for driving case statements. It is a convenient way to associate constants with names. Some of the controls represented in this palette are also available through the Dialog palette.

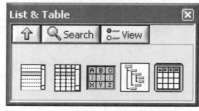

FIGURE 1.15

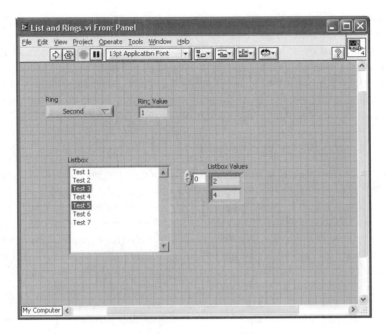

FIGURE 1.16

Figure 1.16 is a simple example that demonstrates how to use the objects in this palette. Shown is the menu ring with a digital indicator next to it, and a multiple selection listbox with a digital indicator array next to it. The menu ring is similar to a pull-down menu that allows the user to select one item among a list. Item one in a menu ring is represented by a numeric value of 0, with the second item being 1, and so on. The second item is selected in this example and its numeric value is shown in the indicator. The menu ring terminal is wired directly to the indicator terminal on the block diagram as shown in Figure 1.17.

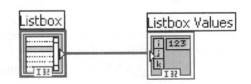

FIGURE 1.17

The multiple selection listbox is not a separate control from a single selection listbox; it's all in the configuration. It's possible to configure listboxes to accept 0 inputs, 1 input, or multiple inputs. It is also possible to allow for the user to modify the text in a listbox. Popping up on a listbox control gives a complete list of the features the control has and how it can be customized. Symbols can also be shown with text in a list box by enabling symbols in the Visible selection. In this example, a multiple selection listbox was configured and is represented by an array of numbers, with 0 corresponding to the first item on the list. In our example, Test 3 and Test 5 are selected and the corresponding array is next to the list box. The array holds two values, 2 and 4, corresponding to the two tests selected from the listbox. Multiple selections are made from the listbox by holding down the Shift key and clicking on the items needed.

1.6.1.5 Array, Cluster, and Matrix

The last palette displayed in Figure 1.15 is Array, Cluster, and Matrix. To create an array, you must first drag the array container onto the front panel of a VI. This will create an array, but does not define the array type. A control or indicator must be dropped inside the array shell. Arrays of any data type can be created using the objects available in the Controls palette, except for charts or graphs. The array index begins at zero and the index display has a control that allows you to scroll to view the elements. A two-dimensional array can be created by either popping up on the array to add a dimension, or by dragging the corner and extending it.

Unlike C++, LabVIEW arrays are always "safe." It is not possible to overwrite the boundaries of an array in LabVIEW; it will automatically resize the array. Languages like C++ do not perform boundary checking, meaning that it is possible to write to the fifth element of a four-element array. This would compile without complaint from the C++ compiler, and you would end up overwriting a piece of memory and possibly crashing your program. LabVIEW will also allow your application to write outside the boundaries of the array, but it will redimension the array to prevent you from overwriting other data. This is a great feature, but is not one that programmers should rely on. For example, if writing to the fifth element was actually a bug in your code, LabVIEW would not complain and it would also not inform you that it changed the array boundaries!

Array controls and indicators have the ability to add a "dimension gap." The dimension gap is a small amount of space between the rows and columns of the control to make it easier for users to read. Another feature of the array is the ability to hide the array indexes. This is useful when users will see only small portions of the array.

A cluster is a data construction that allows grouping of various data types, similar to a structure in C. The classic example of grouping employee information can be used here. A cluster can be used to group an employee's name, Social Security number, and department number. To create a cluster, the container must first be placed on the front panel. Then, you can drop in any type of control or indicator into the shell. However, you cannot combine controls and indicators. You can only drop in all controls or all indicators. You can place arrays and even other clusters inside a cluster shell.

Figure 1.18 shows the array, cluster, and matrix shells as they appear when you first place them on the front panel. When an object is dropped inside the array shell, the border resizes to fit the object. The cluster shell must be modified to the size needed by dragging a corner. Matrix controls appear in a 3x3 of integers. Figure 1.19 shows the array and cluster with objects dropped inside them. A digital control

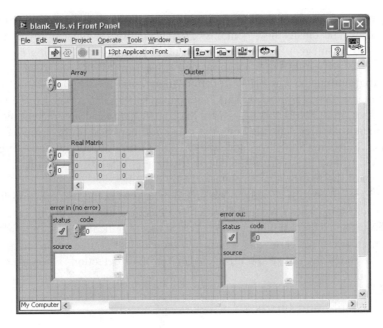

FIGURE 1.18

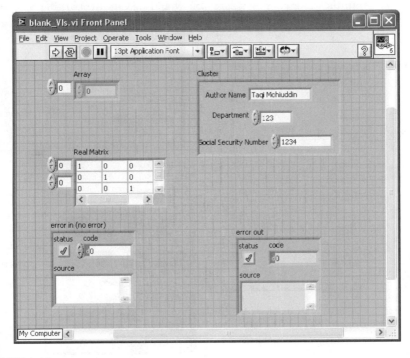

FIGURE 1.19

was dropped in the array shell. The outer display shows the current index number of the array. The cluster now contains a string control for the employee name, a digital control (integer) for the department number, and another string control for the social security number. When only one value from the cluster data is needed when programming, a LabVIEW function allows you to unbundle the cluster to retrieve the piece that is needed. This is explained further in Section 1.6. The matrix control does similar work to an array, but the control's dimensions are kept the same. As the identity matrix was filled out, the x and y dimensions of the array were kept the same. So when a "one" was added to (2,2), the third row and column were filled with zeros. If the cell is part of the matrix it will appear white by default. If the cell is not part of the matrix it will appear gray by default.

The Error In control and Error Out indicator, shown in the two previous figures, are both clusters. These are used for error detection and exception handling in LabVIEW. The clusters hold three objects: a status to indicate the occurrence of an error, a numeric error code, and a string to indicate the source of the error. Many LabVIEW functions utilize the error cluster for error detection. Error handling is discussed in Chapter 6.

1.6.1.6 Graphs and Charts

Figure 1.20 displays the Graphs palette with the built-in graph and chart objects. The Waveform Chart and Waveform Graph are located in the top row, and the

FIGURE 1.20

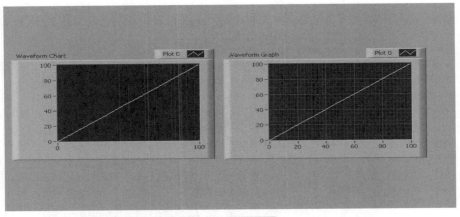

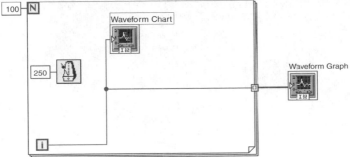

FIGURE 1.21

Intensity Chart and Intensity Graph are in the second row. The XY Graph is also available in the top row of the palette. The graph and chart may look identical at first, but there is a distinction between the two. The graph is used for plotting a set of points at one time by feeding it an array of data values. The chart, on the other hand, is used for plotting one data point or array at a time. A chart also has memory, maintaining a buffer of previous points which are shown in its display.

The example in Figure 1.21 will help to demonstrate the difference between a chart and a graph. A Waveform Chart and Waveform Graph are displayed on the front panel side by side. A For loop is executed 100 times with the index value being passed to the chart. Once the loop is finished executing, the array of index values is passed to the graph. A 250-millisecond delay is placed in the For loop so you can see the chart being updated as the VI executes. Both the chart and graph are used for displaying evenly sampled data.

Graphs and charts have a number of display options enabling programmers to display data in a manner that makes sense. For example, both charts and graphs support a histogram style display. Because histograms plotted with straight lines are awkward to read, interpolation between points and point styles are completely adjustable.

Graph controls and indicators provide a palette for users to adjust the graphs at runtime. The palette allows for auto scaling of both the X and Y axes. Zoom features are available for examining portions of the graph at runtime. Cursors are available

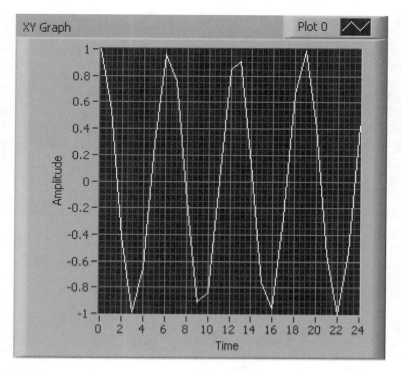

FIGURE 1.22

to measure distances between points. This level of functionality is not very common in graphing packages that come standard with most other languages.

The XY Graph can be used to graph any type of data, similar to a Cartesian graph. Figure 1.22 illustrates the XY Graph with a plot of a cosine wave. Two separate arrays are provided as input to this graph. Several graph and chart attributes can be modified for display purposes. The grid options, mapping (linear or log), scale styles, and marker spacing are some of the items available in the pop-up menu. Their displays can also be resized on the front panel by dragging a corner.

3-D graphs and picture plots are some of the advanced objects available on this palette. The 3-D graphs require three separate arrays of data values for graphing the x, y, and z coordinates. The Polar Plot, Smith Plot, Min-Max Plot, and Distribution Plot are indicators on the Picture subpalette of the Graph palette.

1.6.1.7 String & Path and I/O

The String & Path and I/O palettes are displayed in Figure 1.20. The bottom two objects on the String & Path palette are the File Path Control and File Path Indicator. These are used when performing directory- or file-related operations to enter or display paths.

The I/O palette contains control and indicators for the refnums used in LabVIEW. A refnum is a distinct identifier or reference to a specific communications path. The refnum can point to a file, external device, .NET object, network connection, IVI device, DAQ Card, or VISA communications channel, or to another VI. This identifier is created when a connection is opened to a specific object. When a connection is first opened, the particulars of the connection need to be defined, such as a file path, instrument address, or an IP address. After the connection is opened, a refnum is returned by the open function. This refnum can then be used throughout an application when operations must be performed on the object. The particulars of the connection need not be defined again.

Figure 1.23 demonstrates the refnum through a simple example. In this illustration, a TCP connection is opened to a host computer. The front panel shows controls for the IP address or host computer name and the remote port number that are needed to define the connection. The Network Connection Refnum is an indicator returned by the function that opens the connection. The block diagram shows TCP Open

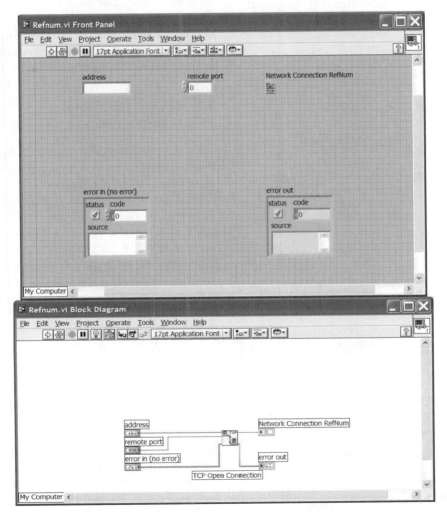

FIGURE 1.23

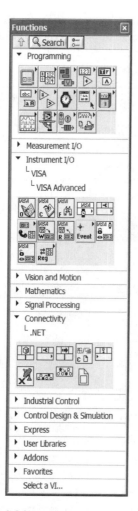

FIGURE 1.24

Connection, a built-in LabVIEW function, with the related data provided. The refnum, or reference, created by this function can then be used to perform other operations. This unique identifier represents the connection, and the specifics do not need to be provided again.

LabVIEW uses refnums to track internally used resources; for example, a file path refnum contains information needed to read or write to a file. This information is using system resources such as memory and must be returned. If the programmer does not close refnums, LabVIEW will leak memory. Over long periods of time, this could degrade the system's performance.

1.7 BLOCK DIAGRAM FUNCTIONS

All coding in LabVIEW is done on the block diagram. Various functions are built in to aid in the development of applications. The Functions palette is displayed in Figure 1.24 and appears when the block diagram window is active. LabVIEW is a programming language and uses the typical programming constructs such as loops, and defines a couple of other structures unique to data flow programming. This section briefly describes some of the tools that are available to LabVIEW programmers.

1.7.1 STRUCTURES

The control structures that are accessible from the Structures palette are shown in Figure 1.25. This palette contains several types of structures including the case, For loop and While loop structures. You will also find several types of variables including the Global and Local Variable on this palette.

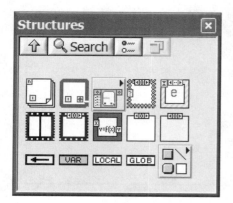

FIGURE 1.25

1.7.1.1 Sequence Structure

There are two types of sequence structure on the structure palette: the Stacked Sequence structure and the Flat Sequence structure. The operation of the two structure types is the same, but the way the sequences are displayed is different. First let us place the stacked sequence structure on the diagram and drag it to the size desired. The structure looks like a frame of film when placed on the diagram. The Sequence structure is used to control the flow or execution order of a VI. In LabVIEW, a node executes when the input data required becomes available to it. Sequence structures can be used to force one node to execute before another, and to ensure that the VI executes in the order intended.

Each frame is basically a subdiagram. The sequence structure will begin executing when the required data becomes available to it, just as any other node. The objects placed inside the first frame (Frame 0) execute first, and the rest of the frames follow sequentially. Within each frame or subdiagram the data flow execution still applies.

The top of Figure 1.26 shows the sequence structure as it appears when first placed on the block diagram. Additional frames are added by popping up anywhere on the border of the structure and selecting Add Frame After (or Before). The second picture depicts the stacked sequence structure after a frame has been added. Only one frame is visible at a time. The display at the top of the frame indicates which frame is currently visible.

The example diagrams in Figure 1.27 will help to define some terms that are related to the Sequence structure. The top window shows Frame 0, and the bottom window shows Frame 1 of the structure. Data can be passed into a Sequence structure by simply wiring it to the border to create a tunnel. The blackened area on the border indicates that a tunnel has been created. Data is passed out of the sequence structure in a similar manner, with the data actually being passed out after all of the frames have been executed. A tunnel is created for each value that needs to be passed in and is available for use in all frames. The same is true for data being passed out of a sequence structure. This point is important because data being passed out of a case structure is handled differently.

Data values can be passed from one frame to the following frames with the use of sequence locals as shown in the top diagram. The sequence local is available in the pop-up menu. The arrow on the local indicates that the data is available for manipulation in the current frame. Note that in Frame 0, the local to the right is not available because the data is passed to it in Frame 1. Frame 2 can use data from both of the sequence locals. The locals can be moved to any location on the inside border of the structure.

Stacked sequence structures can be avoided in most applications. The main problem with stacked sequence structures in LabVIEW programming is readability for other programmers. Controlling the order of execution can be performed with error clusters, or by designing subVIs with dependent inputs. Sequence structures can be a bad habit that is easily developed by some LabVIEW programmers. The authors use sequence diagrams that contain a single frame when working with VIs that do not use a standard error cluster.

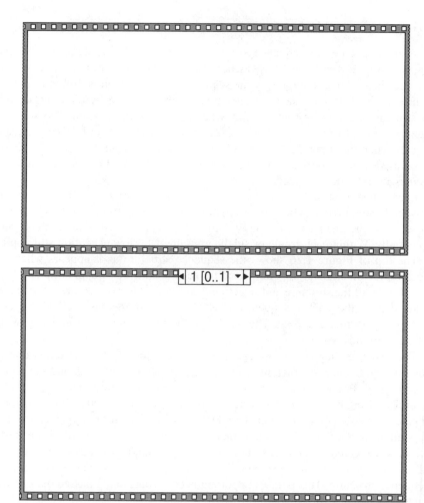

FIGURE 1.26

For years LabVIEW power programmers complained about the use of sequence structures in applications. National Instruments heard the complaints and added a new sequence structure called the flat sequence structure. This structure has the same functionality as its stacked predecessor, but instead of having the sequences stacked on top of one another they are now laid out horizontally. The flat sequence structure truly looks like a section of film now. Figure 1.28 shows a flat sequence structure with two frames. When a flat sequence structure is placed on the code diagram its initial appearance is the same as the stacked. When a second frame is needed the programmer pops up on the structure and selects Add Frame Before (or After) and the new frame will appear adjacent to the existing frame. A definite improvement with respect to readability, but it does result in a loss of valuable diagram real estate.

Sequence structures do not have equivalents to other programming languages; this is a unique structure to dataflow languages. Text-based languages such as Visual

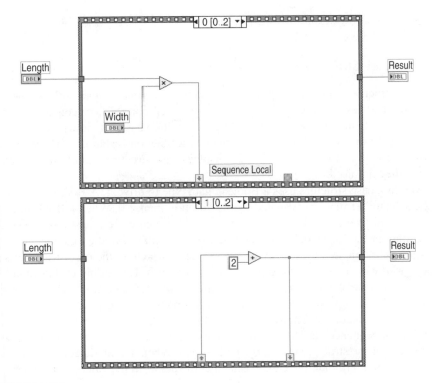

FIGURE 1.27

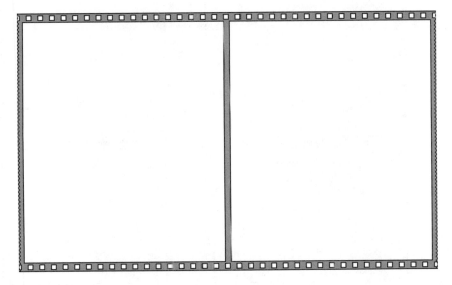

FIGURE 1.28

Basic and C perform operations line-by-line; LabVIEW executes things as they become available.

1.7.1.2 Case Structure

The case structure is placed on the block diagram in the same manner as the sequence structure. The case structure is similar to conditional control flow constructs used in programming languages such as C. The case structure has a bit more responsibility in LabVIEW; in addition to switch statements, it functions as an if-then-else block when used with a Boolean. Figure 1.29 displays case structures and four examples of how they are used.

The first case structure uses a Boolean data type to drive it. A Boolean is wired to the selector terminal represented by the question mark (?). When a Boolean data type is wired to the structure, a true case and a false case are created as shown in the display of the case structure. The false case is displayed in the figure since only one case is visible at a time. As with the sequence structure, the case structure is a subdiagram which allows you to place code inside of it. Depending on the value of the Boolean control, the appropriate case will execute. Of course, the availability of all required data inputs dictates when the case structure will execute.

The flat sequence structure truly looks like a section of film now. Figure 1.28 shows a flat sequence structure with 2 frames. A numerical case structure is shown to the right of the structure driven by the Boolean. When a numeric control is wired to the selection terminal, the case executed corresponds to the value of this control.

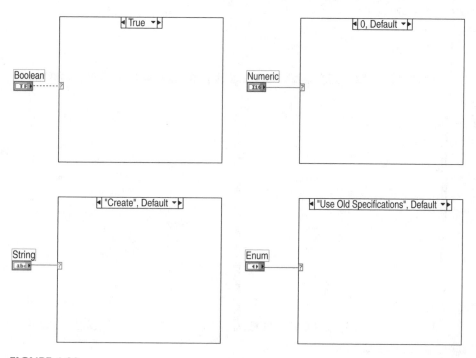

FIGURE 1.29

When the case structure is first placed on the code diagram and the numeric control is wired to the case selector, LabVIEW creates only two cases. You must pop-up on the structure and add as many cases as you need. Normally, Case 0 is the default case, but you can change that to any case you desire. You must specify a default case to account for the different possibilities. If you do not specify a default case, you must create a case for each possibility. You can assign a list or range of numbers to a particular case by editing the display, or case selector label, of the structure with the editing tool. To assign a list to one case, use numbers separated by commas such as 2, 3, 4, 5. To specify a range, separate two numbers by two periods, like 2..5.

You should also be aware that floating point numbers could be wired to the case selection terminal. LabVIEW will round the value to the nearest integer. However, the selector label cannot be edited to a floating point number. The case selector label will display red characters to indicate that it is not valid.

The lower left case structure has a string control wired to the case selector. The case selector display must be edited to the desired string value for each case. The string is displayed in quotes but does not have to be entered that way. The case that matches the string control driving the structure will be executed. LabVIEW allows you to alter the criteria to perform a case-insensitive match to ignore the difference between upper and lower case strings. If there is no match, the default case will execute. Applications with performance requirements should consider using enumerated types to drive the case statements. String parsing and matching can be processor intensive activities.

Finally, an enumerated type is used to drive the case structure in the lower right corner. The text surrounded by the quotes corresponds to the different possible values of the control. When you first wire the enumerated control to the case selector terminal, only two cases are created. You must use the pop-up menu to add the rest of the cases to the structure. Although the enumerated data type is represented by an unsigned integer value, it is more desirable to use than a numeric control. The text associated with the number gives it an advantage. When wired to a case structure, the case selector label displays the text representation of the enumerated control. This allows you to identify the case quickly, and improves readability.

Data is passed in to the case structure by creating a tunnel. Each data value being passed must have a unique tunnel associated with it. This data is made available to all of the cases in the structure. This is similar to the Sequence structure described earlier. However, when data is being passed out of the case, each case must provide output data. Figure 1.30 illustrates this point. The picture shows the code of a VI using an enumerated type to control the execution of the case structure. This VI takes two numeric values as input and performs an operation on them, returning the result as output. Depending on the selection, addition, subtraction, multiplication, or division is performed.

The top window shows the "Subtract" case displayed. Number 2 is subtracted from Number 1 and the result is passed out to Result. Note that the tunnel used to pass the data out is white. This indicates that a data value is not being output by all cases. All of the cases must have a value wired to the tunnel. The bottom window shows the Add case displayed. Now all of the cases have an output wired to the tunnel, making it turn black. This concept holds true for any data type driving the

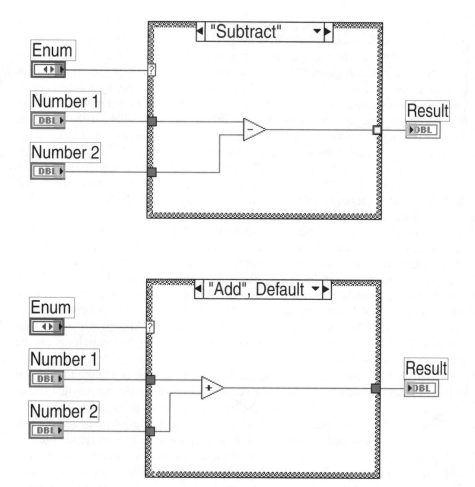

FIGURE 1.30

case structure. An alternative to wiring all the cases to the output in order to remove the error is to right click on the output tunnel and select Use Default if Unwired. This "fills in" the tunnel coloring with just a white dot in the middle indicating that not all the cases are wired, but the structure will output the default value if it is not connected.

1.7.1.3 For Loop

The For loop is used to execute a section of the code, a specified number of iterations. An example of the For loop structure is shown in Figure 1.31. The code that needs to be executed repeatedly is placed inside of the For loop structure. A numeric constant or variable can be wired to the count terminal to specify the number of iterations to perform. If a value of zero is passed to the count terminal, the For loop will not execute. The iteration terminal is an output terminal that holds the number

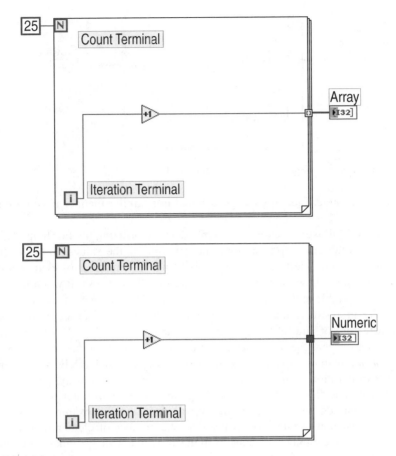

FIGURE 1.31

of iterations the loop has executed. Therefore, the first time the loop executes, the iteration value is 0.

The top block diagram shows a For loop that will execute 25 iterations. A 1 is added to the value of the iteration terminal and passed out to an indicator array via a tunnel. The output of the For loop is actually an array of the 25 values, one for each iteration. Because the loop executed 25 times, LabVIEW passes an array with the 25 elements out of the tunnel. In this case, the array holds values 1 through 25 in indexes 0 through 24, respectively; this is known as auto indexing. Both the For loop and While loop assemble arrays when data is passed out. Auto indexing is the default only for the For loop, however. LabVIEW allows the programmer to disable auto indexing so that only the last value is passed out of the loop. This is shown in the bottom code diagram. Popping up on the tunnel and selecting the appropriate item from the menu disables indexing. The output from the tunnel is wired to a numeric indicator in this diagram. If you observe the wire connecting the indicator and the tunnel, you will notice that the wire is thicker in the top diagram because it is an array. This allows you to quickly distinguish an array

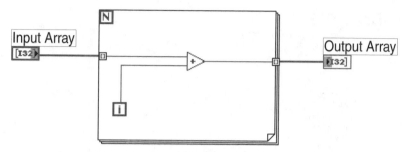

FIGURE 1.32

from a single value. Indexing can be enabled in a similar manner if you are using a While loop.

Figure 1.32 illustrates another example diagram utilizing the For loop. An array is passed into the For loop to perform an operation on the values. In this example, the count terminal is left unwired. LabVIEW uses the number of elements in the array to determine how many iterations to perform. This is useful when the size of the array is variable and not known ahead of time. One element at a time is passed into the For loop structure and the addition is performed. This property of For loops is also a feature of auto indexing and is available by default in For loops. This is the opposite of what the loop does at the output tunnels. Caution needs to be used when working with multiple arrays being fed into a For loop. LabVIEW will perform a number of iterations equal to the array length of the shortest array. Popping up on the terminal and selecting Disable Indexing can disable auto indexing.

What if you do wire a value to the count terminal in this example? If the value passed to the count terminal is greater than the number of elements in the array, LabVIEW uses the number of elements in the array to decide how many iterations to perform. If the value passed to the count terminal is less than the number of elements in the array, LabVIEW will use the count terminal value. This indexing feature on the input side of the For loop can also be disabled by using the pop-up menu. Once indexing is disabled, the whole array is passed in for each iteration of the loop.

The last feature of auto indexing is the ability to handle arrays of multiple dimensions. A two-dimensional array fed into a For loop will iterate the values in one dimension; in other words, a one-dimension array will be fed into the For loop. A nested For loop can be used to iterate through the one-dimension arrays.

Figure 1.33 shows the code diagram of a VI that calculates the factorial of a numerical value. A shift register is utilized to achieve the desired result in this example. The shift register has two terminals, one on the left border and one on the right border of the structure. The shift register is used for passing a data value from the current iteration to the next one. The right terminal holds the data of the current iteration and is retrieved at the left terminal in the next iteration. A shift register pair can be created by popping up on the left or right border of the For loop structure and selecting Add Shift Register. The shift register can hold any LabVIEW data type.

In the example shown, a constant value of 1 is wired to the shift register. This initializes the value of the shift register for the first iteration of the loop. If nothing

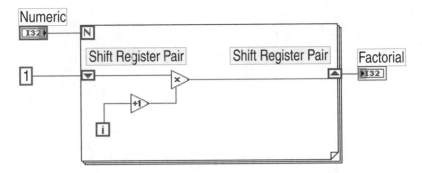

FIGURE 1.33

was wired to the shift register, the first iteration would contain a value of 0. The Numeric control wired to the count terminal contains the value for which the factorial is being calculated. A 1 is added to the iteration terminal and then multiplied to the previous result. This successfully yields the desired factorial result. Shift registers can be configured to remember multiple iterations by popping up and selecting Add Element from either side. A new terminal will appear just below the existing one on the left border of the structure. When you have two terminals, this allows you access to the two previous iteration values. The top terminal always holds the last iteration value.

Care should be used when leaving shift registers uninitialized. When a default value is not wired to a shift register the last stored value will become the initial value. Take the code diagram in Figure 1.34. This VI takes arrays of test name, measured data and status and assembles them into a tab-delimited string that can be written to a text file. Notice that the shift register is not initialized. The first time the VI is run the initial value would be an empty string resulting in the correct assembled string output. If the user were to call this VI a second time the initial value would now be the last assembled string. When the FOR loop executes the second time the old string will get concatenated to the new string. Obviously the

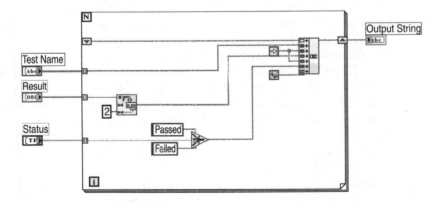

FIGURE 1.34

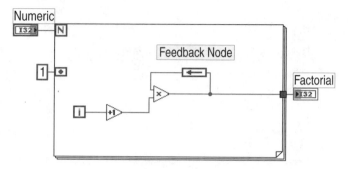

FIGURE 1.35

string could grow very quickly resulting in a memory problem. In some cases the programmer may want this kind of behavior. One application of the unitialized shift register is the Type 2 global, which is discussed in section 2.1.

Shift registers are the only mechanisms available to perform recursive operations in LabVIEW. Recursion is the ability for a function to call itself during execution, and it has frustrated thousands of students learning C and C++. The good news for LabVIEW programmers is that VIs cannot wrap back onto themselves in a wire diagram. There are times when a recursive operation is the best way to solve a problem, and using shift registers simulate recursion. Although not truly recursive, access to the last iterations can be used to perform these ever-popular algorithms in LabVIEW. It is not possible for LabVIEW to overrun a call stack with shift registers, which is very possible with recursive functions in C. One of the problems with recursion is that if exit criteria are not correct, the function will not be able to stop calling itself and will crash the application. Memory usage is also a bit more efficient for shift registers because there is not as much call stack abuse.

A newer option available for passing back values in LabVIEW loops is the feedback node. The feedback node transfers values from one loop to the next in For and While loops. The functionality of the feedback node is similar to the shift register, but is a bit cleaner on the code diagram. Figure 1.35 shows the factorial example discussed above with a feedback node used in place of the shift register. Note in the example that there is an input port on the loop for setting an initial value. The feedback node is on the main level of the Structures palette.

Outputs of a For loop, by default, will be arrays consisting of a collection of outputs for each iteration of the loop. One advantage of the For loop when handling arrays is LabVIEW's efficiency. Because the For loop's iteration count is derived from an iteration count or length of an array, LabVIEW can precompute the number of elements in array outputs. This allows LabVIEW to reserve one contiguous block of memory to write output arrays to. This is important because, as we mentioned earlier, LabVIEW will expand array boundaries, but this involves a performance hit because LabVIEW needs to go to the operating system and reallocate the entire array and perform a duplication of the existing elements. Small arrays will not be a significant performance degradation, but larger arrays can slow things down quite a bit.

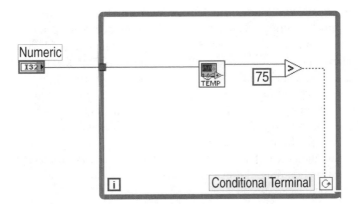

FIGURE 1.36

1.7.1.4 While Loop

The While loop is an iteration construct that executes until a predetermined Boolean value is passed to its conditional terminal. The conditional terminal is located in the lower right corner of the While loop structure, as shown in Figure 1.36. The While loop will execute at least once because the condition is evaluated at the end of the current iteration. When the While loop is placed on the block diagram the conditional terminal is setup to stop if true by default. In this case when a false is passed to the conditional terminal, the loop will execute another iteration before evaluating the value once again. If the terminal is left unwired, the loop will execute infinitely. By right clicking on the conditional terminal the user can change the behavior of the loop to continue if a true is received. If continue if true is selected, and a true value is passed to the conditional terminal, the loop will execute another iteration before evaluating the value once again. If the terminal is left unwired, the loop will execute once before stopping.

Figure 1.36 illustrates the use of the While loop. The output of the subVI is compared to find out if it is greater than 75.0. This evaluation determines whether the loop will execute one more iteration. If the value is greater than 75.0, a true value is passed to the conditional terminal causing it to execute again. If the value is less than or equal to 75.0, a false value causes the loop to terminate.

Automatic indexing is available for the While loop also, but it is not the default. When data is passed in or out of the loop structure, you must use the pop-up menu to enable indexing. Shift registers can be created on the left or right border of the While loop. The shift registers operate in the same manner as described as the For loop.

While loops can be used to perform the functions of a For loop with a little less efficiency. Popping up on the terminals can use auto indexing and array creation. As you will see throughout this book, While loops are used by the authors more often than For loops. This is not a matter of personal preference, but good design decisions. When working with previously collected data, such as reading a file and processing the file contents, For loops will be more efficient and are used in these types of applications. Points read in the form of arrays can be done far more efficiently with For loops because LabVIEW can precompute memory allocations. The problem with For loops is that there is no mechanism to abort execution of the loop; i.e., there is no break command. While loops stop their execution any time the predetermined Boolean value is fed into the condition terminal.

Stopping execution of a loop is important when performing automation, which is the authors' primary use of LabVIEW. One of the inputs to the condition indicator will be the Boolean value of the error cluster, which we feed through a shift register for every iteration. In an automation application, the ability to break execution is more important than the efficiency of array handling. There is a tradeoff of efficiency against exception handling, but in automation it makes more sense to stop execution of troubled code.

1.7.1.5 Event Structure

An *event structure* is a structure containing one or more subdiagrams corresponding to defined events. When an event occurs, the corresponding subdiagram executes. The event structure will wait until an event occurs or a timeout occurs. By default the timeout value is -1 (never times out). An event could be anything from a control value change to a mouse click. The programmer defines what conditions constitute an event when defining the subdiagram. There is also a dynamic event terminal that allows for events to be registered at runtime.

Perhaps the easiest way to describe an event structure is through an example. Figure 1.37 shows a VI with an enumerated control and indicator and a stop button. The event structure is set up to monitor a value change on the stop button and the enumerated control. If the control value is changed then the event structure executes case 1. The new value is written to the enumerated indicator. If the stop button is pushed the event 0 case is executed. The Boolean value is written to the loop conditional terminal to stop execution. Note that the event structure will wait forever because no value is wired to the timeout input at the top left of the structure.

1.7.1.6 Disable Structure

There are two disable structures available on the structure palette. There is a disable structure and a conditional disable structure. The disable structure is similar to a case structure except there is no conditional input for the structure. The disable can have as many subdiagrams as you care to create, but one and only one must be enabled. This structure allows a user to enter several mutually exclusive subdiagrams in one location. When editing, the programmer can select which subdiagram to execute. The programmer can then change the enabled subdiagram without having to change the code as would be needed for a standard case structure.

The conditional disable structure gives the programmer the ability to select what code to operate based on the target platform. This means there can be different functions called if the code is run on a windows machine vs. a PDA or FPGA. Again, the basic look of the structure is similar to the cases structure without the case selector input.

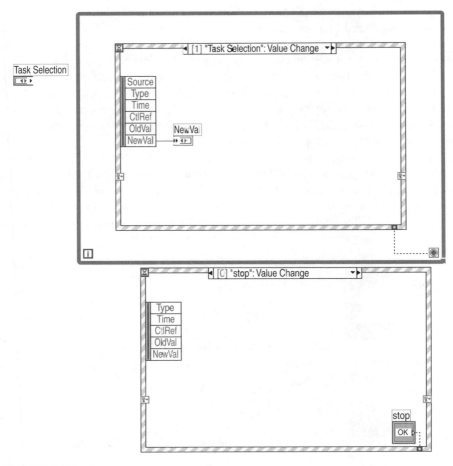

FIGURE 1.37

1.7.1.7 Timed Structure

The timed loop structure is similar to a While loop, but gives the programmer the ability to control the execution rate and run time. The use of the timed structure can allow for exact timing of the loop execution, programmatic change of loop timing, multi-loop timing and loop priorities.

When a timed loop is placed on the code diagram there is an input node on the left of the structure with an input for an error cluster. The input node can be resized to expose additional input parameters. Some of the inputs include the period of the loop execution, the starting offset, a loop timeout and a timing source. When the input node is viewed it starts with an icon representing the input type and the default value next to it. Once a value is wired to the input node the display will change to simply indicate what the parameter is. The programmer can wire values to the input node as they would with a subVI or a timed loop configuration window can be opened by right clicking on the loop structure. By default there are only two inputs for timing source, priority and period. If additional settings are desired, click on the Configure Advanced Settings checkbox.

There is an output node on the right of the structure. Initially there is only one output available for the error cluster. Again, the node can be expanded to expose additional outputs. The loop can output the expected and actual end iteration values. There is a Boolean output indicating if the loop finished late. Finally, there is an output for the end time. By clicking on the timed loop structure you can make visible a right and left data node. This allows the loop to be able to read loop output values as well as to modify the input parameters.

To illustrate a simple example of timed loops the VI in Figure 1.38 uses timed loops to generate a clock display. The VI has 4 timed loops. The first three are loops for the hours, minutes and seconds. The seconds loop has two inputs. The first is the loop name. This input is needed to be able to abort the loop execution when the stop button is pressed. The second input is the execution time of the loop in milliseconds. Obviously for a second timer the period is 1000ms. The display is the remainder of

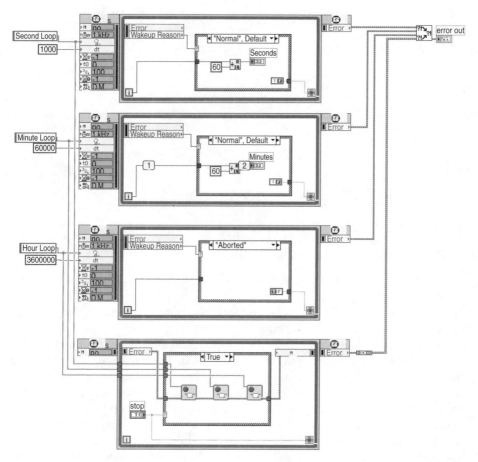

FIGURE 1.38

the loop index divided by 60 in order to reset the value every 60 seconds. The minute and hour loops are exactly the same with a different period setting.

In our example we want to be able to stop execution by clicking on a stop button. In order to perform this action we use the Stop Timed Loop function in the timed loop subpalette. The loop by default executes every 1000ms. If the value of the stop variable is true, the stop timed loop function is called for each of the three named loops. In order to stop a currently executing loop the left data node is needed in each of the three timed loops. The Wakeup Reason input value is Normal by default, but when a stop loop is received the input returns Aborted. This value is used to stop the loops.

1.7.1.8 Formula Node

A formula node is simply a bounded container for math formulas. It allows you to create formula statements, similar to programming in C. Multiple formulas can be enclosed in a single node, and each formula must end with a semicolon.

You can use as many variables as you wish, but you must declare each one as either input or output. Pop-up on the border of the formula node and select either Add Input or Add Output. A terminal is created on the border of the node for which you must enter the name of the variable. An output has a thicker border to help differentiate it from an input terminal. All input terminals must have data wired to them, but output terminals do not have to be used or wired to other terminals. Variables that are not intended for use outside of the Formula Node should be declared as output and left unwired. The input and output terminals can be created on any border of the structure.

The formula node is illustrated in Figure 1.39. The formula node contains a simple formula to demonstrate how it is used. It has one input variable, y, and one output variable, x. The output variable terminal has the thicker border and could have been moved to any location on the structure. The formula Node uses the input variable and calculates the output variable according to the formula created. Consult the Formula Node Syntax topic in Online Help to find out more information on creating formulas and the various operators that are available. You may also find the

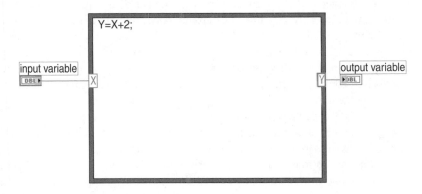

FIGURE 1.39

Formula Node Functions and Operators topic helpful to learn more about the different built-in functions offered.

One advantage of the formula node is that its operation is compiled internally to the node. Long formulas do not take up as much space on your display and can significantly reduce the number of elements in a wire table for the code diagram.

An alternative to using the formula node is the formula express VI. The user can select the formula express VI from the Arithmetic & Comparison palette in the Express palette. This places the express VI on the code diagram. The express VI looks similar to a subVI. Once the VI is placed on the code diagram (depending on your options settings) a configuration screen will display giving you the opportunity to define the formula. This is shown in figure "Express Formula.ai." Express VIs will be discussed further in Section 2.25.

1.7.2 NUMERIC, BOOLEAN, STRING, AND COMPARISON

The Numeric, Boolean, String, and Comparison palettes are displayed in Figure 1.40. The functions shown in the Numeric palette are straightforward and simple to use. The example in Figure 1.33, shown previously, utilized the multiply and increment functions. Most of them can be used for any type of number, including arrays and clusters. The multiply function, for example, requires two inputs and yields the product of the two.

The Numeric palette holds the Conversion, Complex, Data Manipulation and Additional Numeric Constants subpalettes. The functions in the Conversion subpalette are primarily used to convert numerical values to different data types. The Additional Numeric Constants subpalette holds such constants as Pi, Infinity, and e. One issue to note about floating point numbers in LabVIEW is that "not a number" quantities are defined. Values for +/- infinity are defined in floating point numbers, and division by zero will not generate an error but will return NaN (Not a Number). When performing calculations, it is up to the programmer (as always) to validate the inputs before performing calculations. The Data Manipulation subpalette contains functions such as flatten to string, split number and logical shift.

Numbers of various types will be converted when they are involved in a math operation. An integer and complex number will sum to be a complex number. The conversion performed is referred to as Coercion. Any numbers that are coerced will be labeled with a gray dot called a "coercion dot." Coercion is rarely a problem, but it needs to be understood that there is a small performance penalty for coercion between types. Numbers will never be converted "backwards," as a complex number being converted to an integer. Performing this type of conversion requires that you use a conversion method.

A rarely used property of floating point numbers is unit support. It is possible to define quantities with a unit attached. Popping up on any floating-point control, indicator, or constant on the diagram will allow you to expand the display menu. One of the display options is Unit. Once the unit is displayed, popping up on the unit shows the menu of units used by LabVIEW. LabVIEW supports sufficient unit types to make sure every chemistry, electronics, mechanical, and assembly lab has little to ask for, if anything. This feature works very well in simulation, measurement,

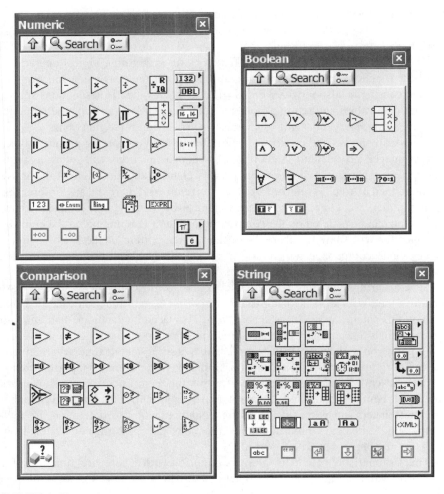

FIGURE 1.40

data display, and educational applications. Unit conversion is also possible, and is done behind the scenes. A floating-point number with a unit of feet can be wired to an indicator with a unit of miles. The display will show in miles; there is no need to perform conversion operations on the results of measurements. In some cases, this represents a possibility for performance enhancement because programmers who perform measurement conversions on their own need to add a few elements to their wire diagrams which will take more time to process. By default, floating-point numbers have no unit dimensions assigned to them.

In LabVIEW 8, there are several changes to the locations of some function palettes as well as some palettes that are available in more than one place. The numeric palette is one example. The Numeric palette is contained in the Programming and the Mathematics palette. Another change to the Numeric palette is the removal of the trigonometric and logarithmic subpalettes. They now are included in the Elementary & Special Functions subpalette under Mathematics. This may be a little confusing at first to long time LabVIEW programmers, but you do adjust over time.

The Boolean palette holds various functions for performing logical operations. All of the functions require Boolean inputs, except for the conversion functions. A Boolean constant is also provided on this palette. The Comparison functions simply compare data values and return a Boolean as the result. You can compare numeric, Boolean, string, array, cluster, and character values using these functions.

Comparing arrays and clusters is a bit different from comparing primitive types such as integers. By default, LabVIEW comparison functions will return a single value for cluster and array comparison. If every element and the length of the arrays are equal, then a "true" is returned. A "false" is returned if there are any differences. If programmers want to compare an array element-by-element, the Compare Aggregate option can be enabled on the comparison operator. Popping up on the comparison operator will show Compare Aggregates at the bottom of the list of options. An aggregate comparison will return an array with Booleans for the result of a comparison of each and every element in the array or cluster.

Several string functions are provided on the Strings subpalette. Figure 1.41 illustrates the use of Concatenate Strings and String Length functions, the first two items on this palette. When Concatenate Strings is placed on the block diagram, two input terminals are normally available. You must pop up on the function and select Add Input if you wish to concatenate more than two strings at one time. Alternatively, you can drag any corner of the function up or down to add more input terminals. You cannot leave any terminal unwired for this function. The example shown has three inputs being concatenated. A control, a string constant, and a line feed character are concatenated and wired to the String Length function to determine the total length. Four subpalettes hold additional functions that perform conversion from strings to numbers, byte arrays, and file paths.

The Comparison palette is pretty straightforward. There are functions for comparing values. Many of these functions are polymorphic. For example the Equal function can compare two numbers, two strings, two arrays of numbers, an array of numbers and a scalar value, etc… The output is either a scalar Boolean value or an array of Boolean values depending on the inputs to the function. There are also functions for determining if a number is in a range, finding the minimum and maximum value and if the number is an empty path. The Select function outputs either the True or False input based on the Boolean input value. This

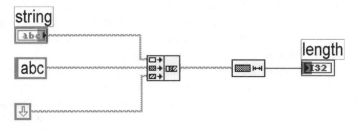

FIGURE 1.41

 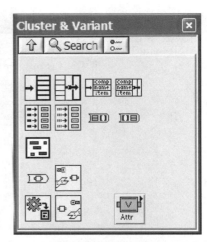

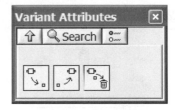

FIGURE 1.42

function is useful when adding conditional execution or exception handling into your application.

1.7.3 ARRAY AND CLUSTER

Array, Cluster & Variant, and Variant Attributes palettes are displayed in Figure 1.42. These palettes contain various functions for performing operations on these data constructs. The array functions provided can be used for multidimensional arrays. You must pop up on the functions and add a dimension if you are working with more than one dimension. Bundle and Unbundle functions are available for manipulation of clusters.

Figure 1.43 displays the front panel and code diagram of an example that uses both array and cluster functions. The front panel shows an array of clusters that contain employee information, similar to the example discussed in Section 1.5.5. This example demonstrates how to change the contents of the cluster for a specific element in the array. The Index Array function returns an element in the array specified by the value of the index wired to it, in this case 0. The cluster at Index 0 is then wired to the Bundle By Name function. This function allows you to modify the contents of the cluster by wiring the new values to the input terminals. Normally, when Bundle By Name is dropped onto the code diagram, only one element of the cluster is created. You can either pop up on the function to add extra items, or drag one of the corners to extend it. The item selection of the cluster can also be changed

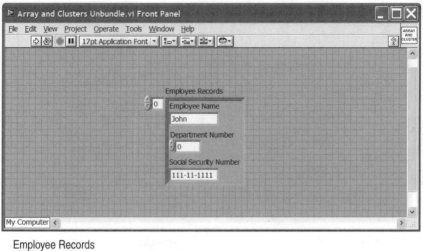

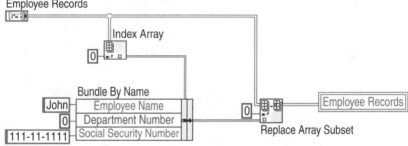

FIGURE 1.43

through the pop-up menu. New values are wired to the function as shown, and are then passed to the Replace Array Element function. This function simply replaces the original data with the values wired to the input terminals at the index specified. The output is then passed to a local variable of the Employee Records control. Local variables can be created by popping up on a control or indicator terminal from the code diagram. Select Local Variable from the Create submenu.

If you work with arrays, one of the array functions with which you should become very familiar is the Dimension array. This function will allow you to set the dimensions on an array. LabVIEW will expand array sizes to prevent users from overwriting the boundaries of an array, but this is bad practice. Each time LabVIEW needs to change the number of elements in a dimension, it must get a memory allocation sufficient to hold the array and copy each and every element into the new array. This is very inefficient, and is a bad programming habit. Pre-dimensioning arrays when you know the length in advance is an efficient habit to develop. The other array function you will become familiar with is the Replace Array element. This function allows you to change the value of an element in an array without duplicating the array.

Other functions in these palettes allow you to perform several other operations on arrays and clusters. The Cluster & Variant palette contains an Unbundle function for retrieving data from a cluster, a function for building cluster arrays, and functions for converting data between clusters and arrays. The Array palette holds functions for detecting array sizes, searching for a specific value in an array, building arrays, decimating arrays, and several other operations. If you are interested in creating easily-read GUIs, the conversion functions between arrays and clusters is something you will want to look into. On occasion, it will be desirable to use array element access in your application, but arrays on the front panel can be difficult to read. Displaying data on the front panel in the form of a cluster and converting the cluster to an array in the code diagram makes both users and programmers happy.

The variant functions have been combined with the cluster functions on the same palette. A variant is a unique data type that can contain string, numeric, date or user-defined data. The user will need to know how the variable was declared in the original function to be able to properly read and write to a function using the variant data type. The variant functions are the To Variant function, Variant to Data, Variant to Flattened String and Flattened String to Variant functions. These functions are all used to convert LabVIEW data types to or from Variant data types.

On the Cluster & Variant palette there is a Variant Attribute subpalette. This subpalette contains three functions used to manipulate variant data. The attribute functions allow the programmer to set, get and delete variant attributes. The Get Variant Attribute function gives the programmer the ability to either get the names and values of all of the variant attributes or the value of a specified attribute.

1.7.4 TIMING

The Timing palette, displayed in Figure 1.44, contains functions for retrieving the system time, wait functions for introducing delays, and functions for formatting time and date strings. The Wait Until Next Multiple function is useful for introducing delays into loop structures. When placed inside a loop, it causes the loop to pause a specified time between iterations of execution. There is a Time Delay Express VI

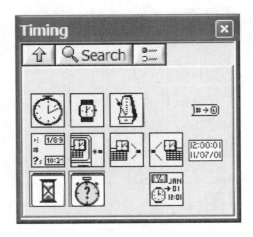

FIGURE 1.44

in this palette that has an error in and out that gives the programmer the ability to control the order of execution. The functions on this palette are simple to use and are self-explanatory.

System dates and times are dependent on the system you run on. Most computers measure the date in the number of seconds that have elapsed since a certain time, for example January 1, 1974, at 12:00am. This number is stored in a 32-bit number and it will be an extremely long time from now before this date rolls over (consider that there are approximately pi * 10^7 seconds in a year). The concern with system dates and times is because of the precision you need. As just mentioned, it is stored in units of seconds. If you need millisecond accuracy, system date and time are not going to be sufficient. Some systems will store hundredths or even tenths of a second, but millisecond accuracy is usually not possible with system times.

1.7.5 DIALOG AND USER INTERFACE

The Dialog & User Interface Palette is shown in Figure 1.45. Dialog boxes are great for informing users that something is happening in the system. Dialog boxes should usually be avoided in automated applications where a user is not monitoring the testing. A dialog box will halt LabVIEW's execution until somebody clicks the "OK" button. If you have an automated system that is expected to run while you are on vacation, it may be a while before you click the button to complete your testing. The dialog functions give the programmer the ability to name the buttons and specify the message displayed.

There are two express dialog functions on this palette. One is an Express Dialog function. This VI acts the same as the standard dialog box, but provides a means for controlling execution flow through the use of the error clusters. The express VI also makes configuring the function easier at the expense of visibility. The second express VI is called Prompt Users for Input. This function displays a dialog box

FIGURE 1.45

with an expandable number of input controls. The programmer can set up each input for numeric, string or checkbox entry. The function outputs the user-provided values.

There are four subpalettes on the Dialog & User Interface palette. The Events subpalette gives the programmer the ability to set up and use user events for program control. The subpalette also contains the Event Structure that is part of the Structures palette. The Menu subpalette contains functions used to add, remove and modify the runtime menus. This can be very useful in an application to give flexibility without creating clutter on the front panel. The Cursor subpalette gives the programmer the ability to control the cursor appearance and disable the mouse on the front panel. The Help subpalette gives control of the help windows and can call up a Web page in the default browser.

The error handler functions are also included in this palette. Chapter 6 covers the topic on exception handling and describes the error handler VIs in more detail. The merge error function merges error I/O clusters from up to four different functions. This function first looks for errors in the four inputs in order and reports the first error found. If the function finds no errors, it will look for warnings and return the first warning found. If the function finds no warnings, it will return no error. This is a helpful function when you have two or more sections of code running in parallel and want to combine the error outputs into a single path.

1.7.6 FILE I/O

Figure 1.46 shows the File I/O palette in addition to one of its subpalettes, the Advanced File Functions. The basic functions allow you to open, close, create, read from, or write to files. These functions will display a dialog box prompting the user to select a file if the file path is not provided. The advanced functions facilitate accessing file and directory information, modifying access privileges, and moving a file to a different directory, among several others.

LabVIEW's file interfaces give programmers as much or as little control over the file operations as desired. If you want to simply write an array to a tab-delimited file, there is a function to do just that. Supplying the array is about all that is necessary. The interface is very simple; you do not have much control over what the file handler will do. Lack

FIGURE 1.46

of control should not be a concern for you if your purpose is to write the tab-delimited string to a file. In fact, the string conversion is done in the function also.

Programmers who are concerned about the amount of space needed by a large set of data can use binary access files. Binary access files will put the bit pattern representing the array directly into the file. The advantages of binary files are the sizes they require. A 32-bit number stored in a binary file takes exactly 32 bits. If the number is stored in a hex format, the number would be 8 digits, requiring 64 bits to store, twice as long. Floating-point numbers have similar storage requirements, and binary files can significantly reduce the amount of disk space required to handle large files.

Binary files also allow programmers to make proprietary file formats. If you do not know the order in which data is stored, it is extremely difficult to read the data back from the file. We are not encouraging developers to make proprietary storage formats — the rest of the engineering community is driving toward open standards — but this is an ability that binary files offer.

Depending on the data being stored in the binary file, the amount of work you need to do is variable. If arrays of numbers are being written to the file, there are binary access VIs to read and convert the numbers automatically. The binary read and write VIs support any data type used in LabVIEW.

If you are trying to write data-like clusters to binary files, there are two options you can use. The first option is to flatten the clusters to a string and write the string to a file. Flattened strings will be binary. File interfaces will be easy to use, but reading back arrays of flattened clusters will be a bit more difficult. You will need to know the length of the flattened string, and be able to parse the file according to the number of bytes each cluster requires. Be sure to provide a robust error handler; the conversion might not work and return all manner of useless data if things go awry. The second option is to use the read and write files directly. Read and write from file is used by all of the higher level file functions, but does not open or close the files directly; you will need to call File Open and Close, in addition to knowing what position in the file to write to.

In general, we do not recommend using binary access files. Binary files can be read only by LabVIEW functions, and a majority of the reasons to use binary files are obsolete. Modern computers rarely have small hard drives to store data; there is ample room to store 1000-element arrays. Other applications, such as spreadsheets, cannot read the data for analysis. Binary files can also be difficult to debug because the contents of the file are not readable by programmers. ASCII files can be opened with standard editors like VI, Notepad, and Simpletext. If parsing or reading file problems show up in your code, it is fairly easy to open up an ASCII file and determine where the problems could be. Binary files will not display correctly in text editors, and you will have to "roll your own" editor to have a chance to see what is happening in the file.

Many programmers use initialization files for use with their applications. Lab-VIEW supplies a set of interfaces to read and write from these types of files. The "platform independent" configuration file handlers construct, read, and write keys to the file that an application can use at startup. Programmers who do not use Windows, or programmers who need to support multiple operating systems, will find this set of functions very useful. There is no need to write your own parsing routines. Data that may be desired in a configuration file is the working directory, display preferences, the last log files saved to, and instrument calibration factors. These types of files are not used often enough in programming. Configuration files allow for flexibility in programs that is persistent. Persistent data is data that is written to the hard disk on shutdown and read back on startup.

The Advanced File Function subpalette contains VIs to perform standard directory functions such as change, create, or delete directories. This subpalette has all the major functions needed to perform standard manipulations, and the interface is much easier to use than standard C.

There is a subpalette for interacting with ZIP files. There are functions for creating a new zip file, adding files to an existing zip file and a close zip file function. These functions can be useful when archiving large data files.

1.7.7 INSTRUMENT I/O, CONNECTIVITY, AND COMMUNICATION

The Instrument I/O and Connectivity palettes contain various built-in functions to simplify communication with external devices. These two palettes along with the Communication subpalette are displayed in Figure 1.47 representing how they appear on a Windows system. The Instrument I/O palette holds VISA, GPIB, Serial, and VXI-related functions. The Connectivity palette contains functions for ActiveX, .NET, Input Devices (keyboard, mouse), Windows Registry Editing, Source Code Control, Communications, Libraries & Executables, and Graphics & Sound. The Communications subpallette contains functions for TCP, UDP, Data

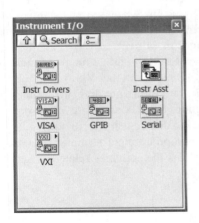

FIGURE 1.47

Socket, Bluetooth, IrDA, SMTP and Port I/O. The specific functions in these palettes will not be discussed in this book; however, Chapter 8 covers ActiveX and .NET in detail.

When designing an application, there may be a few minor details you should consider for communications. Inter-application communications do not involve cables such as GPIB. Windows-specific communications can be done with ActiveX /COM functionality or with .NET. ActiveX and .NET are the current standards for communications in Windows environments.

The only globally available communications protocols are the Unix standards TCP and UDP. Both protocols utilize the Internet protocol (IP). IP-based communications do not need to be between two different computers; applications residing on the same computer can communicate with TCP or UDP connections. TCP or UDP is recommendable because the interfaces are easy to use, standard across all platforms, and will not be obsolete anytime soon.

GPIB, serial, and VXI communications should be performed with the VISA library. VISA is becoming the standard for instrument communications in LabVIEW. The serial support has already been converted to VISA. The serial VIs available in the Instrument I/O palette are built upon VISA subVIs. The IEEE 488 will be supported for some time, but the VISA library is intended to provide a uniform interface for all communications in LabVIEW. Addressing, sending, and receiving from an external device all use the same VISA API, regardless of the communications line. The common API lets programmers focus on talking to the instruments, not on trying to remember how to program serial instruments.

The Instrument I/O palette also contains an express VI called Instrument I/O assistant. This express VI can be used to communicate with instruments and graphically parse the response. This functionality is useful when trying to verify the instrument is connected correctly. This assistant can also be used while verifying the accuracy of a GPIB command before using it in a driver.

LabVIEW VIs are very similar to functions or subroutines in programming languages like C. Once created, VIs can be called inside of other VIs. These subVIs are called simply by placing them on a code diagram, similar to dragging a function from the palettes as discussed in the last section. SubVIs are represented on the block diagram by an icon that you can customize to distinguish it from other subVIs. Once placed on the code diagram, wire the appropriate input terminals to ensure that it will execute correctly. This section explains the activities related in setting up and calling subVIs.

1.7.8 CREATING CONNECTORS

VIs can have inputs and outputs, similar to subroutines. A connector must be defined for a subVI if data is to be exchanged with it. It will be necessary for you to define connectors for most VIs that you create. The process consists of designating a terminal for each of the controls and indicators with which data will need to be exchanged. Once the inputs and outputs have been appointed terminals, data can be exchanged with the VI on a block diagram.

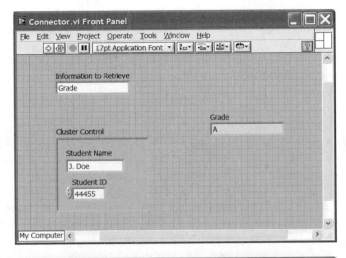

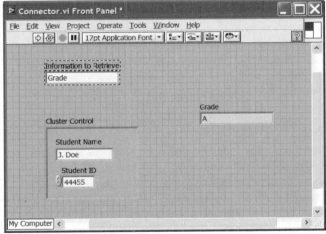

FIGURE 1.48

Figure 1.48 displays the front panel of a VI with the connector pane visible in the top right corner of the window. To display the connector pane on a VI, pop up on the icon that is normally visible and select Show Connector from the menu. Three rectangles or terminals appear in the example, one for each control and indicator. Each control and indicator can be assigned a terminal by using the wiring tool. Click on one of the terminals, then click on a control or indicator to designate the terminal.

The bottom window in Figure 1.48 illustrates how the Information to Retrieve control is assigned the top left terminal on the connector. By default, LabVIEW creates a terminal for each control and indicator on your front panel, but the assignment will be left to the programmer. If the default connector pattern is not appropriate, it can be modified to suit your needs. Once the connector is made visible, use the items in the pop-up menu to select a different pattern, or rotate the current pattern.

Controls and indicators can be assigned to any terminal on the connector. However, controls can only serve as inputs, and indicators can only be used for outputs. You should assign the inputs on the left terminals of the connector and the outputs to the right side, even though you are not required to. All LabVIEW built-in functions follow this convention. This convention also aids the readability of the code. The data flow can be followed easily on a block diagram when subVIs and functions are wired from left to right.

Built-in LabVIEW functions have inputs that are either required, recommended, or optional. If an input is required, a block diagram cannot be executed unless the appropriate data is wired. Correspondingly, LabVIEW allows you to specify whether an input terminal is required. Once you have designated a particular terminal to a control, pop up on that terminal and select This Connection Is from the menu. Then select either Required, Recommended, or Optional. Output indicators have the required option grayed out in the menu. Output data is never required to be wired.

Good programming practice with subVIs is fairly simple. It is a good idea to have a few extra connectors in your VI in case additional inputs or outputs are needed in the future. Default values should be defined for inputs. Defined default values will allow programmers to minimize the number of items on the calling VI's code diagram, making the diagram easier to read. Supplying the same common input to a VI is also tedious; granted, it is not impossible work to do, but it becomes boring. Laziness is a virtue in programming; make yourself and other programmers perform as little work as possible to accomplish tasks.

1.7.9 EDITING ICONS

Icons are modified using the Icon Editor. Either double-click the default icon in the top right corner of the window or pop up on it and select Edit Icon from the menu. Figure 1.49 is an illustration of the Icon Editor containing a default LabVIEW VI icon with a number. This communicates the number of new VIs opened since initiating the LabVIEW program. Each time you start LabVIEW, the VI contains a "1" in the icon as the default.

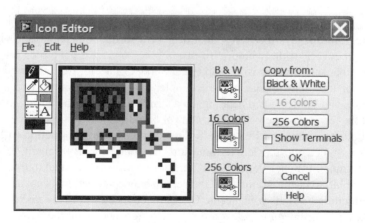

FIGURE 1.49

The Tools palette is located on the left side of the Icon Editor window, and the editing area is in the center. The default foreground color is black, while the background color is white. When you click on the background/foreground color tool, a color palette appears allowing you to select from among 256 colors. You can create different icons for black-and-white, 16-color, and 256-color monitor types. Many people create an icon in color and forget to create in black and white. This is important when you need to print out VI documentation; if you are not using a color printer, the icon will not appear as it should. Try to copy the icon you created from the color area to the black-and-white area.

Figure 1.50 demonstrates the process of customizing an icon. The top window in the figure displays an icon that has been partially customized. First, the contents of the editing area were cleared using the Edit menu. Then, the background color was changed to gray while the foreground was left as black. The Filled Rectangle tool was used to draw a rectangle bordered with a black foreground and filled with a gray background. If you double-click the tool, the rectangle will be drawn for you automatically. The second window displays the finished icon. The Line tool was used to draw two horizontal lines, one near the top of the icon and the other near

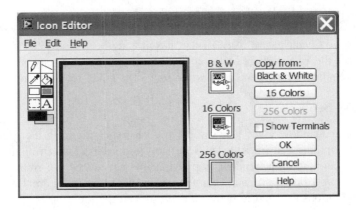

FIGURE 1.50

the bottom. Then, the Text tool was used to write "icon editor" in the editing area. Finally, the same icon was copied into the 16-color and black-and-white icon areas.

Because the icons are graphical representations of the VIs, you can use your imagination and get creative when editing them, if you wish. JPEG- and GIF-formatted picture files can be copied and pasted into the icon editing areas also. Although this can be fun, just remember that the purpose of customizing icons is to allow people to distinguish the VI from other VIs and icons in a program. Try to create icons that are descriptive so that someone looking at the code for the first time can determine its function easily. Using text in the icons often helps achieve this goal. This helps the readability of the code as well as easing its maintenance. Veteran programmers quickly abandon the process of taking an hour to develop an appealing work of art for an icon. We have all had those VIs with the extraordinary icons that were deleted because they became unnecessary in the project.

1.7.10 USING SubVIs

The procedure for using subVIs when building an application is similar to dragging built-in functions from a palette onto the block diagram. The last item on the Functions palette, displayed in Figure 1.24, is used to place subVIs onto block diagrams. When Select a VI is clicked, a dialog box appears prompting you to locate the VI that you want to use. Any VI that has already been saved can be used as a subVI. Place the VI anywhere on the code diagram and treat it as any other function. Once the required inputs have been wired, the VI is ready for execution.

1.7.11 VI SETUP

The VI Setup window gives you several options for configuring the execution of VIs. These options can be adjusted separately for each VI in an application. To access this configuration window, pop up on the icon in the top right corner and select VI Setup from the menu. This window is displayed in Figure 1.51, with the Execution Options selected in the drop down box at the top.

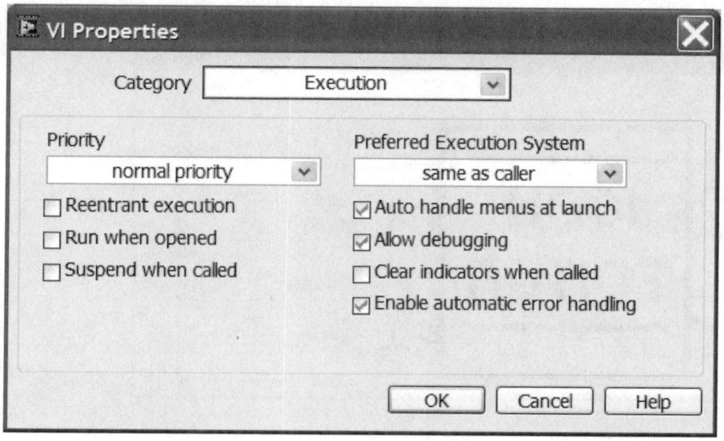

FIGURE 1.51

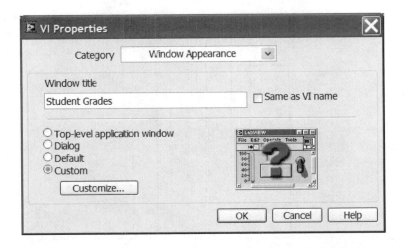

FIGURE 1.52

The items on the execution options panel are used for configuring the VIs execution priority, allowing reentrant execution and setting the preferred execution system. Reentrant execution refers to making multiple calls to the same VI and is covered in the next chapter. VI priority and the execution system selections are used for optimizing the execution of an application. These two topics are discussed further in Chapter 9, which also covers multithreading. We strongly recommend not working with either priority or execution subsystem until you read Chapter 9. This is one of those topics in which not understanding how threads and priorities interact can do more harm than good.

Figure 1.52 displays the VI setup window with Window Appearance Options selected in the drop-down menu. Initially you have the option of Top Level Application Window, Dialog, Default or Custom. These configuration selections allow you to customize the appearance of the VI during execution. The first three options

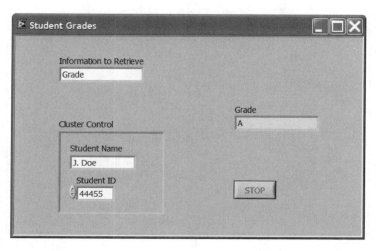

FIGURE 1.53

have predefined settings for appearance and function. The custom option allows the user to setup the VI exactly the way it is needed. In the example shown, Show Scroll Bars, Show Menu Bars, and Show Toolbar have been deselected. These are all enabled by default. There are also checkboxes used to show the front panel when it is called, and to close the panel after it has finished executing. The "Same as VI Name" has also been deselected and the Window Title modified. These alterations cause the VI to appear as shown in Figure 1.53 during execution. When the Stop button is pressed, the front panel returns to its normal appearance. Window options are useful for limiting the actions available to the end user of the program.

Figure 1.54 displays the VI Documentation and Revision History windows. LabVIEW provides some built-in documentation support that can be configured through either VI Setup or Options. A VI history is kept for each VI that is created. This history is used to keep records of changes made to a VI, and serves as documentation for future reference. The "Use the default history settings from the Options dialog box" checkbox has been deselected in the example shown. This informs LabVIEW to use the settings from the VI setup instead of the Options Dialog. The Option settings also allow you to configure the VI history, but this checkbox determines which ones are used.

Also note that two boxes have been checked which configure LabVIEW to add an entry to the VI history every time the VI is saved, and also to prompt the programmer to enter a comment at the same time. The entry LabVIEW adds consists of the time, date, revision number, and the user name. The programmer must enter any comments that will provide information on the nature of the modifications made. Figure 1.55 illustrates the VI history for the VI shown earlier in Figure 1.53. The VI history can be viewed by selecting Show History under the Windows pull-down menu. Chapter 4 discusses the importance of documentation and reveals other documentation methods for LabVIEW applications.

Initially you have the option of Top Level Application Window, Dialog, Default, or Custom. This section describes how VIs, once developed, can be used as subVIs

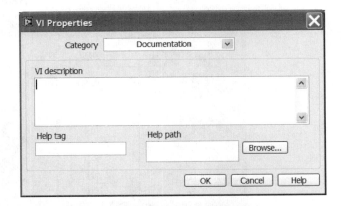

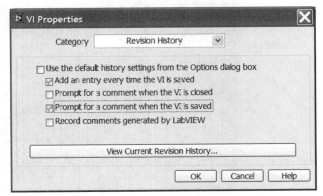

FIGURE 1.54

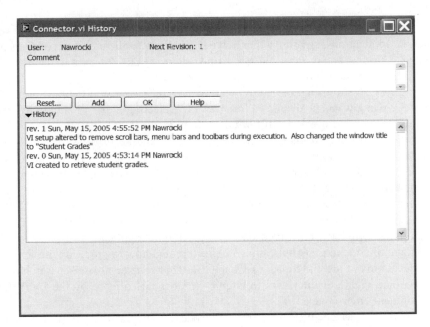

FIGURE 1.55

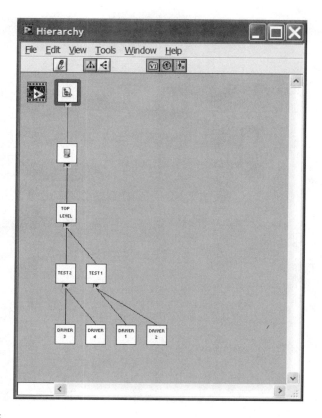

FIGURE 1.56

in larger applications, creating a hierarchy of VIs in an application where layers are created. These layers, or tiers, must be managed during development to increase the readability, maintainability, reuse, and abstraction of code.

Figure 1.56 shows the hierarchy window of a relatively small application. The hierarchy window can be displayed for any VI by selecting Show VI Hierarchy from the Project pull-down menu. This window graphically shows the relationship of a VI to the application. It displays the VI, its callers, and all of the subVI calls that it makes. The hierarchy window shown in the figure corresponds to the main VI at the top. There are two layers of VIs below the main. In this example, the application was developed with three tiers: the main level, the test level, and the driver level.

The inherent structure of LabVIEW allows for reuse of VIs and code. Once a VI is coded, it can be used as a subVI in any application. However, a modular development approach must be used when creating an application in order to take advantage of code reuse. Application architecture and how to proceed with application development are the topics of Chapter 4. This chapter also discusses how to manage and create distinct tiers to amplify the benefits offered by the LabVIEW development environment.

Instrument drivers play a key role in code reuse with LabVIEW. Chapter 5 introduces a formula for the development of drivers to maximize code reuse, based

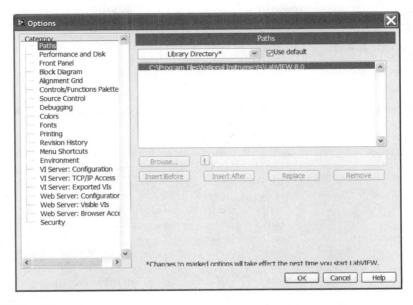

FIGURE 1.57

on National Instruments development method. When this formula is followed, the result is a set of drivers that can be reused in any application while providing abstraction for this lowest tier in the hierarchy.

The intrinsic modularity of LabVIEW can be used to apply an object-oriented methodology to application development. LabVIEW itself is not an object-oriented language; however, it is object-based. The object-oriented approach can be applied to LabVIEW, though in a limited manner. Chapter 10 introduces you to the terminology associated with Object-Oriented Programming, as well as how to apply it in a LabVIEW environment.

1.8   SETTING OPTIONS

This section describes some of LabVIEW's options or preferences that can be configured to suit a programmer's needs. The options selection is available in the Tools pull-down menu. The window that appears is shown in Figure 1.57 along with its default settings. The options shown correspond to the Paths selection from the top drop-down menu. Some of the option selections are self-explanatory and will not be discussed in this section; however, Table 1.3 lists all of the selections and describes the notable settings that can be configured in each.

1.8.1   PATHS

The Paths configurations, shown in Figure 1.57, dictate the directories in which LabVIEW will search when opening or saving libraries, VIs, menus, and other files. The drop-down menu selector allows you to configure the Library, Temporary, Default and Data directories. The last selection in this menu is used to set the VI

TABLE 1.3
Option Headings

Option Selection	Function/Utility
Paths	Configure search directories for opening/saving VIs.
Performance and Disk	Configure to use multithreading and perform check for available disk space prior to launch.
Front Panel	Settings for front panel editing.
Block Diagram	Settings for block diagram programming.
Alignment Grid	Settings for aligning objects on front panel and code diagram to specified grid spacing.
Controls/Functions Palettes	Settings for palettes.
Source Control	Settings for source code control of LabVIEW files.
Debugging	Options that are used for debugging VIs, and execution highlighting during execution.
Colors	Change default colors used by LabVIEW for front panel, block diagram, etc.
Fonts	Settings for Applications, System, and Dialog Font styles.
Printing	Configure print settings.
Revision History	Options for recording revision comments when changes are made to VIs.
Menu Shortcuts	Allows setting key combination shortcuts to menu functions.
Environment	Options for tip-strips, native file dialogs, drop-through clicks, hot menus, auto-constant labels, opening VIs in run mode, and skipping navigation dialog at launch.
VI Server: Configuration	Configure protocols, port numbers, and server resources.
VI Server: TCP/IP Access	Set access privileges to specific list of clients for VI Server.
VI Server: Exported VIs	Specify list of VIs that are accessible to clients using VI Server.
Web Server: Configuration	Enable Web server, configure root directory, set port number and timeout.
Web Server: Browser Access	Set access privileges to specific list of clients for Web server.
Web Server: Visible VIs	Specify list of VIs that are accessible to clients from Web server.
Security	Allows setting of password for LabVIEW login.

Search Path. This informs LabVIEW of the order in which to search directories when opening VIs. When you open a VI containing subVIs that are not part of a library, this search order will be followed to find them. You can configure this to minimize the time it takes to search and find subVIs.

If your group uses a number of common VIs, such as instrument drivers, the directories to the drivers should be added to the VI search path. Current projects should not be added to the search path. The VI search path was intended to allow programmers to easily insert common VIs. The VIs that are written as part of a project and not intended to be part of a reusable library would end up cluttering up the search path, lengthening the time LabVIEW takes to locate VIs.

1.8.2 BLOCK DIAGRAM

Figure 1.58 displays the Block Diagram options window. These options are intended to help you develop code on the block diagram. For the beginning user of LabVIEW,

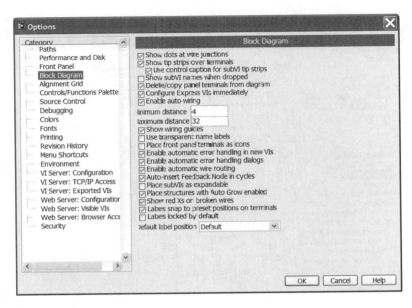

FIGURE 1.58

some of these settings can help you get familiar with the programming environment. Tip-strips, wiring guides, and junction dots are very useful when wiring data to functions and subVIs. Displaying subVI names is also handy because the icons are not always descriptive enough to determine their roles in an application.

1.8.3 ENVIRONMENT

The Environment option window offers the ability to set preferences for the LabVIEW environment. Options for the environment include auto tool select, Just In Time (JIT) advice, undo and abridged menus. Undo and Redo are both available in the Edit pull-down menu. When the option box is unchecked, you can change the default number from 8 to another suitable number. Keep in mind that a higher number will affect the memory usage for your VIs during editing. Since actions are recorded for each VI separately, the number of VIs that you are editing at any one time also affects memory usage. Note that once a VI is saved, the previous actions are removed from memory and cannot be undone. The Environment Option Window is shown in Figure 1.59.

1.8.4 REVISION HISTORY

The Revision History options window is displayed in Figure 1.60. Some of these options are duplicated in the VI History settings under VI Properties. If you compare this to Figure 1.54, you will notice that the first four checkboxes are the same. If you have the Use History Defaults box checked in the VI Property window settings, LabVIEW will use the Revision History options.

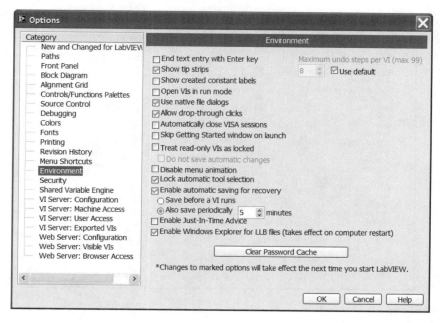

FIGURE 1.59

The radio buttons let you configure the login settings for LabVIEW. These settings will be used to determine the name entered by LabVIEW in the VI History box that records the comments when an entry is made. The second window in Figure 1.60 shows the User Login information. The login name can be modified in this window and is accessed by selecting User Name from the Tools menu.

Using the VI history is simply good programming practice. Listing the change history of a VI allows other programmers to understand what modifications a VI has which can be used to help debug applications. It does not take too long with troubleshooting before you see why an application stopped working because "some-one else" made a modification to code and did not communicate or document the modification. Using history alone is not quite enough. When making comments in the history, note the changes that were made, and, equally important, note why the changes were made. It is fairly common practice to comment code as you write it, but to not keep the comments up to date when modifications are made. Giving other programmers a hint as to why a change was made allows them to see the thought process behind the change.

1.8.5 VI SERVER AND WEB SERVER

The VI Server functionality is a feature that was added to LabVIEW in Version 5.0. It allows you to make calls to LabVIEW and VIs from a remote computer. You can then control them through code that you develop. This also permits you to load and run VIs dynamically. Chapter 8 describes the VI Server in more detail along with the related configurations and some examples.

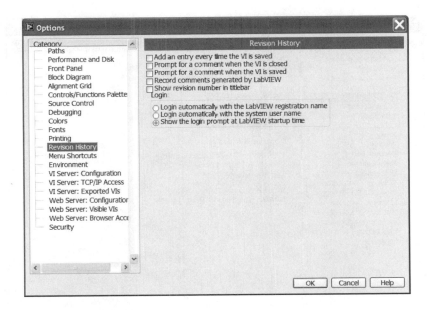

FIGURE 1.60

The Web Server is also an addition to LabVIEW starting in Version 5.0. The built-in Web server must be enabled through the preference settings. The Web server will allow you to view any VIs that are loaded on the same machine using a browser. You can then view the front panel of a VI that may be running from any remote machine. The Web Server and its configurations are discussed further in Chapter 2.

1.8.6 CONTROLS/FUNCTIONS PALETTES

LabVIEW normally displays the default palettes for both Controls (Figure 1.9) and Functions (Figure 1.24). You can change the palette view to match your programming needs by either selecting a new palette set or creating your own palette. The view can be changed easily through the Options menu. The programmer has the option to show the standard icons and text, all icons, all text or tree view by using the Format pull down menu. The Control/Function Palette is shown in Figure 1.61.

Select Tools, Advanced, and then Edit Palette Views to create and customize a new palette set. A window similar to the one shown in Figure 1.62 will appear that will allow you to perform this action. Then select New Setup from the drop-down menu box and enter a name for the new view. A view called "Personalized" was

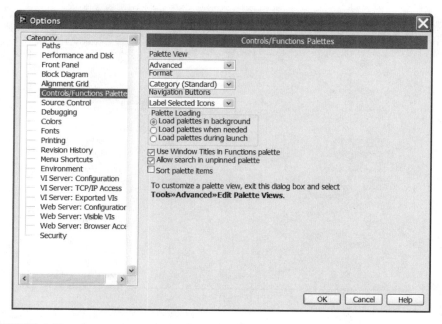

FIGURE 1.61

created for the example in Figure 1.62. The customized Functions palette is also shown, along with the modified User Libraries subpalette. A new setup must be created because LabVIEW does not directly allow you to modify the default palette set. It serves as protection in case the changes a user makes are irreversible.

Once you have created the new setup, the Functions and Controls palettes contain the default subpalettes and icons. The user is allowed to move, delete, and rename items in the palettes as desired. All of the available editing options are accessible through the pop-up menu. Simply pop up on the palette icon or the specific function within a subpalette to perform the desired action. If you compare the Functions palette in Figure 1.62 to the default palette in Figure 1.24, you will notice the changes that were made. Some palettes were deleted while others were moved to new locations. A VI (Data Log.vi) was added to the Users Library displayed in the bottom window. VIs that you have created and may use regularly can be added to a palette in this manner. After a new setup has been created, it will be available to you as an item under the Select Palette Set submenu.

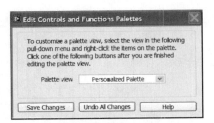

FIGURE 1.62

2 LabVIEW Features

The previous chapter covered many of LabVIEW's basic functions. The functions give a programmer the ability to produce a wide range of applications in a relatively short time. Although the previously discussed functions provide enough of a basis to build an application, there are a number of LabVIEW features that can make an application more flexible and easier to use, and can give your application a professional appearance. Some of these features will be discussed in this chapter.

2.1 GLOBAL AND LOCAL VARIABLES

Global variables are used when a data value needs to be manipulated in several VIs. The advantage of using a global variable is that you only have to define that data type once. It can then be read from or written to multiple VIs. The use of global variables is considered poor programming practice; they hide the data flow of your application and create more overhead. National Instruments suggests that you structure your application to transfer data using a different approach when possible. However, there are instances when global variables are necessary and are the best approach for an application. One example would be updating a display from data being generated in a subVI. The application could have two While loops running in parallel. Data could be generated in a subVI in the top loop while the bottom loop reads the data from the global and writes the information to the user interface. There is no other method for obtaining data from a subVI while it is still running (pending discussion of the Shared Variable).

The global variable must be created and its data types defined before it can be used. To create a global, first drag the icon from the Structures palette and drop it onto a block diagram. Figure 2.1 shows the global as it appears on the diagram. The question mark and black border indicate that it cannot be used programmatically. The global has a front panel to which you can add controls, identical to a VI. A Global does not have a block diagram associated with it. To open the front panel of the global variable, simply double-click on the icon. The front panel of the global is shown in the bottom window of Figure 2.1.

Two controls have been created on the global front panel. A global variable can contain multiple controls on the front panel. Try to logically group related controls and tie them to a single global variable. Once the data types have been defined, save the global as a regular VI. The global can then be accessed in any VI by using the same method you normally follow to place a subVI on the code diagram. If you have more than one control associated with the global variable, pop-up on the icon

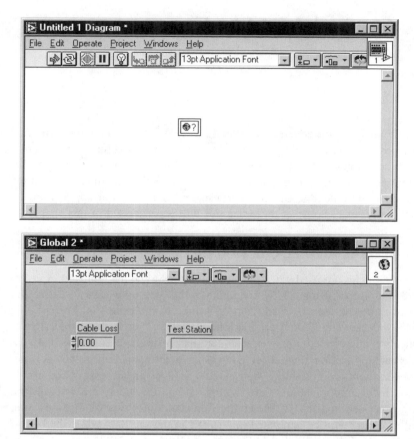

FIGURE 2.1

once you have dropped it onto a block diagram and use the Select Item submenu to
select the appropriate one.

A value can be either written to or read from a global. Use a "read" global to
read data from and a "write" global to write data to a global variable. The first
selection in the pop-up menu allows you to change to either a read or write variable.
Figure 2.2 demonstrates how a global and local variable can be used on the block
diagram. The global created in Figure 2.1 is used in this VI to retrieve the Cable

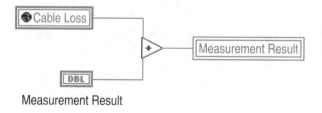

Measurement Result

FIGURE 2.2

Loss parameter. Global variables are easy to distinguish on the block diagram because of the unique icon that contains the name of the variable. The thicker border indicates that it is a read variable.

Measurement Result is a control used in this VI. The result of the addition is being passed to the local variable of Measurement Result. Local variables can be created by popping up on a control or indicator terminal and selecting Local Variable from the Create submenu. Alternatively, drag and drop the local variable from the Structures palette. Then, pop up on it and use the Select Item submenu to choose the name of a control or indicator. A local variable can be created for any control or indicator terminal. As with the global, the local can be used as a read or write variable and toggled using the pop-up menu. In the example shown, the name of the local, Measurement Result, appears in the icon. The icon does not have a thick border, indicating that it is a write variable.

The main difference between local and global variables is access. The local variable is only available on the code diagram it was created on. The global variable can be used in any VI or subVI on the source machine. Due to the fact that the global variable is loaded from a file, any VI has access to this data. While this flexibility seems like a benefit, the result is a loss of data control. If a specified global variable is placed in a number of VIs, one of the VIs could be used by another application. This could result in errant data being written to the global in your main program. With local variables, you know the only place the data can be modified is from within that VI. Data problems become easier to trace.

One alternative to a standard global variable is the use of a *functional global*. A functional global is a VI that contains a While loop with an uninitialized shift register. The VI can have two inputs and one output. The first input would be an Action input. The actions for a simple global would be read and write. The second input would be the data item to store. The output would be the indicator for the item to read back. The case structure would have two states. In the Read state, the program would want to read the global data. The code diagram wires the data from the shift register to the output indicator. The Write case would wire the input control to the output of the shift register. The code diagram is shown in Figure 2.3. The benefit of using this type of global is the prevention of race conditions; an application cannot

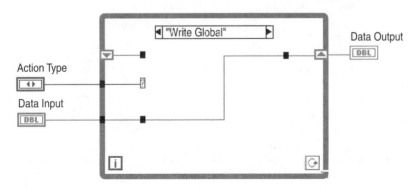

FIGURE 2.3

attempt to write to and read from the global at the same time. Only one action will be performed at a time.

2.2 SHARED VARIABLES

Shared variables are a new type of variable added to LabVIEW 8. The shared variable is similar to a global variable in that you obtain and pass data to other parts of your program or other VIs without having to connect the data by wires. The shared variable has several distinct differences. First, data can be shared not only between one VI and another on your computer, but it can be shared over the network. Another difference is that shared variables can be bound to a source. For example the data could be read in from another shared variable, a front panel control or a LabVIEW RT target. A shared variable can also use data buffering and restrict inputs to one writer at a time. A shared variable node on the code diagram can provide a time stamp to be able to determine the "freshness" of the data. Finally, there is an error in and out for the shared variable, which will help with both exception handling and forcing the order of execution.

To create a shared variable you must first open a LabVIEW project. The shared variable can only be created within a LabVIEW project and must be contained in a project library. If a library does not exist, one will be created for the new variable. Once you have the project open you need to right click in a library and select New, and then Variable. The Shared Variable Properties window will appear. This dialog box is shown in Figure 2.4. Here you will enter the variable type. A shared variable

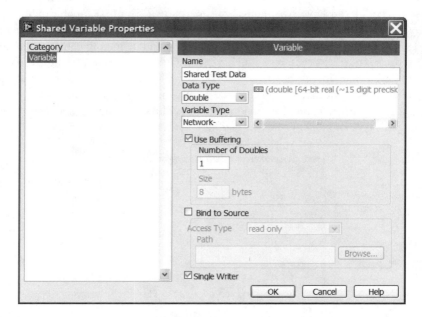

FIGURE 2.4

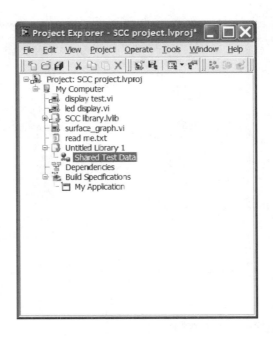

FIGURE 2.5

can be almost any type of variable including numbers, strings, Booleans, waveforms and arrays of the aforementioned datatypes. If those options do not meet your needs you can import a custom data control type.

Once the datatype is selected you need to choose whether the variable will be a network variable. There is a choice of whether to use buffering. If you choose to use buffering make sure you make your buffer large enough. If there is a buffer overflow data will be lost, and there will be no error generated. In the properties window you can select a source to bind the variable to. Finally you have the option to set the variable to only accept changes from one writer at a time. Once you have completed changes to the variable properties you will see the variable has been added to the project. Figure 2.5 shows the new shared variable was added to an empty library since one did not previously exist in this project. Also notice that based on the icon you can see that the variable was setup as a network variable.

Now that the variable is set up you can use the variable in your code. There are two ways to insert the variable in the code. The first is to do a drag and drop from the Project Manager to the VI. The second method is to insert a shared variable from the Structures palette. Once the shared variable node is on the code diagram you can double click on it to select the shared variable to link to. This will only work if the VI you are working with is a part of the project. Figure 2.6 shows a loop that reads a shared variable every 5 sec. The data and time stamp are reported to the front panel. Through this you will be able to see what the current value of the variable is and when it was last changed.

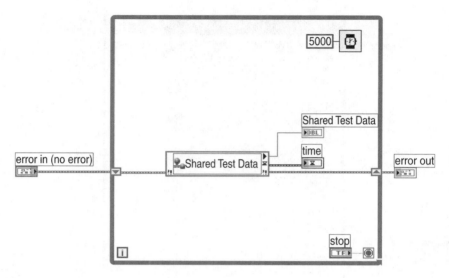

FIGURE 2.6

2.3 CUSTOMIZING CONTROLS

Controls and indicators can be customized, saved, and reused in different VIs. This allows you to modify the built-in controls and indicators to accommodate your applications. This section describes the procedure for creating custom controls and type definitions. A "Type Definition" is a master copy of a control or indicator. When you need to use the same control in several VIs, you can create a type definition and save it. Then, when changes are made to the type definition, they will automatically be applied to all of the VIs that use that control.

2.3.1 CUSTOM CONTROLS

To customize any control on your front panel, select the control and choose Edit Control from the Edit pull-down menu. A new window will appear with the control shown on the panel. This panel has no diagram associated with it and cannot be executed. Figure 2.7 shows this window with a text ring control on the panel. Also note that Control is the current selection in the drop-down menu on the toolbar. The control can be modified in either the Edit mode or the Customize mode; the default is Edit mode when the window first appears. The Edit mode is similar to the Edit mode of the front panel of any VI where alterations can be made. It allows you to make some of the basic changes to a control, such as size and color. The Customize mode lets you make additional changes to specific elements of the control. The top window in Figure 2.7 shows the Edit mode, and the bottom window shows the Customize mode. The first button in the window toolbar allows you to toggle between the two modes of operation.

Each control is a composite of smaller parts. When you modify a control in the Customize mode, you are then able to modify the parts of the control. If you open

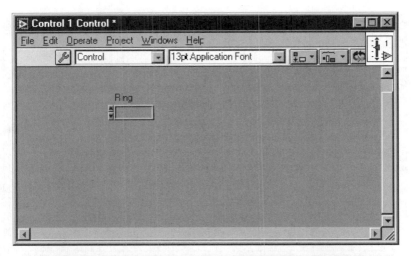

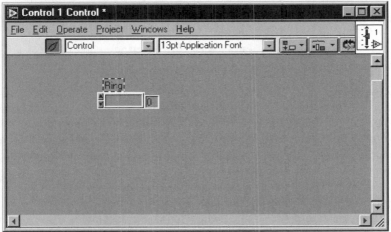

FIGURE 2.7

the Parts Window from the Windows menu, you will be able to see the labels for each part, as well as the position and size of each part. You can scroll through each part via the increment arrow. One of the benefits of this capability is the ability to create custom controls. Text or pictures can be copied and pasted into the control editor. The pictures can become part of the control. This capability makes the creation of filling tanks, pipes, and other user-friendly controls possible.

Figure 2.8 shows the modified ring control on the front panel. The up and down scroll arrows were altered for the ring control. Once the desired modifications are made to a control, you can replace the original control with the modified one without saving the control. Select Apply Changes from the Control Editor window before closing it to use the modified control. Alternatively, you can save the control for use in other VIs. Simply give it a name and save it with a .ctl extension or use Save As

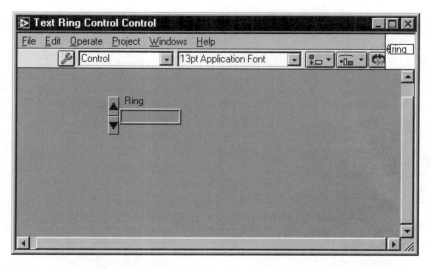

FIGURE 2.8

from the File menu. To use it on another VI front panel, choose Select a Control from the controls palette and use the file dialog box to locate the control.

2.3.2 TYPE DEFINITIONS

A type definition allows you to set the data type of a control and save it for use in other VIs. This may be useful if you change the data type and want that change reflected in several VIs. A type definition allows you to define and control the data type from one location. It can prove to be very practical when using clusters and enumerated types. When items need to be added to these controls, you only have to do it once. Default values cannot be updated from a type definition.

   You create a type definition following a similar procedure as a custom control. Select the control that you wish to create a type definition from and choose Edit Control from the Edit pull-down menu. Figure 2.9 displays the window that appears for an enumerated control. To create a type definition instead of a custom control, select Type Def. from the drop-down menu in the toolbar. The enumerated control has been defined as an unsigned word and three items have been entered into the display. This type definition was saved using the Save As selection from the File menu. The window title tells you the name of the control and that it is a type definition.

   The type definition can be used in multiple VIs, once it has been saved, by using Select a Control from the Controls palette. When you need to modify the type definition, you can open the control using Open on the File menu. You could also select the control from a front panel that uses it and choose Edit Control from the Edit menu, which then opens the window of the saved type definition. A final way to open the control is by double-clicking on it (if this option is selected in your preferences).

   Any changes made to the type definition will be reflected automatically in its instances if they have been set to auto-update. The instances include controls, local

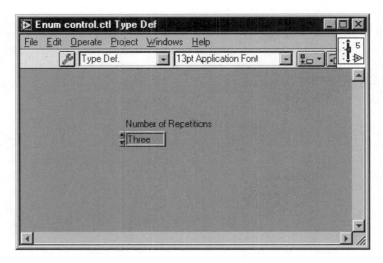

FIGURE 2.9

variables, and constants created from the control. All VIs that use instances of the type definition are set to auto-update by default. When you pop up on an instance of the type definition, you will see a menu similar to the one shown in Figure 2.10. You can then choose to auto-update the control or disconnect it from the type definition. Items can be added to the enumerated control type definition shown in Figure 2.9, and all VIs that use the type definition will be automatically updated. Items cannot be added to the instances unless the auto update feature is disabled.

2.3.3 STRICT TYPE DEFINITIONS

Type definitions cause only the data type of a control to be fixed in its instances. Other attributes of the type definition can be modified within the instances that are used. Size, color, default value, data range, format, precision, description, and name are attributes that can be

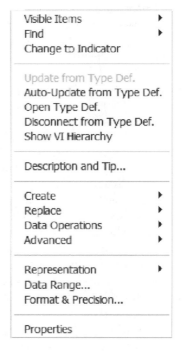

FIGURE 2.10

adjusted. Strict type definitions can be used to fix more of the attributes of a control to create uniformity across its instances. Only the name, description, and default value of a strict type definition can be altered in the instances. This allows you to maintain more control of the type definition. In addition, auto updating cannot be disabled when strict type definitions are used. This forces all changes to be applied to all of the occurrences.

Strict type definitions are created in the same manner as type definitions. The only difference is the drop-down menu in the toolbar should be set to Strict Type Def. After the strict type definition has been saved, changes can be made only to the master copy. The changes made to the master copy are reflected in the VI only when it is open. If a VI is not in memory, the changes are not updated. This could be an issue if a VI is copied to a new location, such as a different PC, without opening the VI between the time the control was modified and the copy was performed. In the absence of the Strict Type Def., the VI would first ask you to find the control. If the control is unavailable, the control will appear grayed out. If you right-click on the control you have the option of disconnecting from the Strict Type Def. If you disconnect, the VI would use the last saved version of the control. In this case, the modifications would not be reflected in the VI on the new PC.

2.4 PROPERTY NODES

Property nodes are a means for getting and setting the properties of a control or indicator during program execution. The properties available will vary depending on the particular control or indicator being used on the front panel of your application. Pop up on either the control from the front panel or the terminal from the code diagram and Select Property Node from the Create submenu. By providing the ability to change the appearance, location, and other properties programmatically, property nodes provide you with a tremendous amount of flexibility while designing and coding your application.

Figure 2.11 illustrates some of the characteristics of property nodes. An enumerated type control, Number of Repetitions, will be used to describe property nodes. A property node was created for the Number of Repetitions control and is placed just below the terminal. In this example, the Visible attribute is being set to "false." When the VI executes, the enumerated control will no longer be visible. All of the properties associated with Number of Repetitions are shown in the property node to the right on the block diagram. Multiple properties can be modified at the same time. Once a property node has been created, simply drag any corner to extend the node or use the pop-up menu and select Add Element.

You can read the current setting or set the value of a control through the property node from the block diagram. Use the pop-up menu to toggle the elements in the node between read and write. An arrow at the beginning of the element denotes that you can set the property, while an arrow after the property name denotes that you can read the property. In the figure shown, the first ten elements have the arrow at the beginning, indicating that they are write elements. When there are multiple properties selected on a property node, they can be selected as either read or write. If both operations need to be performed, separate property nodes need to be created.

The following example demonstrates how to use property nodes in a user interface VI. The front panel of the VI is displayed in Figure 2.12. Depending on the selection made for the Test Sequence Selection, the appropriate cluster will be

Number of Repetitions

Number of Repetitions

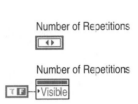

Number of Repetitions

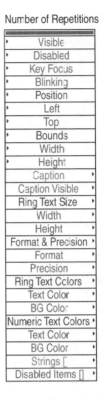

▸	Visible
▸	Disabled
▸	Key Focus
▸	Blinking
▸	Position
▸	Left
▸	Top
▸	Bounds
▸	Width
▸	Height
Caption	▸
Caption Visible	▸
Ring Text Size	▸
Width	▸
Height	▸
Format & Precision	▸
Format	▸
Precision	▸
Ring Text Colors	▸
Text Color	▸
BG Color	▸
Numeric Text Colors	▸
Text Color	▸
BG Color	▸
Strings [	▸
Disabled Items []	▸

FIGURE 2.11

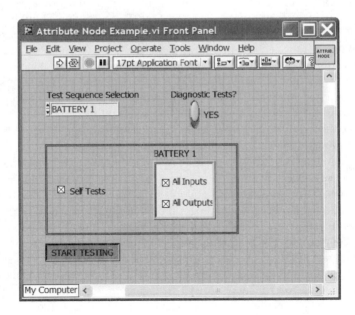

FIGURE 2.12

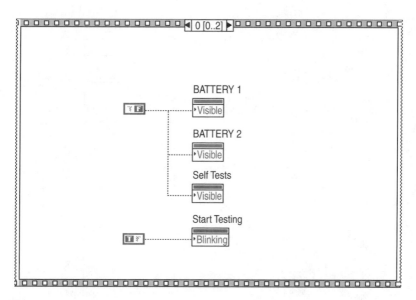

FIGURE 2.13

displayed in the framed border below it. The cluster shown for Battery 1 has two checkboxes that can be manipulated when Battery 1 is selected in the enumerated control. If Diagnostic Tests? is true, the Self Tests checkbox will be visible. When the user is finished making the selections, the Start Testing button is pressed.

Figure 2.13 shows Frame 0 of the sequence structure on the block diagram associated with this VI. The first action taken is to set the visible properties of Battery 1, Battery 2, and Self Tests to false. Also, the Start Testing button's blinking property is set to true. This ensures that the VI begins in a known state in case it was run before. Frame 1 of the sequence is shown in Figure 2.14. The purpose of this frame is to continually monitor the actions the user takes from the front panel. The While loop repeats every 50 milliseconds until Start Testing is pressed. The 50-millisecond delay was inserted so that all of the system resources are not used exclusively for monitoring the front panel. Depending on the selection the user makes, the corresponding controls will be displayed on the front panel for configuring. This example shows that property nodes are practical for various applications.

In the following example, shown in Figure 2.15, more of the properties are used to illustrate the benefits of using the property node. The VI has a number of front panel controls. There is a numeric control for frequency input, a set of Boolean controls to allow the user to select program options, and a stop Boolean to exit the application. To allow the user to begin typing in the frequency without having to select the control with the mouse, the property for key focus was set to "true" in the code diagram. This makes the numeric control active when the VI starts execution. Any numbers typed are put into the control. The second property in the property node for the digital control allows you to set the controls caption. This caption can be changed during execution, allowing you to have different captions based on program results.

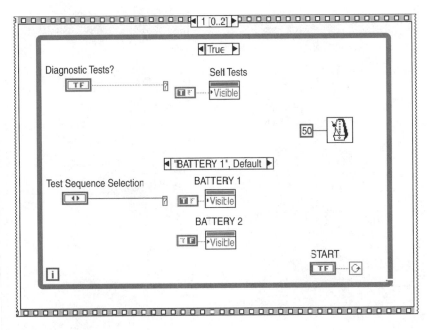

FIGURE 2.14

The second set of controls is the Booleans for setting program options. One of the requirements for this application is to make two additional options available if the second option is set to "true." Setting the Disabled attribute to 2 (which is disabled and grayed out) performs this action when the program starts. If the second option is set to "true," the Disabled attribute is set to 0. This enables the control, allowing the user to change the control value. The other option disables the control, but the control remains visible. The final property node in this example sets the Blinking property of the Stop Boolean to "true." When the VI is run, the Stop button will flash until the program is stopped. The front panel and code diagram are shown in Figure 2.15.

2.5 REENTRANT VIs

Reentrant VIs are VIs configured to allow multiple calls to be made to them at the same time. By default, subVIs are configured to be nonreentrant. When a subVI is nonreentrant, the calls are executed serially. The first subVI call must finish execution before the next one can begin. Calls to the subVI also share the same data space, which can create problems if the subVI does not initialize its variables or start from a known state. Shift registers that are not initialized, for example, can cause a subVI to yield incorrect results when called more than one time.

When a subVI is configured to be reentrant, it can speed up the execution of an application as well as prevent problems caused by sharing the same data space in the system. Each call to the subVI will execute in its own data space. A separate instance is created for each call so that multiple calls can be executed in parallel.

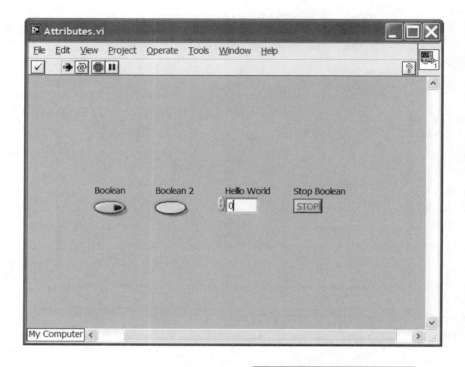

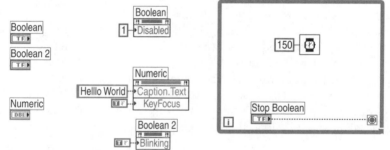

FIGURE 2.15

The calls will, in effect, execute independently of each other. The use of a separate data space for each call will result in more memory usage. Depending on the data types used, and the number of times the VI is executed, this can become an important issue. When creating an application, you should be aware that subVIs of a reentrant VI cannot be reentrant.

A VI can be enabled for reentrant execution through the Execution Options in the VI Properties dialog box. Pop up on the icon in the top right corner of the VI window and select VI Properties from the menu. The Execution options of the VI Properties window are displayed in Figure 2.16. In previous versions of LabVIEW several checkboxes would become disabled when Reentrant Execution is enabled: Show Front Panel When Loaded, Show Front Panel When Called, Run When Opened, Suspend When Opened, and all of the printing options. In addition, Exe-

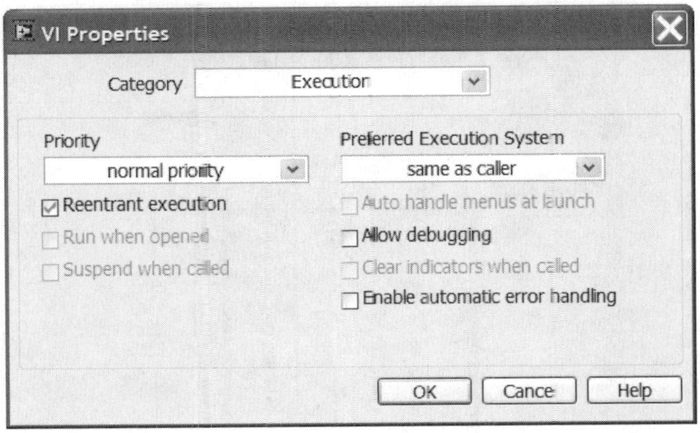

FIGURE 2.16

cution Highlighting, Single-Stepping, and Pausing were no longer available. In LabVIEW 8 reentrant VIs can now be debugged.

When you put a reentrant VI on a code diagram, a copy of the reentrant VI is placed on the code diagram. If you put 2 copies of the reentrant VI on a code diagram or on more than one open code diagram two individual copies of the subVI are created. When you open up a copy of the reentrant subVI, named Reentrant Function SubVI for this example, the front panel opens up with the title reentrant function subVI.vi:1 (clone). When subsequent copies are opened the number after the colon is incremented. Each subVI is a mutually exclusive copy. From the clone copy you cannot modify the original VI. If you want to make changes to the reentrant subVI you must open the original VI from file through the File menu. Changes made to the original VI will be reflected in the cloned copies. You would debug a reentrant VI the same way you would debug a normal VI. In the example in Figure 2.17, the subVI is a function used to perform a select action on 2 numbers. The main level VI has two copies of the subVI. One copy is used to perform addition on two arrays of numbers and one copy of the subVI is used to perform multiplication on the same two arrays of numbers. When you single step through the multiplication version of the subVI you can see that the correct value is seen at the input. You are now able to evaluate the actual numbers going through a specific instance of the reentrant subVI.

2.6 LIBRARIES (.LLB)

The VI library was briefly mentioned in Chapter 1 when saving a VI was discussed. This section will describe the procedure for saving and editing a VI library. There are both advantages and disadvantages to saving a VI inside of a library. National Instruments suggests that VIs be saved as separate files (.vi extension) rather than as libraries unless there is a specific need that must be satisfied. Table 2.1 lists benefits of saving as libraries and separate files; the table is also available through the on-line help VI Libraries topic. These issues need to be considered when saving files.

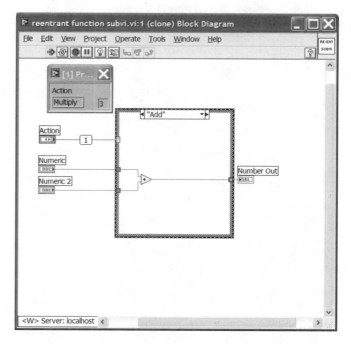

FIGURE 2.17

TABLE 2.1
Library Benefits and Drawbacks

Benefits of Saving as .VI	Benefits of Saving as .LLB
1. You can use your own file system to manage and store VIs.	1. 255 characters can be used for naming files (may be useful for Macintosh, where filenames are limited to 31 characters).
2. Libraries are not hierarchical. Libraries cannot contain subdirectories or sublibraries.	2. Easier for transporting VIs to different platform or to a different computer
3. Loading and saving VIs is faster and requires less disk space for temporary files.	3. Libraries are compressed and require less disk space.
4. More robust than storing entire project in the same file.	4. Can now be viewed in Windows Explorer (starting in LabVIEW 7).
5. There is the possibility of a library becoming corrupt.	5. Can set one or several VIs to start when the library is opened (top level VIs).
6. Source Control cannot operate on individual VIs in a LLB, only the entire LLB.	

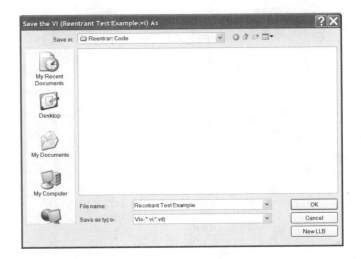

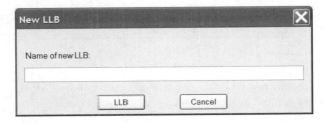

FIGURE 2.18

To save a file as a VI library, you can use Save As from the File pull-down menu. Figure 2.18 shows the file dialog box that appears when you select Save As and then select to make a copy of the VI. One of the buttons in the dialog box window lets you create a new VI library. When you press New VI Library, the window shown below the dialog box in Figure 2.18 appears. Simply enter a name for the library and press VI Library; the extension will be added for you. If you want to add a VI to an existing library, you can perform a Save As option and find the library you wish to save it in. The library is treated as a folder or directory and you can save VIs inside it. The second option available (on Windows platforms) is to open the library in Windows Explorer. To Windows Explorer, the library looks just like another folder so that you can add and delete files from the library.

If you do not want to use the Windows Explorer or if you are running on another OS, you can edit a library by using the LLB Manager. The LLB Manager dialog box is shown in Figure 2.19. In the LLB manager you have the ability to add, remove and copy VIs in the LLB. You can also convert the LLB to a folder or a folder to a LLB. While editing a LLB you can designate a VI in the LLB as a top level VI. A top level VI is opened when the LLB is opened automatically. This allows the programmer to launch a user interface VI that obscures how the underlying code is arranged or used.

FIGURE 2.19

LLB files should not be confused with Project Libraries. In LabVIEW 8, a project library is defined as a collection of VIs, shared variables, type definitions, and other files. When you create a project library a file containing the properties and references for the project library is generated. The project library file will have a .lvlib extension. There are some similarities and differences between a LLB and a project library. Both functions give you the ability to group files for an application. You cannot make a top level VI in a project library though. When using a project library, VIs from the library can be dragged and dropped on the code you are editing.

2.7 WEB SERVER

National Instruments incorporates a Web server with versions of LabVIEW 5.1 or later. Once enabled and configured, the Web server allows users to view the front panels of applications (VIs that have been loaded) from a remote machine using a browser. Both static and dynamic images of a front panel VI can be viewed remotely. Not only will the Web server allow you to view the front panel of an application, but you have the ability to control the VI as well. You can control the application or front panel remotely using a browser. You have the ability to interact with the application through the Web server, but you cannot modify the code. This section will go through the steps for configuring the Web server, and will show an example of how the front panel will appear on an Internet browser. This will be followed by an explanation of controlling the application using remote front panels.

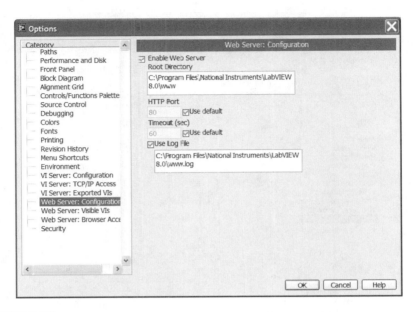

FIGURE 2.20

The Web server is set up through the Options selection on the Tools pull-down menu. This window is shown in Figure 2.20 with Web Server: Configuration selected. The Web server is not running by default, so you must first check Enable Web Server. The default Root Directory is the folder that holds the HTML files published to the Web. The HTTP Port is the TCP port the Web server uses to publish the pages. This port number may have to be changed depending on whether another service is utilizing this default on your machine. Finally, the log file saves information on the Internet connections to the Web server. Once this is done, front panels of loaded VIs can be viewed from a remote browser. However, you may want to perform additional configurations to set access privileges, as well as determine which VIs can be viewed through the server. Figure 2.21 displays the Option window for the Web Server: Browser Access selection. By default, all computers have access to the published Web pages. This window allows you to allow or deny specified computers access to the Web pages. You can enter IP addresses or domain names of computers into the browser access list. The browser list allows you to use the wildcard character (*) so that you do not have to list every computer name or IP address individually. The "X" in front of an entry indicates that a computer is denied access while a √ indicates that it is allowed access. If a diamond appears in front, then this is a signal that the item is not valid.

Figure 2.22 illustrates the window when Web Server: Visible VIs is selected in the left pane. This window is similar to the browser access settings but applies to VIs. You can add names of VIs into the listbox and select whether you want to allow or deny access to them via a browser. Again, the wildcard character is valid when making entries. The X and / indicate whether the VI can be accessed, and the

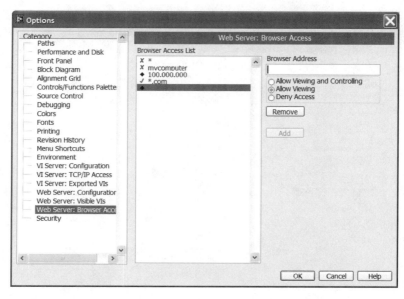

FIGURE 2.21

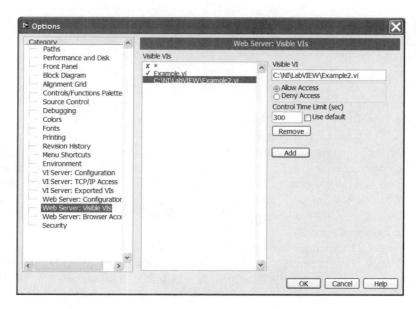

FIGURE 2.22

diamond signifies that the syntax may be incorrect. You can enter the name of a VI or designate the specific path if you have multiple VIs with the same name.

Viewing the front panel of a VI with a Web browser is a straightforward procedure. Before you can view the front panel, make sure the Web server has been enabled and the VI has been opened. The VI must be loaded in memory before anyone can look at it from the Web. Then, from any machine that is allowed access to the Web server, type the URL of the machine running the Web server into the browser. To the end of the URL add the VI name with the html extension. Insert a plus sign or % 20 in the URL to represent the spaces in your VI name. For example, to monitor a VI named "test program.vi" you would enter in the following URL into your browser: http://111.111.111.11/test program.html. This assumes the IP address of the Web server is 111.111.111.11. If you are trying to view a VI on the local machine you could use "localhost" instead of the IP address. Depending on the company you are working for, and your network specifics, the system administrator may need to set permissions or perform other system administrator intervention.

As mentioned earlier you can control a LabVIEW Application or Front Panel Remotely Using a Browser. The first step is to set up the Web server. Once the Web server is set up for allowing viewing and control access you must load or run the application or VI. Once in memory you would navigate to the proper html page in your browser. Now you should see what you saw above. To control the VI you would right click on the VI panel in the browser and select Request Control of VI. If there is no other client controlling the VI a message appears indicating that you have control of the front panel. If another client is already controlling the VI, the Web server queues the request until the other client releases control. To disconnect you can right click on the panel and select Release Control of VI from the Remote Panel Client window or you can close the browser window.

From the server you have the ability to regain control of a VI that is being controlled remotely. If you right click on the VI panel a Remote Panel Client subwindow comes up. You can select Switch Controller, which gives the local VI control again. The browser that was in control will see a message stating that the server has regained control. The person connecting through the browser can request control again. In order to prevent the remote control, the person at the server can select Lock Control from the Remote Panel Client subwindow. Now when the remote browser tries to take control a message will state that either the server has locked control or another client has control. When the server is unlocked the control will transfer to the remote host and the remote host will be notified that control has been granted.

2.8 WEB PUBLISHING TOOL

The Web Publishing Tool can be used to help you customize the way your Web page appears in a browser. Normally, only the front panels of the VIs that you have loaded are displayed in the browser when accessed remotely. This tool lets you add a title with additional text, which is then saved as an HTML file. When you want to view the front panel of a VI, you use the URL of the HTML page you have saved. The

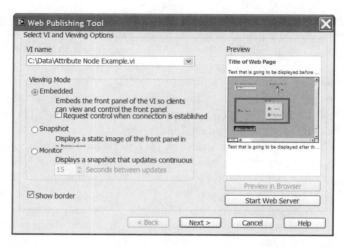

FIGURE 2.23

page then loads the picture of the front panel for you to view along with the title and additional text.

To access this feature, select Web Document Tool from the Project pull-down menu. The Web Document Tool window is displayed in Figure 2.23. Select a VI using Browse from the VI Name drop-down menu, or type in a VI name for which you want to create a page. A dialog box will ask you if you wish to open the VI if it is not already opened. Then you can enter in a title that you want to appear on the page, text before the image, and text after the image. The buttons available allow you to save the changes to an HTML file, start the Web server if it has not already been enabled, and preview the Web page. When you select Save To Disk, a dialog box will tell you the URL that is to be used to access the page from a browser. Remember that if the VI has not been loaded, the Web page will appear without the image of the front panel. Figure 2.24 shows an example of a page created using the Web Publishing Tool in a browser window.

2.9 INSTRUMENT DRIVER TOOLS

There are tools in LabVIEW 8 that help you manage both the external devices connected to your computer and the instrument drivers that are available from National Instruments. An instrument driver is a collection of VIs used to control instruments. LabVIEW drivers abstract the low-level commands that programmable instruments respond to. Drivers allow you to control instruments without having to learn the programming syntax for each instrument. Instrument drivers are discussed in more detail in Chapter 5, including the recommended style to follow when developing them. LabVIEW ships with a second CD-ROM that contains instrument drivers for numerous instruments from various manufacturers.

Under the tools menu there is an Instrumentation submenu. This submenu contains two tools for working with instrument drivers. The first tool is the Find Instrument Drivers function. When this function is selected, a dialog box comes up

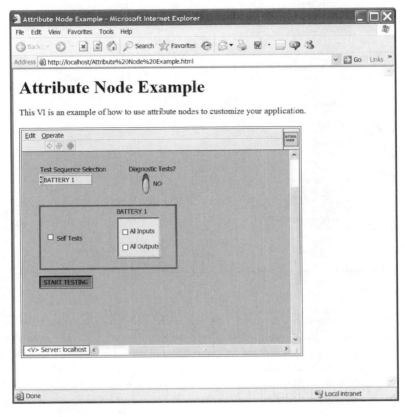

FIGURE 2.24

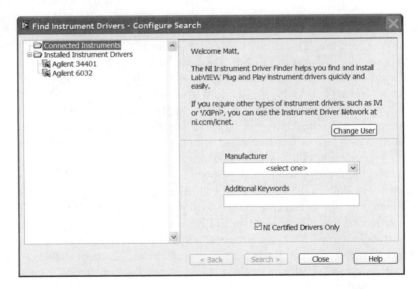

FIGURE 2.25

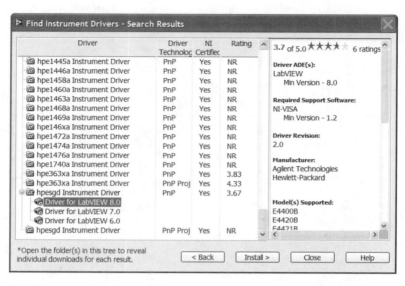

FIGURE 2.26

giving the programmer a list of currently installed drivers as well as a list of any connected instruments. The dialog box is shown in Figure 2.25. If the driver you are looking for is not installed you can search the National Instruments Website for the driver you need. At the top right of the dialog box you can log into the National Instruments Website. Once logged in you can select what manufacturer you are searching for as well as any key words to narrow the search. The result of the search is shown in Figure 2.26.

Now that you have a list of drivers you can search the list for the one that meets your needs. Platform support, LabVIEW version, required support software and ratings are just a few pieces of information available to you. Some drivers also have the option of installing as a standard driver or as a driver project. Once you find the driver you need you can select Install. The driver will install on your computer and will be available to use through the functions palette in the Instrument I/O subpalette.

The second tool available is Create Instrument Driver Project. This tool is a starting point for creating your own driver for an instrument. When the tool is launched it will open the dialog box shown in Figure 2.27. Here you have the choice of creating a new driver form template or creating a new driver from an existing driver. If you select to copy an existing driver, a list of installed drivers will appear in the second pull-down window. If you want to start from a template the second pull-down will have a list of the available driver templates. Once you select the closest matching instrument type the tool will generate a driver project. The project manager view of the generated driver project is shown in Figure 2.28. As you can see, the standard array of VIs are generated such as the VI tree, initialize, configure instrument, reset and close. From here you would modify each of the VIs to work with your instrument. In this example, when you open the Initialize VI, you see that it was made general enough to work in many cases. The initialize VI window is shown in Figure 2.29.

FIGURE 2.27

FIGURE 2.28

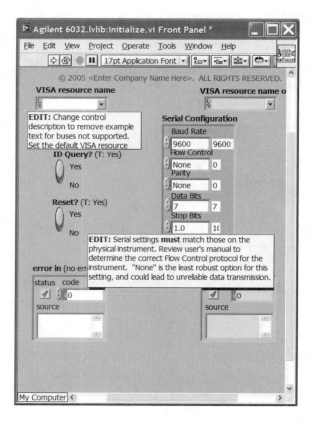

FIGURE 2.29

There is also a selection in the Instrumentation submenu for going directly to the Instrument Driver Network on National Instruments Website. In the Instrument Driver Network you have numerous search options available for finding the driver you need.

2.10 PROFILE FUNCTIONS

In order to make more efficient applications it is often helpful to be able to take a step back and look at the application from a statistical standpoint. This can give the user the ability to see where the weak link in the chain is so as to be able to optimize the code based on the needs of the application. The user can look for VIs that require the most time to execute, use memory inefficiently or where code may be too complex for the application. There are three profile functions in LabVIEW 8 that will help with these issues.

2.10.1 VI PROFILER

LabVIEW has a built-in tool, the VI Profiler, to help you optimize the execution of your applications. It can reveal beneficial information such as how long VIs in the

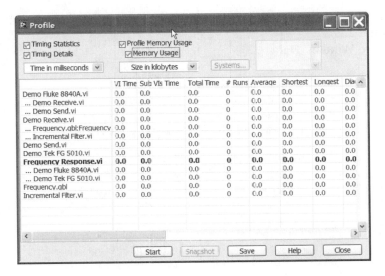

FIGURE 2.30

application take to execute, and how much memory each VI is using. These statistics can then be used to identify critical areas in your program that can be improved to decrease the overall execution time of your application. I/O operations, user interface display updates, and large arrays being passed back and forth are some of the more common sources that can slow down your application. This tool is part of the Full and Professional versions of LabVIEW.

Selecting Performance and Memory from the profile submenu located in the Tools pull-down menu allows you to access the VI Profiler. The profile window is shown in Figure 2.30 as it appears before it is started for capturing the data. The name of the VIs and subVIs, are displayed in the first column, with various pieces of data exhibited in the columns that follow.

Before you start the profiler, you can configure it to capture the information you wish to see by using the checkboxes. VI Time, SubVIs Time, and Total Time are the basic data provided if the other boxes are not selected. The memory usage data is optional because this acquisition can impact the execution time of your VIs. This can result in less accurate data on the timing of your application. Once you start the profiler, the memory options cannot be changed midstream.

Table 2.2 describes the information provided in the various columns in the profile window. It includes items that appear when all of the checkboxes are selected in the profiler window. If memory usage is selected, statistics will be provided for bytes of memory, as well as blocks of memory used. The memory size used is calculated at the end of the execution of a VI and may not be a precise representation of the actual usage during execution.

Frequency Response.vi will be used to demonstrate some actual data taken by the profile window. The Instrument I/O demonstration is an example program that is part of the LabVIEW installation when you choose the recommended install. Figure 2.31 displays the profile window for Frequency Response.vi with the option

TABLE 2.2
Profile Window Data

Statistic	Description
VI time	Total time taken to execute VI. Includes time spent displaying data and user interaction with front panel
SubVIs time	Total time taken to execute all subVIs of this VI. Includes all VIs under its hierarchy
Total time	VI time + subVIs time = total time
Number of runs	Number of times the VI executed
Average	Average time VI took to execute calculated by VI time number of runs
Shortest	Run that took least amount of time to execute
Longest	Run that took longest amount of time to execute
Diagram	Time elapsed to execute code diagram
Display	Amount of time spent updating front panel values from code
Draw	Time taken to draw the front panel
Tracking	Time taken to follow mouse movements and actions taken by the user
Locals	Time taken to pass data to or from local variables on the block diagram
Average bytes	Average bytes used by this VI's data space per run
Minimum. bytes	Minimum bytes used by this VI's data space in a run
Maximum bytes	Maximum bytes used by this VI's data space in a run
Average blocks	Average blocks used by this VI's data space per run
Minimum blocks	Minimum blocks used by this VI's data space in a run
Maximum blocks	Maximum blocks used by this VI's data space in a run

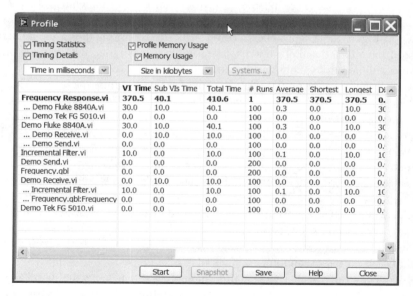

FIGURE 2.31

for Timing Statistics selected. The data shown in the profile window represents a single run, after which the profiler was stopped. This data will vary depending on the configuration of the system from which this VI is executed.

Information on subVIs is normally hidden in the profile window. When you double-click on a cell containing a VI name, its subVIs will appear below. You can show or hide subVIs by double-clicking. Clicking on any of the column headings causes the table to sort the data by descending order. The time information can be displayed in microseconds, milliseconds, or seconds by making the appropriate selection in the drop-down menu. Memory information can be displayed in bytes or kilobytes.

2.10.2 BUFFER ALLOCATIONS

There are two additional options under the profile submenu. The first is to show Buffer Allocations. This function shows the user where buffers are created (memory is allocated) for data operations on the code diagram. This can be a useful tool for optimizing your code for efficient use of memory as well as speed. There may be a case where you are doing manipulation on arrays that could cause an exponential increase in memory needed. If you catch this before pumping a lot of data through your VI you could change the way you are handling the arrays such as predefining the array and then inserting the values into the defined array instead of doing an append which makes a new copy each time. Sometimes problems like this can be missed when the code is debugged with a small data set, but when it is on a live system with large amounts of data the system can come crashing down.

When you select Show Buffer Allocations from the profile submenu, a dialog box will come up giving you the option of what data types to analyze. You can select any combination of data types that you want to look at. Once you have selected the appropriate types you select Refresh. This will put a small black box everywhere a buffer is created.

2.10.3 VI METRICS

The final Profile option is VI Metrics. This function gives the user a means for providing a view into the actual code used in a VI or Project. The dialog that comes up gives the user options for what portion of the VI or VIs you want to analyze such as Code Diagram, User Interface and Globals/Locals. Once all the inputs are set you generate a window containing the statistics. Depending on your selections you will have statistics on parameters such as number of nodes, structures, wire sources and maximum diagram depth. From the window you can select to save the information to a tab-delimited text file. An example of the metrics window is shown in Figure 2.32.

The VI metrics can be a useful tool for code optimization and for project planning. Sections like the maximum diagram depth can alert the developer that there may be too many nested structures leading to code that will be hard to understand and maintain. The information for number of structures, nodes, subVIs,

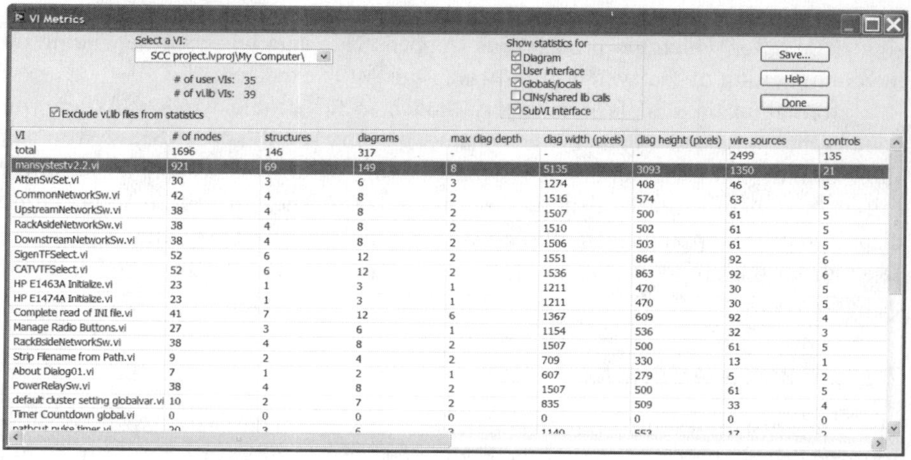

VI Metrics

Select a VI:
SCC project.lvproj\My Computer\

\# of user VIs: 35
\# of vi.lib VIs: 39

Exclude vi.lib files from statistics

Show statistics for
Diagram
User interface
Globals/locals
CINs/shared lib cals
SubVI interface

Save...
Help
Done

VI	# of nodes	structures	diagrams	max diag depth	diag width (pixels)	diag height (pixels)	wire sources	controls
total	1696	146	317	-	-	-	2499	135
mansystestv2.2.vi	921	69	149	8	5135	3093	1350	21
AttenSwSet.vi	30	3	6	3	1274	408	46	5
CommonNetworkSw.vi	42	4	8	2	1516	574	63	5
UpstreamNetworkSw.vi	38	4	8	2	1507	500	61	5
RackAsideNetworkSw.vi	38	4	8	2	1510	502	61	5
DownstreamNetworkSw.vi	38	4	8	2	1506	503	61	5
SigenTFSelect.vi	52	6	12	2	1551	864	92	6
CATVTFSelect.vi	52	6	12	2	1536	863	92	6
HP E1463A Initialze.vi	23	1	3	1	1211	470	30	5
HP E1474A Initialze.vi	23	1	3	1	1211	470	30	5
Complete read of INI file.vi	41	7	12	6	1367	609	92	4
Manage Radio Buttons.vi	27	3	6	1	1154	536	32	3
RackBsideNetworkSw.vi	38	4	8	2	1507	500	61	5
Strip Filename from Path.vi	9	2	4	2	709	330	13	1
About Dialog01.vi	7	1	2	1	607	279	5	2
PowerRelaySw.vi	38	4	8	2	1507	500	61	5
default cluster setting globalvar.vi	10	2	7	2	835	509	33	4
Timer Countdown global.vi	0	0	0	0	0	0	0	0
pathcut pulse timer.vi	20	3	6	2	1140	552	17	2

FIGURE 2.32

and wire sources can help give some view into the complexity of the code. This can later be used for planning future projects and for measuring overall efficiency.

2.11 AUTO SUBVI CREATION

Creating and calling subVIs in an application was discussed in the introductory chapter in Section 1.6.9. Connector terminals, icons, VI setup, and LabVIEW's hierarchical nature were some of the topics that were presented to give you enough information to create and use subVIs. There is another way to create a subVI from a section of code on a VI diagram. Use the selection tool to highlight a segment of the code diagram that you would like to place in a subVI and choose Create SubVI from the Edit pull-down menu. LabVIEW then takes the selected code and places it in a subVI that is automatically wired to your current code diagram.

Figure 2.33 will be used to illustrate how Create SubVI is used. The code diagram shown is used to open a Microsoft Excel file, write the column headers, and then write data to the cells. A section of the code has been selected for placing inside of a subVI using the menu selection. The result of this operation is shown in Figure 2.34, where the selected segment of the code has been replaced with a subVI. The subVI created by LabVIEW is untitled and unsaved, an activity left for the programmer. When the subVI has been saved, however, the action cannot be undone from the Edit pull-down menu. Be cautious when using this feature so as to prevent additional time spent reworking your code. Also, a default icon is used to represent the subVI, and the programmer is again left to customize it.

Notice that although the Test Data control was part of the code selected for creating the subVI, it is left on the code diagram. These terminals are never removed from the code diagram, but are wired into the subVI as input. Then, the appropriate controls and indicators are created in the subVI, as shown in Figure 2.35. The

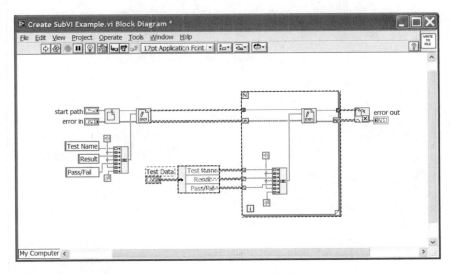

FIGURE 2.33

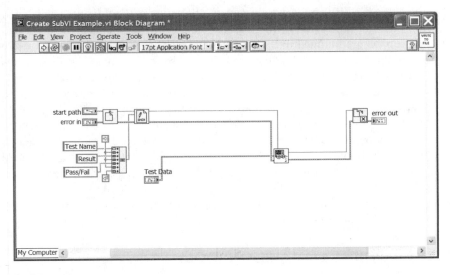

FIGURE 2.34

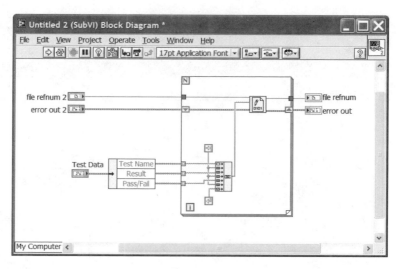

FIGURE 2.35

refnums, error clusters, and the Test Data cluster terminals appear on the code diagram, with their counterparts on the front panel.

There are instances when creating a subVI using the menu option is illegal. When there are real or potential problems for creating a subVI from the selected code, LabVIEW will not perform the action automatically. For potential problems, a dialog box will notify you of the issue and ask you whether you want to go ahead with the procedure. If there is a real problem, the action will not be performed, and a dialog box will notify you of the reason. Some of the problems associated with creating subVIs are outlined in the on-line help under the "Cycles" topic. A cycle is data initiating from a subVI output and being fed back to its input. Attribute nodes within loops, illogical selections, local variables inside loops, front panel terminals within loops, and case structures containing attribute nodes, local variables, or front panel terminals are causes of the cycles described in the help.

This tool should not be used to create all of your subVIs. It is a feature intended to save time when modifications to VIs are needed. National Instruments suggests that you use this tool with caution. Follow the rules and recommendations provided with on-line help when using this feature. Planning is needed before you embark on the writing of an application. There are several things that should be considered to get maximum benefit from code that you develop. Chapter 4, Application Structure, discusses the various tasks involved in a software project, including software design.

2.12   GRAPHICAL COMPARISON TOOLS

Keeping track of different versions of VIs is not always an easy task. Documenting changes and utilizing version control for files can help if the practices are adhered to strictly. When the size of an application grows, or if there is a team involved in the development, it becomes more difficult to follow the guidelines. In LabVIEW 5.0, graphical comparison tools were introduced to help manage the different ver-

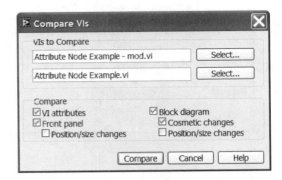

FIGURE 2.36

sions of VIs. These tools are available only with the Professional Development
System and Professional Developers Toolkit. You can compare VIs, compare VI
hierarchies, or compare files from the source code control tool.

2.12.1 COMPARE VIs

To compare two VIs, you must select Compare VIs from the Compare submenu in
the Tools pull-down menu. This tool graphically compares two VIs and compiles a
list of the differences between them. You then have the option of selecting one of
the items to have it highlighted for viewing. Figure 2.36 displays the Compare VIs
window as it appears.

Use the Select buttons to choose the two VIs for comparison. Only VIs that have
already been opened or loaded into memory can be selected as shown in Figure
2.37. The listbox displays VIs, Globals, and Type Definitions that are currently in

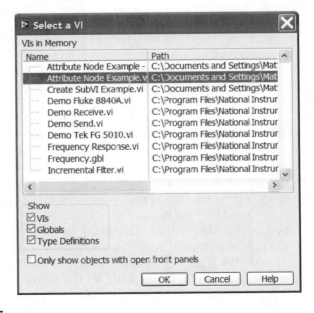

FIGURE 2.37

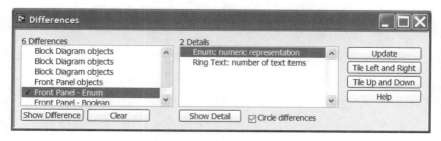

FIGURE 2.38

memory. You should keep in mind that LabVIEW does not allow you to open two VIs with identical names at the same time. If you want to compare two VIs that have identical names, you must rename one of them so they both can be loaded at the same time. Once renamed, both will appear in the listbox.

Once both VIs have been selected, simply press Compare to allow LabVIEW to begin compiling the list of differences. Figure 2.38 shows the window that lists the differences found during the comparison. The first box lists the difference and the second lists the details of the difference. There are two details associated with the difference selected in the figure shown. You can view the differences or the details by clicking the appropriate button. The comparison tool can tile the two VIs' windows and circle the differences graphically to make them convenient for viewing. A checkmark is placed next to the items that have already been viewed in both listboxes. The selection of differences in the list can sometimes be very long. Small differences in objects' locations, as well as cosmetic differences, will be listed even though they may not effect the execution of your VI. To graphically show the changes select Show Detail from the list of differences. Figure 2.39 shows one of the changes displayed.

2.12.2 COMPARE VI HIERARCHIES

You can compare two VI hierarchies by selecting Compare VI Hierarchies from the Compare submenu in the Tools pull-down menu. This is to be used for differentiating two versions of a top-level or main-level VI. Figure 2.40 displays the Compare VI Hierarchies window. Use the buttons next to file paths to select any VI for comparison through the file dialog box. With this tool, two VIs with the same name can be selected for comparison of hierarchies, unlike the tool for comparing VI differences. LabVIEW takes the second VI selected, renames it, and places it in a temporary directory for comparison purposes. This saves you the trouble of having to rename the VI yourself when you want to use the tool to find differences.

When Compare Hierarchies is clicked, descriptions of the differences are displayed along with a list of all of the VIs. All of the descriptions provided are relative to the first VI selected. For example, if a VI is present in the first hierarchy and not in the second, the description will tell you that a VI has been added to the first hierarchy. If a VI is present in the second hierarchy and not in the first, the description will tell you that a VI has been deleted from the first hierarchy. Shared VIs are

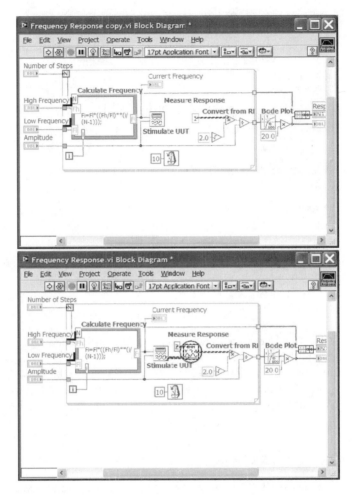

FIGURE 2.39

present in both hierarchies. A symbol guide is provided in the window to assist you while reviewing the list. An option is also available to allow you to view these differences graphically by having the two window displays tiled for convenience. This is similar to the highlighted differences shown when comparing two VIs. The variance is circled in red on both VI hierarchies for quick identification.

2.12.3 SCC COMPARE FILES

Comparing files through the Source Code Control (SCC) Tool is similar to the procedures for comparing VIs and VI hierarchies. The use of the third party Source Code Control applications is described further in Section 2.18. Select Show Differences from the Source Control submenu from the Tools pull-down menu or by right clicking on the VI in the Project Explorer and selecting Show Differences.

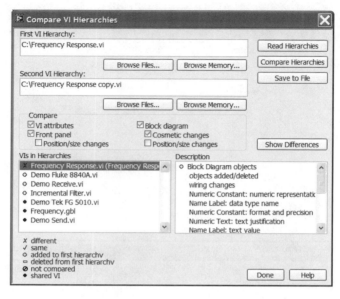

FIGURE 2.40

The SCC Compare Files tool allows you to select a project that you have already created from the pull-down menu. You also have the option of comparing files and having the differences shown by clicking on the respective buttons. Unlike the previous compare VIs function, as the VIs are in SCC, the VIs still have the same names. The SCC utilities keep track of which VI is which, making the function easier to use.

2.13   REPORT GENERATION PALETTE

The Report Generation VIs allow users to programmatically send text reports to a system printer. The Report Generation palette and the Report Layout subpalette are shown in Figure 2.41. These VIs are only available for Windows 2000 and XP because they only work on Win32 systems. The VIs use an ActiveX server, NI-Reports Version 1.1, to perform all of the functions provided in the palette. The NI-Reports ActiveX object is installed on your system at the same time you install LabVIEW.

Chapters 7 and 8 explain ActiveX, the related terminology, and show several examples to help you get started using this powerful programming tool. You can view the source code of the Report Generation VIs by opening them as you would any other VIs. These VIs use the built-in functions from LabVIEW's ActiveX subpalette. National Instruments developed this ActiveX object, as well as the VIs that use the server, to offer a simplified method for generating reports. Once you become familiar with using the ActiveX functions, you will be able to create your own report-generation VIs using the NI-Reports object, and utilize other ActiveX servers that are available.

FIGURE 2.41

The first VI available on the palette is Easy Text Report.vi. This is a high-level VI that performs all of the formatting as well as sending the report to the designated printer. Use this VI if you just want to send something to the printer without concern for detailed control or formatting of the report. Easy Text Report performs all of the actions you would normally have to execute if it were not available. It calls other VIs on this palette to perform these actions. All you have to provide is the relevant information and it is ready to go. Simply wire the printer name and the text you want printed. You can optionally set margins, orientation, font, and header and footer information with this VI.

Figure 2.42 illustrates a simple example on the use of Easy Text Report.vi. The VI shown is the same one shown in Figure 2.33, where test data is written to a file for storage. The VI has been modified slightly to send the same cluster information to the printer using the report generation feature. The printer name, desired text, and header and footer information is wired to Easy Text Report.

The other VIs on the palette let you dictate how the report is printed in more detail. In order to print a report, you must first use New Report.vi to create a new report. This VI opens an Automation Refnum to NI-Reports server object whose methods will be used to prepare and print the report. Once the report is created, you can use Set Report Font.vi or any of the other VIs on the Report Layout subpalette for formatting. Append Numeric Table to Report.vi lets you add numeric data,

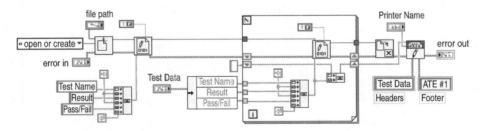

FIGURE 2.42

through a two-dimensional array, to your printout. You can then print your information with the Print Report VI. Finally, remember to release the resources being used for printing by executing Dispose Report.vi. You can open and view the source code for Easy Text Report.vi to get a better understanding of how to use the Report Generation VIs. This VI performs the sequence of actions described in using the NI-Reports ActiveX object.

2.14 APPLICATION BUILDER

The Application Builder allows you to create standalone applications, shared libraries and source distributions. The Application Builder is an add-on package that must be purchased separately, normally shipping only with the Professional Development System of LabVIEW. The tool is accessible through the project explorer. The application builder enables you to generate an executable version of your LabVIEW code, create an installer for your application and install any support files using one dialog box.

The first step to create a standalone application is to create a build specification. By selecting a file in your project explorer and selecting Build Application from the tools menu you will get the dialog box shown in Figure 2.43. In this window you have several sections that can be configured including Application Information, Destinations and Source Files. The Application Information category gives you the ability to set the build name, target name, destination directory, version information and miscellaneous other parameters. The Source Files category gives you the ability to select the VIs that will launch with the application as well as to include files that

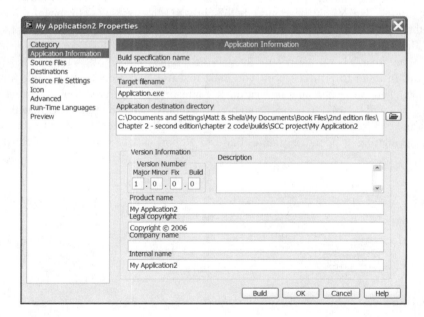

FIGURE 2.43

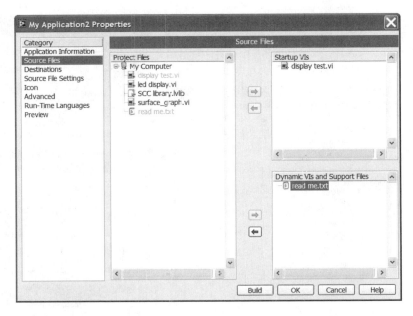

FIGURE 2.44

are not part of the application such as VIs that are dynamically loaded or documentation files. The builder will include all of the subVIs in the top level VI's hierarchy so that you do not have to create a library. An example of this is shown in Figure 2.44.

The Source File Settings category gives you the ability to select display and function options for each VI in the build. You can also set whether a VI is a startup VI, dynamic VI or Include only if referenced. In the Advanced category there are some additional options like the ability to enable debugging.

Once you are done updating all the settings in the dialog box you will see the new build specification listed in the project explorer. This can bee seen in Figure 2.45. To generate the executable you simply right-click on the build specification and select Build. The option is also available under the Project menu in the project explorer.

After you have created the executable, you can place it on any machine for use as a distributed program. When you choose to create an installer, LabVIEW's runtime library gets installed automatically on the target computer. This allows you to run executable LabVIEW programs on computers that do not have LabVIEW installed on them. LabVIEW's runtime library consists of the execution engine and support functions that are combined into a dynamic link library (DLL).

2.15 SOUND VIs

Sound VIs are found on the Sound subpalette of the Graphics and Sound palette. These built-in VIs permit you to easily incorporate sound and the manipulation of sound files into your applications. This section is only a brief overview of the sound VIs, and detailed information on their use will not be provided. The Sound palette and its associated subpalettes are displayed in Figure 2.46.

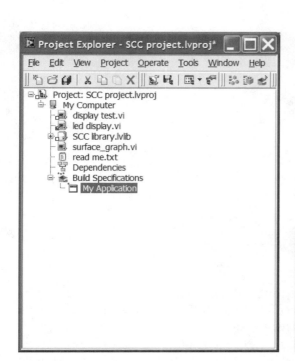

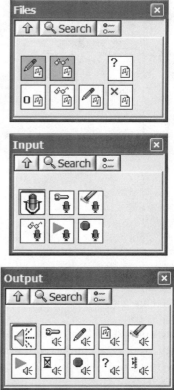

FIGURE 2.45   **FIGURE 2.46**

You can read and write waveform files to an output device using Snd Read or Write Waveform VIs. These are both general VIs that simplify the use of the sound VIs. You should ensure that your multimedia playback and recording audio devices and drivers are installed properly on your system before using the sound VIs.

For more control, use the VIs located on the Input, Output and Files subpalettes. The Sound Input VIs allow you to configure and capture data from an input device. The Sound Output VIs let you configure, send data, and control volume for your output device. The File VIs give you the ability to read and write sound files. Use Sound Input Configure or Sound Output Configure functions first when sending or retrieving data from the devices, and pass the Task ID when performing other functions. The Task ID serves as a reference to your device that is then used by the other VIs.

Two VIs that you may find useful in your applications are Beep.vi and Play Sound File.vi. The first VI simply makes a beep sound on Windows, Macintosh, and Unix. It can be used to send an audio signal to the user when an action has occurred. For example, you may want to have your application beep to let the user know that it needs attention when a test has completed its run or when an error has occurred. Play Sound File can be used in a similar manner. It is capable of playing

back any waveform audio file (*.wav). Just provide the file path of the audio file
you wish to play.

2.16 APPLICATION CONTROL

There are several LabVIEW functions that give the developer added control of the
way a LabVIEW program executes. In older versions of LabVIEW these functions
were grouped in one Palette (Application Control), but they are now located in
several palettes. These application-control functions will be discussed below.

2.16.1 VI SERVER VIs

The VI server functions are part of the Application Control Palette. The Application
Control Palette is shown in Figure 2.47. The VI server functions allow the Lab-
VIEW programmer control over a VI's properties, subVI calls, and applications.
Through the use of the VI server functions, you can set the properties of the VI
user interface through the code diagram. You can set whether to show the run
button, show scrollbars, set the size of the panel, and even the title that appears at
the top of the window. VI Server functions also allow you to dynamically load VIs
into memory, which can speed up your application by not having VIs in memory
unless they are being used. This is especially useful when you have seldom used
utility VIs that can be loaded when needed and then released from memory to free
up system resources.

Remote access is another capability made available through the VI server. A VI
can be opened and controlled on a remote computer with relative ease. Once the
connection is made to the VI, the interaction is the same as with a local VI using
the server VIs. A brief description of the VI server VIs is provided below, followed
by some examples.

Open VI Reference allows the user to obtain a reference to the VI specified in
the VI Path input. You also have the option of opening this VI on the local machine

FIGURE 2.47

or a remote machine. If you want to open the VI on a remote computer, the Open Application Reference VI needs to be used. The input to this VI is the machine name. The machine name is the TCP/IP address of the desired computer. The address can be in IP address notation or domain name notation. If the input is left blank, the VI will assume you want to make calls to LabVIEW on the current machine. The output of this VI is an application reference. This reference can then be wired to the Open VI Reference VI. Similarly, the application reference input to the Open VI Reference VI can be left unwired, forcing LabVIEW to assume you want the LabVIEW application on the current computer.

Once the application reference has been wired to the Open VI Reference, there are a few inputs that still need to be wired. First, a VI path needs to be wired to the VI. This is the path of the VI to which you want to open a session. This session is similar to an instrument session. This session can then be passed to the remainder of the application when performing other operations. If only the VI name is wired to this input, the VI will use relative addressing (the VI will look in the same directory as the calling VI). The next input will depend on whether you want to use the Call by Reference Node VI, which allows the programmer to pass data to and from a subVI. You can wire inputs and outputs of the specified VI through actual terminal inputs. The Call by Reference Node VI will show the connector pane of the chosen VI in its diagram. To be able to use this function, the Type Specifier VI Refnum needs to be wired in the Open VI Reference VI. To create the type specifier, you can right-click on the terminal and select Create Control. A type specifier refnum control appears on the front panel. If you go to the front panel, right-click on the control, and choose Select VI Server Class, there will be a number of options available. If you select Browse, you can select the VI for which you want to open a type specifier refnum. The Browse selection brings up a typical open VI window. When the desired VI has been selected, the connector pane of the VI appears in the refnum control as well as in the Call By Reference Node VI. There is also a Close Application or VI Reference VI. This VI is used to close any refnums created by the above VIs.

An example of these functions would be helpful to see how they fit together. We created a VI that has two digital control inputs and a digital indicator output. The code diagram simply adds the two controls and wires the result to the indicator. The VI in this example is appropriately named "Add.vi." In our VI reference example, we first open an application reference. Since we are using the local LabVIEW application, we do not need to wire a machine name. Actually, we do not need to use this VI, but if we want to be able to run this application on other computers in the future, this VI would already be capable of performing that task. The values of the two inputs are wired to the connector block of the Call by Reference Node VI. The Output terminal is wired to an Indicator. The code diagram of this VI is shown in Figure 2.48.

There is a second way to perform the same function as the above example. The example could be executed using the Invoke node instead of the Call by Reference Node VI. The Invoke node and the Property node are both in the Application Control Palette. These nodes can be used for complete control over a subVI. The Property node allows the developer to programmatically set the VI's settings such as the

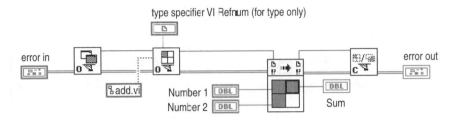

type specifier VI Refnum (for type only)

error in   type specifier VI Refnum (for type only)   error out

add.vi   Number 1 [DBL]
                Number 2 [DBL]   Sum [DBL]

FIGURE 2.48

execution options, history options, front panel window options, and toolbar options. The settings can be read from the VI or written to the VI. The Invoke node allows the developer to perform actions like set control values, read control values, run the VI, and print the VI. The previous example of calling the Add VI to add the two numbers was rewritten to use the Invoke node instead of the Call by Reference Node VI. Figure 2.49 shows the code diagram of this VI.

As can be seen from the example, more code is necessary to perform the task of executing a subVI. The true power in the use of the Invoke and Property nodes is the control the programmer has over the VI. By using these nodes, you can configure the VI in the required manner for the application. For example, you may want to use the same VI for a number of applications but need the front panel to be displayed in only one of the applications. The application that needs the front panel to be displayed can use the Property node to set that value to "true." This allows the code to be reused while still being flexible.

We will now provide an example of using the Property node to control a VI's property programmatically. This example VI will call a subVI front panel to allow the user to input data. The data will then be retrieved by the example and displayed. The desired subVI front panel that needs to be displayed during execution is shown in Figure 2.50. This VI will first open a connection to a subVI through the Open VI Reference function previously described. Once the reference to the VI is opened, we will set the subVI front panel properties through the Property node. By placing the Property node on the code diagram, and left-clicking on the property selection with the operator tool, a list of available properties appears. From this menu we will select our first property. First, we will select Title from the Front Panel Window section. This will allow us to change the title that appears on the title bar when the VI is visible.

The next step is to change additional properties. By using the position tool, you can increase the number of property selections available in the Property node. By dragging the bottom corner, you can resize this node to as many inputs or outputs as are needed. The properties will execute from top to bottom. For our user interface, we will set the visibility property of the ScrollBar, MenuBar, and ToolBar to "false." The final property to add is the front panel visibility property. We will set this property to "true" in order to display the front panel of the subVI.

Next, we will use the Invoke node to select the Run VI method. This will perform the same action as clicking on the Run button. The Invoke node only allows one method to be invoked per node. A second Invoke node is placed on the code diagram

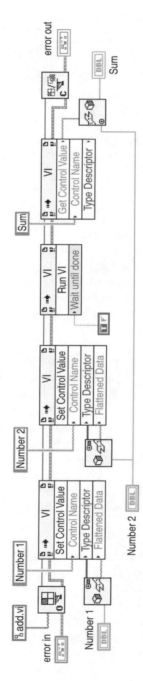

FIGURE 2.49

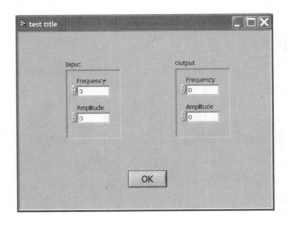

FIGURE 2.50

in order to obtain the subVI output. The name of the output control is wired into the Invoke node. The output of the node is wired to an Unflatten from String function, with the resulting data wired to the front panel. A Property node is then placed on the code diagram to set the subVI front panel visibility attribute to "false." The final step is to wire the VI reference to the Close function. The code diagram is shown in Figure 2.51.

2.16.2 MENU VIs

The Menu subpalette can be found in the Dialog & User Interface palette. The Menu VIs allow a programmer to create a menu structure for the application user to interact with. The menus are similar to the typical window menus at the top of the toolbar. The first VI is the current VI's menu. This function provides the current VI's menu refnum to be use by the other functions. The next VI is the Get Menu Selection VI. This VI returns a string that contains the menu item tag of the last selected menu item. The function has two primary inputs. The programmer can set the timeout value for the VI. The VI will read the menu structure for an input until the timeout value has been exceeded. There is also an option to block the menu after an item is read. This can be used to prevent the user from entering any input until the program has performed the action necessary to handle the previous entry. The Enable Menu Tracking VI can be used to re-enable menu tracking after the desired menu has been read.

There are two VIs in the palette for specifying the menu contents. The Insert Menu Items VI allows the programmer to add items to the menu using the item names or item tags inputs. There are a number of other options available and are described in the on-line help. The Delete Menu Items VI allows the programmer to remove items from the menu. There are also functions available for setting or getting menu item information. These VIs manipulate the menu item attributes. These attributes include item name, enabled, checked, and shortcut value. Finally, there is a VI that will retrieve the item tag and path from the shortcut value.

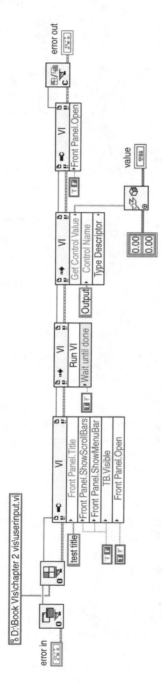

FIGURE 2.51

Before continuing with an example of using the menu VIs, we will discuss the Menu Editor. If you open the Edit menu from either the front panel or code diagram of a VI, there will be a selection for Run-Time Menu. You will need to make this selection to launch the menu editor. The first choice is a menu ring control that allows you to select the default menu structure, the minimal menu structure, or a custom menu structure. If you look at the remainder of the panel while the default menu structure is selected, you will see that there is a preview section showing the menu names, as they would appear on the front panel during execution. If you click on the menu, the submenus will be displayed. When you are creating your own menu, this preview section will allow you to see how the menu will appear during runtime. Below the preview area is a display area. This area is used to create the new menu. To the right of this display is a section to set the item properties.

To create your own menu, you must select Custom. This will clear the list of menus from the display. If you want to use some of the items from the default menu structure, you can select Item Type from the Property section. This will allow you to copy one item, the entire default structure, or something in between. The items that appear in the window can be deleted, added, moved, or indented. In addition, menu separators can be added to make the menu easier to read. Finally, in the Properties section there is an option to set up shortcut keys for your menu selection. These will work the same way as Ctl-C works for Copy in most menu structures.

We will now add a simple menu to a VI through the menu editor. This VI will contain two menus: a File menu and an Options menu. Under each menu heading there will be two submenus. They will be named "First" and "Second" for convenience. Once we open the menu editor and select Custom, we must select either Insert User Item from the menu or click the Plus button on the editor window. This will add a line in the display with question marks. You will now be able to add information to the Properties section. In the Item Name section we will type in "File." The underscore is to highlight a specific letter in the label for use with the ALT key. If the user presses ALT-F, the file menu will be selected. Notice that the name placed in the Item Name section is also written into the Item Tag section. These can be made different, if you choose.

The next step is to add another entry. After pressing the Add button, another group of question marks is placed below the File entry. You can make this entry a submenu by clicking on the right arrow button on the editor. Using the arrow keys will allow you to change an item's depth in the menu structure, as well as move items up and down the list. The menu editor display is shown in Figure 2.52. When you have completed the desired modifications, you must save the settings in an .mnu file. This file will be loaded with the VI to specify the menu settings.

The same menu structure can be created at run time by using the Insert Menu Items function. The first step is to create the menu types. This is done by wiring the insert menu items function with the menu names wired to the item tags input, and nothing wired to the menu tag input. These can be done one at a time, or all at once through an array. You can then call the insert menu items VI for each menu, inserting the appropriate submenu items as necessary. The different options can be modified through the Set Menu Item Info function. Figure 2.53 shows the code used to create the same menu as described above.

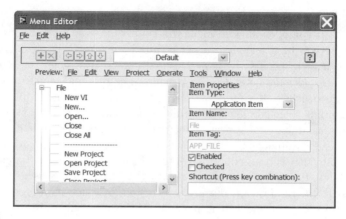

FIGURE 2.52

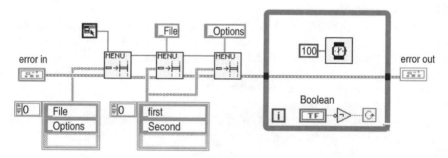

FIGURE 2.53

The following example will illustrate how to use the menu in an application. The VI menus created in the previous example will be modified to have one of the File items display Close. The example will require the use of a state machine. State Machines are covered in Chapter 3. The Get Menu Selection VI is inside the While loop. The output of the VI is wired to the selector of the case structure. The timeout value of the VI is set for 250ms. This is to prevent the VI from using all of the processor time. The Block Menu input is set to "true" to allow the state machine to process the inputs without additional inputs being selected. If the user does not select a menu, an empty string is returned from the VI. A case for the empty string will need to be created to do nothing. This will allow the state machine to continually poll the menus for input. A case matching the close input must be created (_File:_Close). This case will exit the state machine by wiring a "false" to the conditional terminal of the While loop. Cases for the options will need to be created to account for those inputs. The code that you want to run when the menu item is selected should be placed in cases corresponding to the menu items. After the case structure executes, the Enable Menu Tracking VI will need to be wired since menu tracking was disabled when the Get Menu Selection VI was executed. Finally, one of the cases will need to be made the default. A default case is always required when

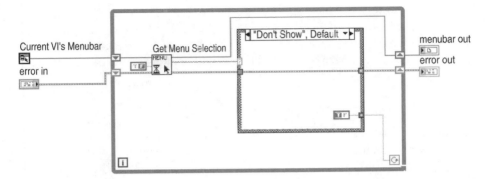

FIGURE 2.54

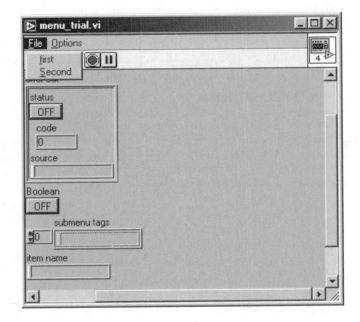

FIGURE 2.55

a case structure is driven by strings. The code diagram for this VI is shown in Figure 2.54. The front panel is shown in Figure 2.55.

2.16.3 HELP VIs

The Help VIs are located in the Dialog and User Interface palette. There are four VIs in the Help subpalette. The first VI, Control Help Window, allows the programmer to select the visibility of the help window as well as its location. The second VI, Get Help Window Status, reads the status of the help window (visibility and position). The third VI, Control Online Help, allows the programmer to display the online help window. The inputs allow the programmer to display the contents, the index, or jump to a specific section of the online help. The VI can also close the online help window. The final VI, Open URL in Default Browser, can be used to display Web-based help in the browser window. These VIs are especially useful when creating custom menus. The menu selections can include items to query the online help or display the help window.

2.16.4 OTHER APPLICATION CONTROL VIs

There a couple of additional functions in the application control palette that should be discussed. The first is the Quit LabVIEW VI. The VI will stop all executing VIs and close the LabVIEW application. It will not execute if a "false" is wired to the input. Next is the Stop VI. This VI stops execution of the current application. It performs the same function as the Stop button on the toolbar. Finally, there is the Call Chain VI. This VI returns an array of VI names, from the VI in which the function is executing to the top level VI in its hierarchy. This function is especially useful when performing error handling. It will allow you to locate a user-defined error quickly and perform the necessary actions to deal with it.

2.17 ADVANCED FUNCTIONS

There are several functions in LabVIEW that are intended for the advanced Lab-VIEW user. They give the user the ability to call external code modules, manipulate data and synchronize code execution. The functions were part of the Advanced palette in older versions of LabVIEW, but currently have several locations. These VIs and their uses will be discussed in the following sections.

2.17.1 DATA MANIPULATION

There are some functions that give the programmer the ability to manipulate data. These VIs are grouped together in a subpalette in the Numeric palette. The first function in the Data Manipulation subpalette is the Type Cast function. This function converts the data input to the data type specified. For instance, the Type Cast function can convert an unsigned word input to an enumerated type. The resulting output would be the enumerated value at the index of the unsigned word. Another function on this palette allows you to split a number into its mantissa and exponent.

There are a number of low-level data functions. There is both a Left and Right Rotate with Carry function, which rotate the bits of the input. There is a Logical Shift function to shift the specified number of bits in one direction depending on the sign of the number of bits input. A Rotate function is available to wrap the bits to the other end during the data shift. This function also derives direction from the sign of the number of bits input. Moving on to a slightly higher level, there are functions to split a 16- or 32-bit number and to join an 8- or 16-bit number. Finally, there are functions to swap bytes or words of the designated inputs.

The final functions in this palette are the Flatten to String and Unflatten from String functions. The Flatten to String function converts the input to a binary string. The input can be any data type. In addition to the binary string, there is a type string output to help reconstruct the original data at a later time. The Unflatten from String function converts the binary string to the data type specified at the input. If the conversion is not successful, there is a Boolean output that will indicate the error. Appendix A of the G Programming Reference Manual discusses the data storage formats. This information can also be found in your on-line help.

2.17.2 CALLING EXTERNAL CODE

LabVIEW has the ability to execute code written in C, as well as execute functions saved in a DLL. There are two methods for calling outside code. The programmer can call code written in a text-based language like C using a Code Interface Node (CIN). The programmer also has the ability to call a function in a DLL or shared library through the use of the Call Library function. The CIN and the Call Library functions reside in the Libraries and Executables subpalette. This subpalette is located in the Connectivity palette. Further descriptions of the CIN and Call Library functions are in Chapter 5.

2.17.3 SYNCHRONIZATION

The Synchronization palette contains five categories of VIs: semaphores, occurrences, notifications, rendezvous, and queues. The semaphore is also known as a "mutex." The purpose of a semaphore is to control access to shared resources. One example would be a section of code that was responsible for adding or removing items from an array of data. If this section of code is in the program in multiple locations, there needs to be a way to ensure that one section of the code is not adding data in front of data that is being removed. The use of semaphores would prevent this from being an issue. The semaphore is initially created. During the creation, the number of tasks is set. The default number of simultaneous accesses is one. The next step is to place the Acquire Semaphore VI before the code that you want to protect. The Boolean output of the VI can be used to drive a case structure. A "true" output will indicate that the semaphore was not acquired. This should result in the specified code not being executed. If the semaphore is acquired, the size of the semaphore is reduced, preventing another semaphore from executing until the Release Semaphore VI is executed. When the release is executed, the next available semaphore is cleared to execute.

The Occurrence function provides a means for making a section of code halt execution until a Set Occurrence function is executed. The Generate Occurrence function is used to create an occurrence reference. This reference is wired to the Wait on Occurrence function. The Wait function will pass a "false" Boolean out when the Occurrence has been generated. If a timeout value has been set for the Wait function, a "true" Boolean will be wired out at the end of the specified time if no occurrence has been generated. This is a nice feature when you need to perform polling. For example, you may want to display a screen update until a section of code completes execution. A While loop running in parallel with the test code can be created. Inside the While loop is the code required for updating the display. The Wait on Occurrence function is wired to the conditional terminal. The timeout is set

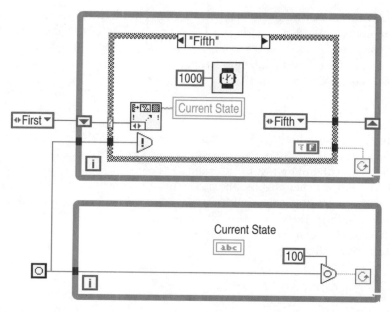

FIGURE 2.56

for the amount of time desired between While loop iterations. If the occurrence is not generated, a "true" is passed to the conditional terminal, and the While loop executes again. If the occurrence is generated, a "false" is sent to the conditional terminal, completing execution of the While loop. An example code diagram using occurrences is shown in Figure 2.56.

The Notification VI is similar to occurrences. Both functions wait for a message to be generated in order to execute. There are two main differences between occurrences and notifications. The Send Notification VI sends a text message as well as instructing the Wait on Notification VI to continue. This is true of one or multiple Wait on Notification VIs. The second difference is the ability to cancel a notification. In addition to these differences you also have the ability to obtain the status from the Get Notifier Status VI.

The Rendezvous VIs are used to synchronize multiple parallel tasks. When the rendezvous is created, the number of items or instances is entered. In order for code at the Wait at Rendezvous VI to execute, the specified number of Wait at Rendezvous VIs must be waiting. The programmer has the option of setting a timeout on the wait VI. There is also a function that can increase or decrease the number of rendezvous required for the code to continue. When all of the required Wait on Rendezvous VIs execute, the tasks all continue at the same time.

The final set of functions is the Queue functions. The queue is similar to a stack. Items are placed on the stack and removed from the stack. The first step in using the queue functions is to create a queue. The create queue function allows the programmer to set the size of the queue. If the maximum size is reached, no additional items will be able to be put into the queue until an item is removed. The Insert Queue Element and Remove Queue Element VIs perform as advertised. There is one option for both of these VIs that can be of use: the Insert Queue Element has the option of inserting the element in the front (default) or back. The Remove Queue Element allows the data to be taken from the end (default) or beginning. The Queue functions are beneficial when creating a list of states to execute in a state machine. More information and examples of the queue used with state machines is included in the state machines chapter.

There is a lone function in the Synchronization Palette that does not belong in one of the subpalettes. This function is the First Call function. The function has no inputs and one output. The output is a Boolean value indicating if this is the first time that this section of code or subVI has executed. This can be a useful function when there are tasks that need to be performed the first time through a loop or the first time a call has been made to a subVI. Instead of having a shift register running through a loop to track if you have been to a certain case or not can be simplified with this function.

2.18 SOURCE CODE CONTROL

Source Code Control (SCC) is a means for managing projects. The use of SCC encourages file sharing, provide a means for centralized file storage, prevents multiple people from changing the same VI at the same time, and provides a way to track changes in software. The Professional Developers version of LabVIEW and the Full Development version of LabVIEW with the Professional G Developers Toolkit provide means for source code control.

In earlier versions of LabVIEW there was support for Microsoft Visual SourceSafe, Rational ClearCase for Unix and a built-in source code control application. In LabVIEW 8 you have numerous options for source code control implementation including PVCS (Serena) Version Manager, MKS Source Integrity, and CVS with the Push OK Windows client software. If you are using Linux or Mac, the only source code control software that will work in the LabVIEW environment is Perforce. The built-in source code control application is no longer available and will not be discussed here. Some general setup and usage examples will be discussed here, but will not be discussed in depth. Additional information for SourceSafe and Perforce should be obtained from the Professional G Developers Tools Reference Guide and the specific vendor user manuals.

2.18.1 CONFIGURATION

To start using source code control you first need to install the application software. Once installed you can configure LabVIEW to work with the third party application by going to the Tools menu and selecting Options and navigating to the Source Control section, or by selecting Source Control and selecting Configure Source Control. The dialog box that displays is shown in Figure 2.57. The Source Control Provider Name pull down menu initially comes up as < None >. If you have installed your SCC application properly you should see the name in the pull down menu when you click on it. You then have the ability to select where the database (Source Control Project) is located. The database can be installed either on a server or

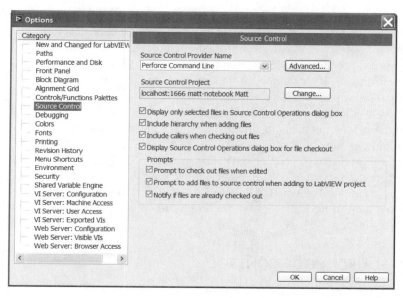

FIGURE 2.57

locally. Depending on which SCC application is installed, the available options may be different.

2.18.2 ADDING AND MODIFYING FILES

The first step to adding files to SCC is to create a VI or project. Once created the time can be added to source code control by either right clicking on the item in the Project Manager window and selecting Add to Source Control or by Selecting Source Control from the Tools Menu and choosing Add to Source Control. If you add a VI to a project and it is not in SCC you will get a dialog box asking if you want to add it to source control.

You can remove files from SCC by going to the Tools menu in the Project Manager and selecting Remove from Source Control in the Source Control submenu. You will get a warning message that the file will be removed from the SCC database and may be deleted from disk depending on the SCC setup.

To check out a file you can right click on the name in the Project manager window and selecting Check Out. To check the file back in or to undo the checkout you can right click on the file name in the Project Manager and select the appropriate item from the list. Figure 2.58 shows the Project Manager with files stored in source control. When you check in a file you have the ability to add in a comment for the history. Adding comments cannot be encouraged enough. This is the best way to make sure you can find out to what version you would need to rollback in the event major problems are found after several modifications are made. The history is a valuable tool that should be used every time.

Most of the time, when the files are stored on a separate server, you will want to get the latest version of your files in order to have the most up-to-date code before

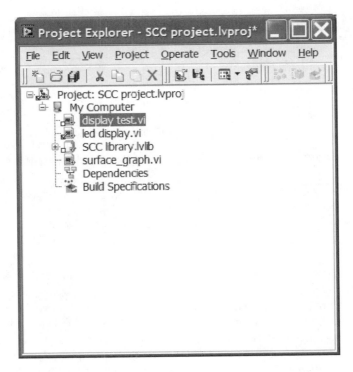

FIGURE 2.58

making modifications. This might also be useful if you have been changing a VI and then decide you want to start over. You will be replacing the modified VI with the one stored in the SCC application. To get the latest version you would go to the Source Code selection from the Tools menu and click on Get Latest Version. This will get the version of the code that the SCC application is presenting as the latest version. Depending on the settings in the SCC application one can set the latest version to a specific version label or VI version based on what code is desired. This is a topic for how to use the SCC application and will not be discussed further here.

2.18.3 ADVANCED FEATURES

There are a number of advanced features available including Show History and Show Differences. Often when modifying a VI or debugging your code you may notice something had been modified from what you remember. This is especially true in a multiple developer environment. By selecting Show History from the Source Control submenu you can get a list of changes for the VI or project. The list should include who saved the VI, when it was saved and any comments that were entered. This is where the history really pays off. You will be able to see what changes were made each time the VI was saved in order to see what had been modified and hopefully why. The LabVIEW function calls up the native file history in the SCC application. If you are running Perforce through LabVIEW the Show History selection will look like Figure 2.59.

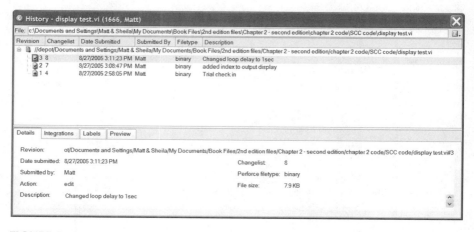

FIGURE 2.59

There are times when the history does not give you enough information on changes that have been made (or maybe even no comments at all). This is where the LabVIEW differencing tool comes in handy. In the past you would have to save the old version of the VI under a new name and then call the LabVIEW differencing tool. Now the process is integrated in LabVIEW so that you can just select Show Differences in the Source Control submenu or by right clicking on the VI in the project manager and the software will compare the versions without having to change names.

2.19   GRAPHS

There are a number of graphing controls available in the Functions palette on the front panel. In addition to these functions on the front panel, there are a number of VIs available in the Graphics and Sounds subpalette on the code diagram. The three main types of graphs will be discussed below.

2.19.1   STANDARD GRAPHS

The standard graphs include the waveform graph and chart, the intensity graph and chart, and the XY graph. These graphs have been standard LabVIEW controls for a long time. The waveform graphs and charts are simply displays of 2-D data. A graph is created by collecting the data into an array and then wiring the resulting data to the graph. Each time the graph is updated, the graph is refreshed. The chart is a continual display of data. Each data item is written to the chart, with the previous data remaining. The XY graph is capable of displaying any arrangement of points. Each point is plotted individually, without any relation to time or distribution.

The intensity chart and graph are slightly different than the waveform counter-parts. The intensity chart was created to display 3-D data in a 2-D graph. This is achieved by using color over the standard display. The values of the X and Y parameters correspond to a Z parameter. This Z parameter is a number that represents the color to be displayed.

2.19.2 3-D Graphs

Three graphs were added to the Graph palette with LabVIEW 5.1. These graphs are the 3-D surface graph, the 3-D parametric graph, and the 3-D curve graph. These graphs utilize ActiveX technology, and therefore are only available on the Windows versions of the full development and professional development systems. The 3-D surface graph displays the surface plot of the Z axis data. The 1-D X and Y data are optional inputs that displace the surface plot with respect to the X and Y planes. The 3-D parametric graph is a surface graph with 2-D surface inputs for the X, Y, and Z planes. The 3-D curve graph is a line drawn from 1-D X, Y, and Z arrays.

When the 3-D graphs are placed on the front panel, the corresponding VI is placed on the code diagram with the plot refnum wired to the connector. Since these graphs are designed using ActiveX controls, the properties and methods can be modified through the code diagram using the Property and Invoke nodes. These nodes can be found in the ActiveX subpalette in the Communication palette. The nodes can also be found in the Application Control palette. Properties and Methods will be discussed in depth in the Active X chapters (7 and 8). Using the Property node, the programmer is able to set values like the background color, the lighting, and the projection style.

A VI that generates data for the 3-D Surface graph is presented as an example. The VI modifies a number of the attributes through the Property and Invoke nodes. The resulting front panel and code diagram are shown in Figure 2.60.

In addition to the graph VIs, there are additional 3-D graph utility VIs in the Graphics & Sound palette. There are VIs for setting the Basic, Action, Grid, and

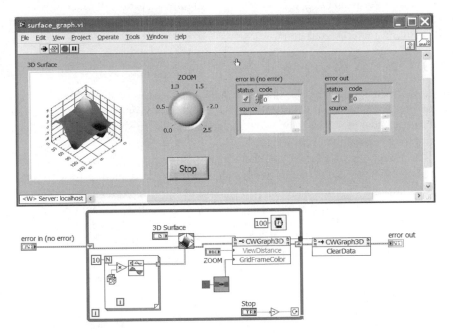

FIGURE 2.60

Projection properties. These VIs simply call the property node internally, allowing the user to simply input the desired properties without having to do as much wiring. These VIs are essentially drivers for the graph properties. There are two additional VIs in the palette. There is a VI to set the number of plots available on a given 3-D graph. And finally, the Convert OLE Colors VI is used to convert LabVIEW colors to OLE colors.

2.19.3 DIGITAL AND MIXED SIGNAL GRAPHS

In LabVIEW 7, National Instruments added support for Digital graphs. Someone taking data from a data acquisition card can now graph the outputs on a digital graph. With the digital graph the user can now combine timing and digital signal information in a waveform graph. There is also a palette of functions for building digital waveforms, generating a digital waveform and for manipulating the information in the waveform. In the digital waveform palette there is a subpalette containing functions for converting Boolean arrays and spreadsheet strings to or from digital waveforms. The digital waveform is simply a cluster of information relevant to a digital signal.

In LabVIEW 8, National Instruments introduced a Mixed Signal Graph. The mixed signal graph gives the user the ability to plot analog and digital waveforms on the same graph. Waveforms of any type and number can be bundled together and wired to a mixed signal graph. The graph will group the analog and digital waveforms and plot them together on the graph. All graphs on the mixed signal graph will share the same x-axis. Figure 2.61 shows an example of a mixed signal graph. The Generate Digital Waveform function was used to generate 3 waveforms. A function was also used to generate a sine wave. These waveforms are then bundled together to produce the resulting mixed signal graph.

2.19.4 PICTURE GRAPHS

In the Graph palette, there is a subpalette that contains additional graphing functions. The Picture subpalette contains a Polar plot, Smith plot, Min-Max plot, Distribution plot, and a Radar plot. This subpalette also contains a picture control and a subpalette of graphing related data types.

The Functions palette contains additional picture graph VIs in the Graphics & Sounds subpalette in the programming palette. There is a Picture Plot subpalette that contains the VIs corresponding to the above mentioned plots, as well as functions for building different waveforms and parameter inputs. There is a Picture Functions subpalette that provides a variety of functions such as draw rectangle, draw line, and draw text in rectangle. Finally, there is a subpalette that reads and writes data in the different graphics formats.

2.20 DATA LOGGING

One feature available for application debugging and data storage is front panel data logging. This feature stores the values of all front panel controls with a

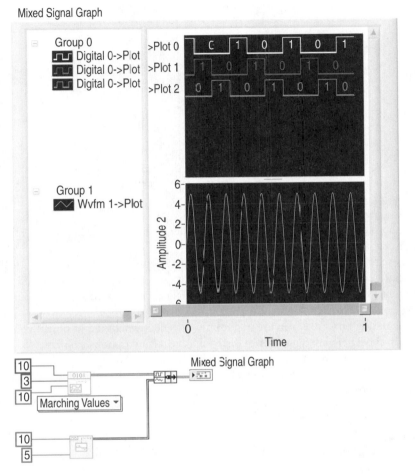

FIGURE 2.61

corresponding time and date stamp to a file. The data from this file can be retrieved and displayed. The VI from which the data is stored can retrieve and display the data through an interactive display. The data can also be retrieved interactively using the file I/O functions. This function is described in more detail in the Exception Handling chapter.

2.21 FIND AND REPLACE

There are times when developing an application that you want to find instances of an object. For example, when wiring a VI with Local variables, you want to make sure that there is no opportunity for a race condition to occur. In order to prevent any possible race conditions you will have to find all occurrences of the Local variable. By finding each instance of the Local, you can ensure there is no possibility

FIGURE 2.62

of another section of the code changing the value of the Local before the specific section has made its modifications.

There is a function in LabVIEW that gives the programmer the ability to find the location of a selected object. The Find and Replace function can be found in the Edit menu. By selecting Find and Replace, a dialog box appears on the screen. The dialog box is shown in Figure 2.62.

The dialog box gives you the opportunity to select what kind of object you want to search for. There are a number of different objects and options available. To select a specific type of object, you need to click on the Add button. A pull-down menu will appear giving you a list of objects to search for. A list of the object categories, and a brief description, is shown in Table 2.3.

TABLE 2.3
Find and Replace Function Objects

Menu Items	Description
Functions	This selection brings up the Functions palette. Any object available in the Functions palette can be selected for the search.
VIs	This option lets you select from any VIs in memory except VIs contained in the vi.lib.
Type Defs	This choice allows you to select any Type Definitions that are in memory and not contained in the vi.lib.
Globals	This selection allows you to choose any Global variables that are in memory and not contained in the vi.lib.
Objects in vi.lib	Choosing this option displays all VIs, Global variables, and Type Definitions located in vi.lib.
Others	This option allows you to select Attribute Nodes, Break Points, and Front Panel terminals. This search does not let you look for specific items, only for all instances of the given object.
VIs by name	This selection provides the user with a dialog box to input the desired VI to search for. This display contains a list of all VIs in memory in alphabetical order. You have the option of selecting whether to show any or all of the VIs, Globals, and Type Definitions.

After selecting the object you want to search for, you will need to specify the search parameters. There is a pull-down menu that allows you to select all VIs in memory, selected VIs, or the active VI. If you choose selected VIs, you can select any number of VIs currently in memory by clicking the Select button. You can also choose to search VIs in vi.lib and the hierarchy window.

At the top of the Find dialog box, there is a choice of what to search for. The default selection is Objects, but you can also search for text. If you select Text, the dialog box will show choices for what to search for, and the search scope. To choose what to search for, the Find What section allows you to type in the text string you are attempting to find. You can select whether to find an exact match including case (Match Case), match the whole word, or to match regular expression. By clicking on the More Options button, you are able to limit your search. You will have the option to select or deselect items including the diagram, history, labels, and descriptions.

Some objects also have the ability to launch their own search by right-clicking on the object and selecting Find. For example, if you select a Local variable and select Find from the pop-up menu, you will have the option of finding the indicator/control, the terminal, or the Local variables. If you select Local variables, a list of all instances of the Local, including the one you selected, will be displayed. You have the option to clear the display, go to the selected Local in the list, or launch the Find function. A checkmark will be displayed next to any Local variables you have already selected.

2.22 PRINT DOCUMENTATION

Many LabVIEW programmers print out VIs and documentation only when they need to create a manual-type document for their application, or need to view some code while away from their computers. There are a number of options when printing that can save time and provide you with documentation that you can use to create your own on-line references. By using the RTF functions, you have the ability to create help-file source code.

The first step to creating your own documentation is selecting Print Documentation from the File menu. A Print Documentation dialog box will pop up, prompting you to enter a format for your documentation. You have options from printing everything, just the panel, or custom prints that allow you to configure the options you want to select. Once you select what you want in your documentation, you then have the option of what style to print.

By selecting the Destination, you have the choice of printing to a printer, creating an HTML file, creating an RTF file, or creating a text file. If you select HTML file, you will be prompted for additional selections. These selections include image format and depth. Your image format options are PNG (lossless), JPEG (lossy), or GIF (uncompressed). In the Depth field you have a choice of black and white, 16 colors, 256 colors, or true color (24 bit). The RTF file options include the depth field and a checkbox for Help compiler source. The option for printing to a text file is the number of characters per line.

When you select Print, you need to be careful. If you are printing HTML documents, the print window will ask you for a name and location for the file. Each of the images will be saved in separate picture files. If you save a large VI in this manner, you could get a lot of files spread around your drive. You need to make sure to create a folder for your storage space to ensure that your files are in one place and are not difficult to find. The same holds true for RTF file generation.

2.23 VI HISTORY

The history function provides a means of documenting code revisions. This function is useful if no form of source code control is being used. This history will allow you to record what modifications have been done to a VI, and why the changes were made. Along with the text comment you enter, the date, time, user, and revision number are stored. Selecting VI Revision History from the Edit menu can access the VI History window. The window that appears is shown in Figure 2.63.

If you want to use the History function, you will first need to set up the VI's preferences. There are three methods for modifying a VI's history preferences. You can select History from the Preferences found in the Edit pull-down menu. The history preferences accessed here allow you to define when the history is saved. You can choose to have an entry created when the VI is saved, prompt for comments when the VI is closed, prompt for comments when the VI is saved, or allow LabVIEW to record its generated comments. Other options involve different login selections.

You can also select preferences from the VI Setup. In the documentation section of the VI Setup, there is an option to use the default settings from the history preferences. If this option is not selected, a number of options for creating the history become enabled. The final method for accessing history preferences is through the VI server. In the Property node there is a selection for History. A number of the

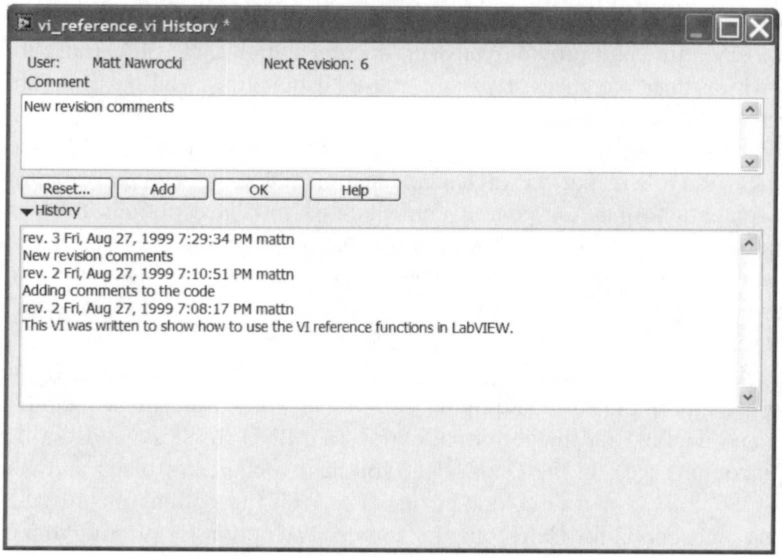

FIGURE 2.63

options available in the preferences are available. The use of the VI server allows you to set the preferences for any VI that you can open with the VI server. If you have a large number of VIs in which to change preferences, you could create a VI to set the preferences programmatically through the VI server.

If you want to add a history entry, you can enter your comments into the comment section of the history window. When all of your comments have been made, you can select Add in the history window. This will add the comments to the VIs history. Clicking on the Reset button will delete the history. The Reset option will also give you the option of resetting the revision number. You have the ability to print the history information when you select Print Complete Documentation or History from the Custom Print settings.

2.24   KEY NAVIGATION

The Key Navigation function provides a means to associate keyboard selections with front panel controls. An example of key navigation is when you have to enter information into a form; pressing the tab key cycles through the valid inputs. Another example of Key Navigation is the ability to press a function key to select a desired control. If you want the user to be able to stop execution by pressing the Escape key, the Stop Boolean can be associated with the Escape key. If the Escape key is pressed while the program is running, the Boolean control is selected, causing the value to change.

To open the Key Navigation dialog box, you need to right click on the control you want to configure, select Advanced and select Key Navigation. The dialog box that is opened is shown in Figure 2.64. The input section on the top left allows you

FIGURE 2.64

to define the key assignment for the control. You can combine specific keys like the Escape key, or a function key with the Shift or Control key to expand the possible setting combinations. The listbox on the right displays the currently assigned key combinations. Since you cannot assign the same key combinations to multiple controls, this window can be used to avoid previously defined assignments. If you do select a previously defined combination, the previous setting will be removed.

Below the windows is a selection for skipping the control while tabbing. If this checkbox is selected, the user will not be able to tab to this control. The changes will take effect when you click on the OK button. There are a few issues to keep in mind when using Key Navigation. If you assign the Return (Enter) key to a control, you will not be able to use the Return key to enter the data in any controls. The exception to this rule is if you are tabbing to the controls. If you tab to a control and press Return, the value of the control will be switched. A final consideration is the ability to use your application on different computers or platforms. Not all keyboards have all of the keys available for assignment. This will prevent the user from being able to use keyboard navigation. In addition, there are a few different key assignments among Windows, Sun, and Macintosh computers. If an application will be used on multiple platforms, try to use common keys to make the program cross-platform compatible.

2.25 EXPRESS VIs

An Express VI is similar to a subVI or function. The main difference is with express VIs you have the ability to configure the operation through a dialog box instead of through wiring. This can minimize the wiring needed on your code diagram, but will obscure settings making the code more difficult to debug. The inputs and outputs that are displayed on an Express VI node will change depending on how the Express VI is configured. The express VI will look like a box with input and output connections (National Instruments calls this format expandable nodes). The Express VI will always have a blue border as is shown in Figure 2.65. A subVI with expandable nodes will have a yellow border.

If you wanted to save a configured Express VI as a subVI for use in other applications you would right click on the Express VI and select Open Front Panel.

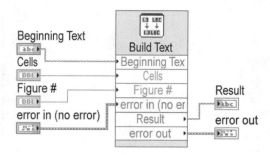

FIGURE 2.65

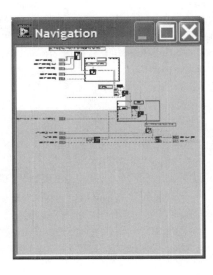

FIGURE 2.66

You will receive a warning that you will not be able to configure this VI as an Express VI anymore. Once you dismiss the message, the Express VI will be converted to a subVI. From here you can save the subVI and edit as needed.

2.26 NAVIGATION WINDOW

All of us at one time or another have had to work with code diagrams that would not fit on three monitors. Now for the sake of argument I will assume that we are talking about editing code written by someone else. It can be a laborious task to go from one end of the diagram to the other trying to find pieces of the code you are modifying. One LabVIEW tool that will make the work a little easier is the Navigation Window. This tool is located in the View pull down menu. When this tool is launched a window appears that allows the user to see a zoomed-out view of the entire code diagram. There is a white box that shows the portion of the code diagram that is visible in the current window. By dragging this white box around you can move the displayed portion of your window to the section of the code you want to view. The navigation window is shown in Figure 2.66.

2.27 SPLITTER BAR

Splitter Bars are a tool used for partitioning the front panel into sections. This tool is similar to splitting a worksheet in Microsoft Excel. The splitter bar can be used to make the diagram look more professional, make navigation easier, freeze portions of the panel from moving or even for creating a tool bar.

   To insert a splitter bar on your front panel you would select a vertical or horizontal splitter bar from the Containers palette. Once the splitter bar has been added you can right click on the bar and select from several settings. You can select whether there is a scrollbar (frozen window or unfrozen). You can select whether the items in the window scale with changes to the panel size and if the items in the panel are locked to a location in the panel. There are also a couple different styles of splitter bars to choose from.

BIBLIOGRAPHY

G Programming Reference, National Instruments, Austin, TX, 1999.
LabVIEW On-line Reference, National Instruments.
Professional G Developers Tools Reference Manual, National Instruments.
LabVIEW 5.1 Addendum, National Instruments.

3 State Machines

State machines, as a programming concept to provide an intelligent control structure, have gained a large degree of acceptance from the LabVIEW programming community. LabVIEW itself supports state machines in the form of template code. LabVIEW has also changed the types of inputs to a case structure to further support State Machines. This chapter will discuss state machine fundamentals and provide examples of how to implement different versions of state machine.

3.1 INTRODUCTION

State machines revolve around three concepts: the state, the event, and the action. No state machine operates effectively without all three components. This section will define all three terms and help you identify meaningful states, events, and actions in your programming. Major mistakes programmers make when working with state machines is not defining appropriate states. We will begin with the concept of state.

"State" is an abstract term, and programmers often misuse it. When naming a state, the word "waiting" should be applied to the name of the state. For example, a state may be waiting for acknowledgment. This name defines that the state machine is pending a response from an external object. States describe the status of a piece of programming and are subject to change over time. Choosing states wisely will make the development of the state machine easier, and the robustness of the resulting code much stronger. Relevant states allow for additional flexibility in the state machine because more states allow for additional actions to be taken when events occur.

Events are occurrences in time that have significant meaning to the piece of code controlled by the state machine. An event that is of significance for our previous example is the event "Acknowledgment Received." This external occurrence will inform the state machine that the correct event has occurred and a transition from states is now appropriate. Events can be generated internally by code controlled by the state machine.

Actions are responses to events, which may or may not impact external code to the state machine. The state machine determines which actions need to be taken when a given event occurs. This decision of what action needs to be taken is derived from two pieces of information: the current state and the event that has occurred. This pair of data is used to reference a matrix. The elements of this matrix contain the action to perform and the next state the machine should use. It is possible, and often desirable, for the next state to be equal to the current state. Examples in this section will demonstrate that it is desirable to have state changes occur only when

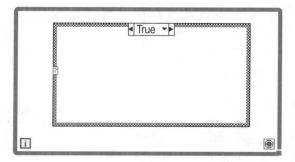

FIGURE 3.1

specific actions occur. This type of behavior is fairly typical in communications control. Unless a specific sequence of characters arrives, the state should not change, or perhaps the state machine should generate, an error.

The state machine itself always makes state changes. The current state is not normally given to code external to the state machine. Under no circumstances should external code be allowed to change the current state. The only information external code should give to the state machine is an event that has occurred. Changing state and dictating actions to perform is the responsibility of the state machine.

3.1.1 STATE MACHINES IN LABVIEW

A state machine, in simple terms, is a case structure inside a While loop, as shown in Figure 3.1. The While loop provides the ability to continuously execute until the conditional operator is set "false." The case statement allows for variations in the code to be run. The case that is selected to run can be, and usually is, determined in the previous iteration of the While loop. This allows for a relatively simple block of code to make decisions and perform elegant tasks. In its simplest form, a state machine can be a replacement for a sequence structure. In more complex forms of the state machine, the resulting structure could be used to perform the operations of a test executive. The topic of state machines is covered in a number of places, including the National Instruments LabVIEW training courses; however, there is not much depth to the discussions. This chapter will describe the types of state machines, the uses, the pitfalls, and numerous examples.

When used properly, the state machine can be one of the best tools available to a LabVIEW programmer. The decision to use a state machine, as well as which type to use, should be made at an early stage of application development. During the design or architecting phase, you can determine whether the use of a state machine is appropriate for the situation. Chapter 4 discusses how to approach application development including the various phases of a development life cycle.

3.1.2 WHEN TO USE A STATE MACHINE

There are a number of instances where a state machine can be used in LabVIEW programming. The ability to make a program respond intelligently to a stimulus is the most powerful aspect of using state machines. The program no longer needs to be linear. The program can begin execution in a specified order, then choose the next section of code to execute based on the inputs or results of the current execution. This can allow the program to perform error handling, user-selected testing, conditional-based execution, and many other variations. If the programmer does not always want the code to be executed in the same order for the same number of iterations, a state machine should be considered.

An example of when a state machine is useful to a programmer is in describing the response to a communications line. An automated test application that uses UDP to accept commands from another application should be controlled with a state machine. Likely states are "waiting for command," "processing command," and "generating report." The "waiting for command" state indicates that the application is idle until the remote control issues a command to take action. "Processing command" indicates that the application is actively working on a command that was issued by the remote application. "Generating report" notes that the state for command has completed processing and output is pending.

Events that may occur when the code is executing are "command received," "abort received," "error occurred," and "status requested." All of these possibilities have a different action that corresponds to their occurrence. Each time an event occurs, the state machine will take a corresponding action. State machines are predictable; the matrix of events, actions, and states is not subject to change.

The purpose of the state machine is to provide defined responses to all events that can occur. This mechanism for control is easily implemented, is scalable for additional events, and always provides the same response mechanism to events. Implementing code to respond to multiple events without the state machine control leads to piles of "spaghetti code" that typically neglect a few events. Events that are not covered tend to lead to software defects.

3.1.3 TYPES OF STATE MACHINES

There are a number of different styles of state machines. To this point there is no defined convention for naming the style of a state machine. In an effort to standardize the use of state machines for ease of discussion, we propose our own names for what we feel are the four most common forms of state machines in use. The four styles are the Sequence, the Test Executive, the Classical, and the Queued style state machines. Discussions of the four types of state machines, as well as examples of each type, follow in Sections 3.3 to 3.6.

3.2 ENUMERATED TYPES AND TYPE DEFINITIONS

In the introductory chapters on LabVIEW programming, we stated that an "enumerated type control" is similar to a text ring. The enumerated type control is basically a list of text values associated with a numeric value. The main difference for the enumerated control is that the string is considered part of the data type. The enumerated control can be found in the "List & Ring" section of the Control palette. The default data representation of the enumerated control is an unsigned word.

However, its representation can be changed to unsigned byte or unsigned long by popping up on the control and selecting Representation. When an enumerated type control is copied to the back panel, the result is an enumerated constant. Using the "numeric" section of the Function palette can also create an enumerated constant.

Enumerated constants can make state machines easier to navigate and control. When an enumerated type is connected to a case structure, the case indicator becomes the enumerated text instead of a numeric index. When a programmer is using a case structure with a number of cases, navigation of the case structure becomes difficult if the input is a numeric constant. When the user clicks on the case structure selector, a list of case numbers is shown. It is difficult for the user to determine which case does what, requiring the user to go through a number of cases to find the one that is desired. A benefit of using an enumerated constant with the case structure is readability. When someone clicks on the selector of a case structure controlled by an enumerated-type input, the lists of cases by name are shown. If the enumerated constant values are used to describe the action of the case, the user can easily find the desired case without searching through multiple cases. Similarly, you can use strings to drive state machine structures to enhance readability. However, string matching functions can be time consuming for performance critical applications. When enumerated or string constants are used with a state machine, there are a few added advantages. When the state machine passes the next state to execute to the case structure, the state to be executed becomes obvious. When a state branches to subsequent states, the user can see by the constant which state will execute next. If numerics are used, the person going through the code will have to go through the states to see what is next.

A second advantage involves the maintenance of existing code. When numeric inputs are used to control the state machine, a numeric constant will point to whichever state corresponds to the defined index. A better way to aid in modifications is the use of enumerated types. When the order of states is changed, the enumerated constants will still be pointing to the state with the matching name. This is important when you change the order of, add, or remove states. The enumerated constants will still point to the correct state. It should be noted that in the event a state is added, the state needs to be added to the enumerated constant in order to make the program executable; however, the order of the enumerated constant does not have to match the order of the case structure. This problem does not exist when using string constants to drive the case structure. This leads into the next topic, type definitions used with state machines.

3.2.1 TYPE DEFINITIONS USED WITH STATE MACHINES

A "type definition" is a special type of control. The control is loaded from a separate file. This separate file is the master copy of the control. The default values of the control are taken from this separate file. By using the type definition, the user can use the same control in multiple VIs. The type definition allows the user to modify the same control in multiple VIs from one location.

The benefit of using type definitions with state machines is the flexibility allowed in terms of modifications. When the user adds a case in the state machine, each enumerated type constant will need to have the name of the new case added to it. In a large state machine, this could be a very large and tedious task — not to mention fairly error prone if one enumerated type is not updated. If the enumerated constant is created from a type definition, the only place the enumerated type needs to be modified is in the control editor. Once the type definition is updated, the remaining enumerated constants are automatically updated. No matter how hard we all try, modifications are sometimes necessary; the use of type definitions can make the process easier.

3.2.2 Creating Enumerated Constants and Type Definitions

Selecting Enumerated Type from the List & Ring section of the Tools pallete creates the enumerated control on the front panel. The user can add items using the Edit Text tool. Additional items can be added by either selecting Add Item Before or After from the popup menu, or pressing Shift + Enter while editing the previous item. The enumerated constant on the code diagram can be created by copying the control to the code diagram via "copy and paste," or by dragging the control to the code diagram. Alternative methods of creating enumerated constants include choosing Create Constant while the control is selected, and selecting the enumerated constant from the Function pallete.

To create an enumerated-type definition, the user must first create the enumerated-type control on the front panel. The user can then edit the control by either double-clicking on the control or selecting Edit Control from the Edit menu. When the items have been added to the enumerated control, the controls should be saved as either a Type Definition or a Strict Type Definition. The strict type definition forces all attributes of the control, including size and color, to be identical. Once the type definition is created, any enumerated constants created from this control are automatically updated when the control is modified, unless the user elects not to auto-update the control.

3.2.3 Converting between Enumerated Types and Strings

If you need to convert a string to an enumerated type, the Scan from String function can accomplish this for you. The string to convert is wired to the string input. The enumerated type constant or control is wired to the default input. The output becomes the enumerated type constant. The use of this method requires the string to match the enumerated type constant exactly, except for case. The function is not case-sensitive. If the match is not found, the enumerated constant wired to the default input is the output of the function. There is no automatic way to check to see if a match was found. Converting enumerated types into strings and vice versa is helpful when application settings are being stored in initialization files. Also, application logging for defect tracking can be made easier when you are converting enumerated types into strings for output to your log files.

There are two methods that can be used to ensure only the desired enumerated type is recovered from the string input. One option is to convert the enumerated output back to a string and perform a compare with the original string. The method needed to convert an enumerated type to a string is discussed next. A second option is to use the Search 1-D Array function to match the string to an array of strings. Then you would use the index of the matched string to typecast the number to the enumerated type. This assumes that the array of strings exactly matches the order of the items in the enumerated type. The benefit of this method is that the Search 1-D Array function returns a –1 if no match is found.

If the programmer wants to convert the enumerated-type to a string value, there is a method to accomplish this task. The programmer can wire the enumerated type to the input of the format into string function in the Function palette. This will give the string value of the selected enumerated type.

3.2.4 DRAWBACKS TO USING TYPE DEFINITIONS AND ENUMERATED CONTROLS

The first problem was mentioned in Section 3.2.2. If the user does not use type definitions with the enumerated type constant, and the constant needs to be modified, each instance of the constant must be modified when used with a state machine. In a large state machine, there can be a large number of enumerated constants that will need to be modified. The result would be one of two situations when changes to the code have to be made: either the programmer will have to spend time modifying or replacing each enumerated control, or the programmer will abandon the changes. The programmer may decide the benefit of the changes or additions does not outweigh the effort and time necessary to make the modifications. This drawback limits the effectiveness of the state machine because one of the greatest benefits is ease of modification and flexibility.

The programmer needs to be careful when trying to typecast a number to the enumerated type. The data types need to match. One example is when an enumerated type is used with a sequence-style state machine. If the programmer is typecasting an index from a While or For loop to an enumerated type constant, either the index needs to be converted to an unsigned word integer, or the enumerated type to a long integer data type. The enumerated data type can be changed in two ways. The programmer can either select the representation by right-clicking on the enumerated constant and selecting the representation, or by selecting the proper conversion function from the Function palette.

If the programmer needs to increment an enumerated data type on the code diagram, special attention needs to be paid to the upper and lower bounds of the enumerated type. The enumerated values can wrap when reaching the boundaries. When using the increment function with an enumerated constant, if the current value is the last item, the result is the first value in the enumerated type. The reverse is also true; the decrement of the first value becomes the last value.

3.3 SEQUENCE-STYLE STATE MACHINE

The first style of state machine is the sequence style. This version of the state machine is, in essence, a sequence structure. This version of the state machine executes the states (cases) in order until a false value is wired to the conditional terminal. There

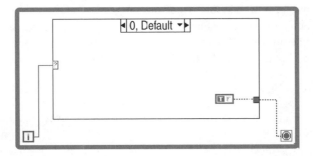

FIGURE 3.2

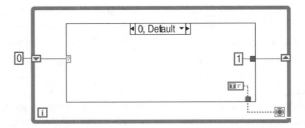

FIGURE 3.3

are a couple ways in which to implement this style of state machine. The first and simplest way is to wire the index of the While loop to the case statement selector. Inside each case, a Boolean constant of "true" is wired to the While loop conditional terminal. The final case passes a false Boolean to the conditional terminal to end execution of the loop. Figure 3.2 shows a sequence machine using the index of a While loop. This type of state machine is more or less "brain dead" because, regardless of event, the machine will simply go to the next state in the sequence.

A second way to implement the sequence-style state machine is to use a shift register to control the case structure. The shift register is initialized to the first case, and inside each case the number wired to the shift register is incremented by one. Again, the state machine will execute the states in order until a false Boolean is wired to the conditional terminal of the While loop. This implementation of a state machine is modified for all of the other versions of state machines described in this chapter. Figure 3.3 shows a simple sequence machine using a shift register.

3.3.1 WHEN TO USE A SEQUENCE-STYLE STATE MACHINE

This style of state machine should be used when the order of execution of the tasks to be performed is predefined, and it will always execute from beginning to end in order. This state machine is little more than a replacement for a sequence structure. We generally prefer to use this type of structure instead of sequences because it is far easier to read code that uses shift registers than sequence locals. The single biggest problem with the sequence structure is the readability problems caused by sequence locals. Most programmers can relate to this readability issue.

The biggest benefit is gained when the sequence-style state machine is implemented with enumerated-type constants. When enumerated types are used, the code becomes self-documenting (assuming descriptive state names are used). This allows someone to see the function of each action at a glance.

3.3.2 EXAMPLE

When writing test automation software there is often a need for configuring a system for the test. Generally, there are a number of setup procedures that need to be performed in a defined order. The code diagram in Figure 3.4 shows a test that performs the setup in the basic coding procedure. This version of the VI performs all of the steps in order on the code diagram. The code becomes difficult to read with all of the additional VIs cluttering the diagram. This type of application can be efficiently coded through the use of the sequence-style state machine.

There are a number of distinct steps shown in the block diagram in Figure 3.4. These steps can be used to create the states in the state machine. As a general rule, a state can be defined with a one-sentence action: set up power supply, set system time, write data to global/local variables, etc. In our example, the following states can be identified: open instrument communications, configure spectrum analyzer, configure signal generator, configure power supply, set display attributes and variables to default settings, and set RF switch settings.

Once the states have been identified, the enumerated control should be created. The enumerated control is selected from the List & Ring group of the Control pallete. Each of the above states should be put into the enumerated control. The label on the case statement will go as wide as the case statement structure.

There are two main factors to consider when creating state names. The first is readability. The name should be descriptive of the state to execute. This helps someone to see at a glance what the states do by selecting the Case Statement selector. The list of all of the states will be shown. The second factor to consider is diagram clutter or size. If enumerated constants are used to go to the next state, or are used for other purposes in the code, the size of the constant will show the entire state name. This can be quite an obstacle when trying to make the code diagram small and readable. In the end, compromises will need to be made based on the specific needs of the application.

After the enumerated control has been created, the state machine structure should be wired. A While loop should be selected from the Function palette and placed on the diagram with the desired "footprint." Next, a case structure should be placed inside the While loop. For our example we will be using the index to control the state machine. This will require typecasting the index to the enumerated type to make the Case Selector show the enumerated values. The typecast function can be found in the Data Manipulation section of the advanced portion of the Function pallete. The index value is wired to the left portion of the typecast function. The enumerated control is wired to the middle portion of the function. The output of the typecast is then wired to the case structure. To ensure no issues with data representations, either the representation of the enumerated control or the index should be changed. We prefer to change the index to make sure someone reading the code will

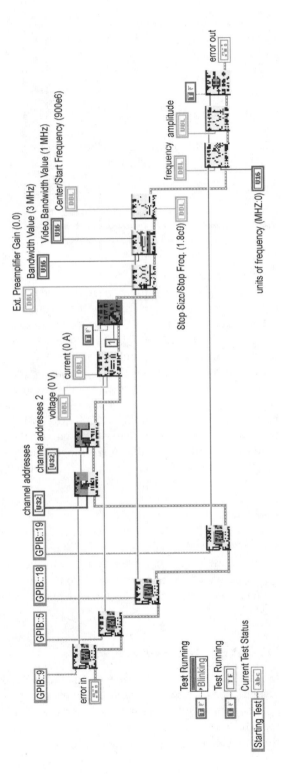

FIGURE 3.4

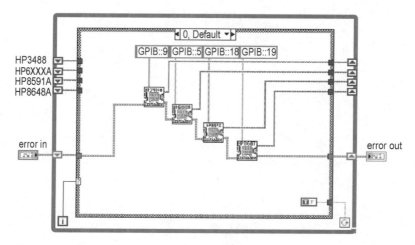

FIGURE 3.5

see what is being done. As the index is a long integer, it will need to be converted to an unsigned word to match the default representation of the enumerated control. The Conversion functions are part of the numeric section of the function pallete.

Now that the enumerated control has been wired to the case structure, the additional states can be added to match the number of states required. With the structure in place, the code required to perform each state should be placed into the appropriate case. Any data, such as instrument handles and the error cluster, can be passed between states using shift registers. The final and possibly most important step is to take care of the conditional terminal of the While loop. A Boolean constant can be placed in each state. The Boolean constant can then be wired to the conditional terminal. Because the While loop will exit only on a false input, the false constant can be placed in the last state to allow the state machine to exit. If you forget to wire the false Boolean to the conditional terminal, the default case of the case statement will execute until the application is exited.

At this point, the state machine is complete. The diagram in Figure 3.5 shows the resulting code. When compared to the previous diagram, some of the benefits of state machines become obvious. Additionally, if modifications or additional steps need to be added, the effort required is minimal. For example, to add an additional state, the item will have to be added to the enumerated control and to the case structure. That's it! As a bonus, all of the inputs available to the other states are now available to the new state.

3.4 TEST EXECUTIVE-STYLE STATE MACHINE

The test executive-style state machine adds flexibility to the sequence-style state machine. This state machine makes a decision based on inputs either fed into the machine from sections of code such as the user interface, or calculated in the state being executed to decide which state to execute next. This state machine uses an initialized shift register to provide an input to the case statement. Inside each case,

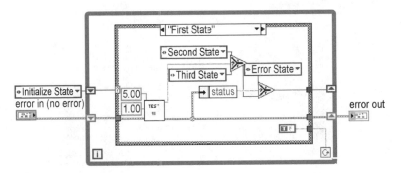

FIGURE 3.6

the next state to execute is decided on. An example of this state machine is shown in Figure 3.6.

3.4.1 THE LABVIEW TEMPLATE STANDARD STATE MACHINE

The template state machine provided by LabVIEW is a test executive styled machine. In order to add one to an application, select "New …" under the file menu. The state machine template is listed in the Design Patterns set as shown in Figure 3.7. The template state machine is shown in Figure 3.8. This stock state machine covers all

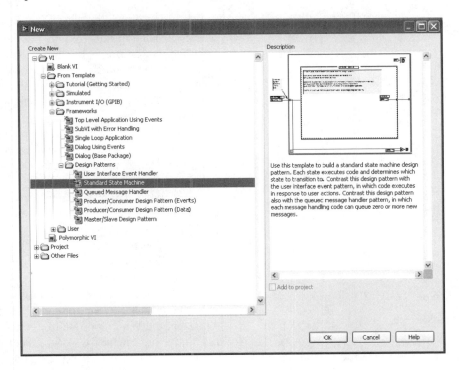

FIGURE 3.7

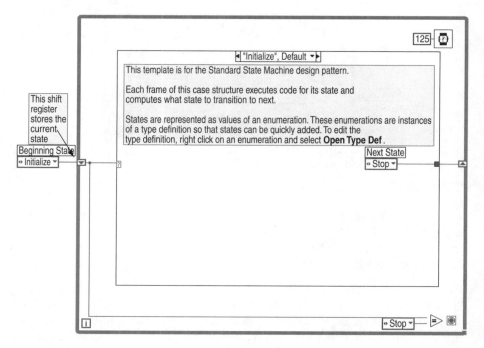

This shift
register
stores the
current
state

FIGURE 3.8

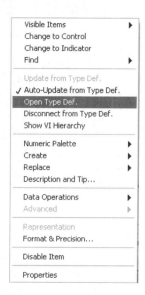

FIGURE 3.9

the basics but typically requires some modifications before being used. Key features of the template state machine are the use of enumerated types for state definitions, a delay, and defined initialize and stop states.

Right clicking on the enumerated type is shown in Figure 3.9. This shows we were in error; the template state machine isn't based on type definitions, not enumerated types. As stated above, editing the type definition for the state machine will cleanly propagate any changes in the state list through the state machine. The state machine's feedback register is also initialized.

The state machine compares the current state to "stop." If the state machine is in Stop, the While loop will be fed a true and exit, otherwise it will keep going. The next state is determined inside the case structure. It is also possible to eliminate this comparison and feed a constant Boolean true or false from inside the case statement. The template implementation is a bit easier to implement as opposed to adding and wiring a Boolean to each state. The only disadvantage is there is an additional comparison every iteration. The state is compared to stop, and the Boolean result is compared to true. In general, the performance is not going to be impacted by this.

The 125-ms wait located in the upper right hand corner of the state machine is there to provide some pacing. There are scenarios where delaying action between states is needed, such as processing user interface input. In general, we tend to delete this part of the template machine. If there is a need for particular states to have delays, put the delays inside the relevant states.

The template state machine implements all the features of a test executive styled machine that are needed.

3.4.2 WHEN TO USE A TEST EXECUTIVE-STYLE STATE MACHINE

There are a number of advantages to this style of state machine. The most important benefit is the ability to perform error handling. In each state, the next state to execute is determined in the current state. If actions were completed successfully, the state machine will determine what state to execute next. In the event that problems arise, the state machine can decide to branch to its exception-handling state. The next state to execute may be ambiguous; there is no reason for a state machine to execute one state at a time in a given order. If we wanted that type of operation, a sequence state machine or a sequence diagram could be used. A test executive state machine allows for the code to determine the next state to execute given data generated in the current state. For example, if a test running in the current state determined that the Device Under Test (DUT) marginally makes spec, then the state machine may determine that additional tests should be performed. If the DUT passes the specified test with a considerable margin, the state machine may conclude that additional testing is not necessary.

The user can make one of the cases perform dedicated exception handling. By unbundling the status portion of the error cluster, the program can select between going to the next state to execute or branching off to the Error State. The Error State should be a state dedicated to handling errors. This state can determine if the error is recoverable. If the error is recoverable, settings can be modified prior to sending the state machine back to the appropriate state to retry execution. If the error is not recoverable, the Error State, in conjunction with the Close State, can perform the cleanup tasks involved with ending the execution. These tasks can include writing data to files, closing instrument communications, restoring original settings, etc. Chapter 6 discusses the implementation of an exception handler in the context of a state machine.

3.4.3 RECOMMENDED STATES FOR A TEST EXECUTIVE-STYLE
STATE MACHINE

Test executive state machines should always have three states defined: Open, Close, and Error. The Open state allows for the machine to provide a consistent startup and initialization point. Initialization is usually necessary for local variables, instrument communications, and log files. The existence of the Open state allows the state machine to have a defined location to perform these initialization tasks.

A Close state is required for the opposite reason of that of the Open state. Close allows for an orderly shutdown of the state machine's resources. VISA, ActiveX, TCP, and file refnums should be closed off when the state machine stops using them so that the resources of the machine are not leaked away.

When this type of state machine is developed using a While loop, only one state should be able to wire a false value to the conditional terminal — in the case of the template state machine, only one state in the comparison should end execution of the state machine. The Close state's job is to provide the orderly shutdown of the structure, and should be the only state that can bring down the state machine's operation. This will guarantee that any activities that must be done to stop execution in an orderly way are performed.

The Error state allows for a defined exception-handling mechanism private to the state machine. This is one of the biggest advantages of the test executive style over "brain dead" sequence-style machines. At any point, the machine can conclude that an exception has occurred and branch execution to the exception handling state to record or resolve problems that have been encountered. A trick of the trade with this type of state machine is to have the shift register containing the next state use two elements. This allows for the Error state to identify the previous state and potentially return to that state if the exception can be resolved.

If the error condition is resolvable, the error state can set the error code to 0 and the error indication to false. Sometimes it is advantageous to put the error information into the error string so it can be fed out of the state machine. A good error handler state may make it impossible to tell an error has occurred from outside the state machine.

The Error state should not be capable of terminating execution of the state machine; this is the responsibility of the Close state. If your exception-handling code determines that execution needs to be halted, the Error state should branch the state machine to the Close state. If necessary, the error state can close off any resources related to the error so they do not cause an additional error in the close state. This will allow for the state machine to shut down any resources it can in an orderly manner before stopping execution.

3.4.4 DETERMINING STATES FOR TEST EXECUTIVE-STYLE STATE MACHINES

When working with a test executive machine, state names correlate to an action that the state machine will perform. Each name should be representative of a simple sentence that describes what the state will do. This is a guideline to maximize the flexibility of the state machine. Using complex or compound sentences to describe the activity to perform means that every time the state is executed, all actions must be performed. For example, a good state description is, "This state sets the voltage of the power supply." A short, simple sentence encapsulates what this state is going to do. The state is very reusable and can be called by other states to perform this activity. A state that is described with the sentence, "This state sets the power supply voltage and the signal generator's output level, and sends an email to the operator stating that we have done this activity," is not going to be productive. If another state determines that it needs to change the power supply voltage, it might just issue the command itself because it does not need the other tasks to be performed. Keeping state purposes short allows for each state to be reused by other states, and will minimize the amount of code that needs to be written.

3.4.5 EXAMPLE

This example of the state machine will perform the function of calculating a threshold value measurement. The program will apply an input to a device and measure the resulting output. The user wants to know what level of input is necessary to obtain an output in a defined range. Although this is a basic function, it shows the flexibility of the text executive-style state machine.

The first step should be performed before the mouse is even picked up. In order to code efficiently, a plan should already be in place for what needs to be done. A flowchart of the process should be created. This is especially true with coding state machines. A flowchart will help identify what states need to be created, as well as how the state machine will need to be wired to go to the appropriate states. A flowchart of the example is shown in Figure 3.10.

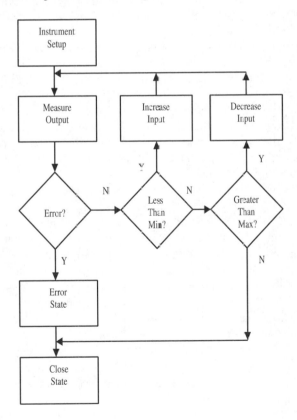

FIGURE 3.10

Once the test has been laid out, the skeleton of the state machine should be created. Again, the While loop and case statement need to be placed on the code diagram. An enumerated control will need to be created with the list of states to be executed. Based on the tasks identified in the flowchart, the following states are necessary: Instrument Setup, Measure Output, Compare to Threshold, Increase Input, Decrease Input, Error, and Close. A better approach is to combine the Increase and Decrease Input states into a Modify Input state that will change the input based on the measurement relationship to the desired output. However, this method makes a better example of state machine program flow and is used for demonstration purposes.

Once the enumerated control is created, an enumerated constant should be made. Right-clicking on the control and selecting create constant can do this. The Instrument Setup state should be selected from the enumerated list. This is the initial state to execute. The user needs to create a shift register on the While loop. The input of the shift register is the enumerated constant with the Instrument Setup state selected. The shift register should then be wired from the While loop boundary to the Case Statement selector. Inside each state an enumerated constant needs to be wired to the output of the shift register. This tells the state machine which state to execute next. Once the structure and inputs have been built, the code for each state can be implemented.

The Instrument Setup state is responsible for opening instrument communications, setting default values for front panel controls, and setting the initial state for the instruments. One way to implement the different tasks would be to either break these tasks into individual states or use a sequence-style state machine in the Initialize state. We prefer the second method. This prevents the main state machine from becoming too difficult to read. The user will know where to look to find what steps are being done at the beginning of the test. In addition, the Initialize state becomes easier to reuse by putting the components in one place.

After initializing the test, the program will measure the output of the device. The value of the measurement will be passed to the remainder of the application through a shift register. The program then goes to the next state to compare the measurement to a threshold value. Actually, a range should be used to prevent the program from trying to match a specific value with all of the significant digits. Not using a range can cause problems, especially when comparing an integer value to a real number. Due to the accuracy of the integer, an exact match cannot always be reached, which could cause a program to provide unexpected results or run endlessly.

Based on the comparison to the threshold value, the state machine will either branch to the Increase Input state, Decrease Input state, or the Close state (if a match is found). Depending on the application, the Increase or Decrease state can modify the input by a defined value, or by a value determined by how far away from the threshold the measurement is. The Increase and Decrease states branch back to the Measure Output state.

Although not mentioned previously, each state where errors can be encountered should check the status of the error Boolean. If an error has occurred, the state machine should branch to the Error state. What error handling is performed in this state is dependent on the application being performed. As a minimum, the Error state should branch to the Close state in order to close the instrument communications.

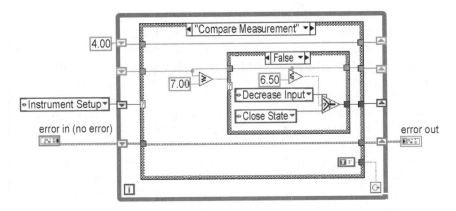

FIGURE 3.11

Finally, there should be a way to stop execution. You should never assume a program will complete properly. There should be a way for the program to "time out." In this example, the test will only execute up to 1000 iterations of the Measure Input state. One way to implement this requirement is to do a comparison of the While loop index. As the Initialize state is only executed once, the state is negligible. That leaves three states executing per measurement (Measure, Compare, and the Change). The Measure Output state can compare the loop index to 3000 to verify the number of times the application has executed. If the index reaches 3000, the program can either branch to the Close state directly or set an error in the error cluster. By using the bundling tools, the program can set the error Boolean to "true," set a user-defined code, and place a string into the description. The program can indicate that the test timed out or give some other descriptive error message to let the user know that the value was never found. Another way to implement this "time out" is to use shift registers. A shift register can be initialized to zero. Inside the Measurement state, the program can increment the value from the shift register. This value can be compared to the desired number of cycles to determine when the program should terminate execution. Figure 3.11 shows the completed state machine. The code is also included on the CD accompanying this book.

3.5 CLASSICAL-STYLE STATE MACHINE

The classical state machine is taught to computer programming students, and is the most generic of state machine styles. Programmers should use this type of state machine most frequently, and we do not see them often enough in LabVIEW code. The first step to using the classical state machine is to define the relevant states, events, and actions. Once the triad of elements is defined, their interactions can be specified. This concludes the design of the state machine, and coding may begin to implement the design. This section will conclude with a design of a state machine for use with an SMTP mail VI collection. This design will be used to implement a simple mail-sending utility for use with LabVIEW applications.

Step One is to define the states of the machine. States need to be relevant and in most scenarios should be defined with the word "waiting." Using "waiting" helps frame the states correctly. State machines are not proactive; they do not predict events about to happen. The word "waiting" in the state name appropriately describes the state's purpose.

Once the states for the machine are defined, then the events that are to be handled need to be defined. The events are typically driven by external elements to the state machine. The states that are defined for the state machine should not be considered when it comes time to determine what events are going to occur. It will be common for a state to handle one, maybe two events. Anytime an event other than the one or two it was designed to handle arrives it becomes an error condition.

3.5.1 WHEN TO USE A CLASSICAL-STYLE STATE MACHINE

Classical state machines are a good design decision when events that occur are coming from outside the application itself. User mouse-clicks, messages coming from a communications port, and .NET event handling are three examples. As these events may come into the application at any moment, it is necessary to have a dedicated control structure to process them. LabVIEW 5.0 introduced menu customization for LabVIEW applications. Classical style state machines used to be the best solution for user events. LabVIEW 7 introduced the event handling structure. The event structure would appear to easily outperform the classical state machine. The event structure itself is lacking in terms of an error state, but it is easy enough to add one. Add a hidden error cluster to the front panel, you can trigger an event by setting the error cluster's values. The event structure can have a frame dedicated to handling a set value in the error cluster.

As an example, if a user menu selection would make other menu selections not meaningful, the state machine, now an event structure, could be used to determine what menu items need to be disabled, if your file menu had the option for application logging and a selection to determine the level of logging. Typically, application logging is defined in "steps": one step might log everything the application does for debugging purposes, one level might only track state changes, and one will only log critical errors. When the user determines that he did not want the application to generate a log file, then setting the level of logging detail is no longer meaningful. The event structure handling the menu events would make the determination that logging detail is not useful and disable the item in the menu.

3.5.2 EXAMPLE

One of the requirements of this example is to read information from a serial port searching for either user inputs or information returned from another application or instrument. This example will receive commands from a user connected through a serial port on either the same PC or another PC. Based on the command read in from the serial port, the application will perform a specific task and return the appropriate data or message. This program could be a simulation for a piece of equipment connected through serial communications. The VI will return the expected inputs, allowing the user to test the code without the instrument being present. The user can also perform range checking by adjusting what data is returned when the program requests a measurement.

For this style of state machine the states are fairly obvious. There needs to be an Initialize state that takes care of the instrument communication and any additional setup required. The next state is the Input state. This state polls the serial port until a recognized command is read in. There needs to be at least one state to perform the application tasks for the matched input. When a command is matched, the state machine branches to the state developed to handle the task. If more than one state is necessary, the first state can branch to additional test states until the task is complete. When the task is completed, the state machine returns to the Input state. Finally, there needs to be an Error state and a Close state to perform those defined tasks.

The first step is to identify what commands need to be supported. If the purpose of the test is to do simulation work, only the commands that are going to be used need to be implemented. Additional commands can always be added when necessary. For our example, the VI will support the following commands: Identity (ID?), Measurement (Meas), Status (Status), Configure (Config), and Reset (RST). For our example, only one state per command will be created.

Once the commands are identified, the state machine can be created. As in the previous example, the input of the case statement is wired from an initialized shift register. Inside each state, the next state to execute is wired to the output of the shift register. This continues until a false Boolean is wired to the conditional terminal of the While loop.

The most important state in this style of state machine is the Input state. In our example, the list of commands is wired to a subVI. This subVI reads the serial port until a match is found in the list. When a match is found, the index of the matched command is wired out. This index is then wired to an Index Array function. The other input to this function is an array of enumerated type constants. The list is a matching list to the command list. The first state to execute for a given command should be in the array corresponding to the given command. A quick programming tip: when using this method, the index of the match should be increased by one. Then in the match array of enumerated constants, the first input should be the Error state. Because the match pattern function returns a −1 when no match is found, the index would point to the zero index of the array. This can allow the program to branch to the Error state if no match is found. Then, each command in order is just one place above the original array. The code for this state is shown in Figure 3.12.

The VI will continue to cycle through reading the serial port for commands and executing the selected states until the program is finished executing. There should be a way to stop the VI from the front panel to allow the VI to close the serial communications.

In this example, we are simulating an instrument for testing purposes. Using the Random Number Generator function and setting the upper and lower limits can use the measurement outputs to perform range checking. The state can be set up to output invalid data to check the error-handling capabilities of the code as well. This is a nice application for testing code without having the instrument available.

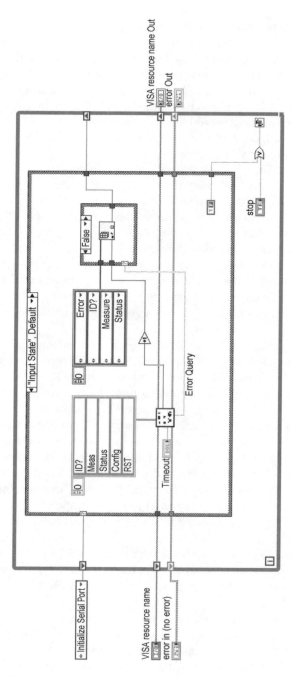

FIGURE 3.12

This next example will focus on developing a Simple Mail Transfer Protocol (SMTP) VI. Communications with a mail server are best handled through a state machine. The possibilities of errors and different responses from the server can make development of robust code very difficult. A state machine will provide the needed control mechanism so that responding to the various events that occur during a mail transfer conversation can be handled completely.

Before we begin the VI development, we need to get an understanding of how SMTP works. Typically, we learn that protocols containing the word "simple" are anything but simple. SMTP is not very difficult to work with, but we need to know the commands and responses that are going to present themselves. SMTP is defined in Request For Comments (RFC) 811, which is an Internet standard. Basically, each command we send will cause the server to generate a response. Responses from the server consist of a three-digit number and text response. We are most concerned with the first digit, which has a range from two to five.

The server responses that begin with the digit two are positive responses. Basically, we did something correctly, and the server is allowing us to continue. A response with a leading three indicates that we performed an accepted action, but the action is not completed.

Before we can design the state machine, we need to review the order in which communications should occur and design the states, events, and actions around the way things happen. When designing state machines of any kind, the simplest route to take is to thoroughly understand what is supposed to happen and design a set of states around the sequence of events. Exception handling is fairly easy to add once the correct combinations are understood.

Figure 3.13 shows the sequence of events we are expecting to happen. First, we are going to create a TCP connection to the server. The server should respond with a "220," indicating that we have successfully connected. Once we are connected, we are going to send the Mail From command. This command identifies which user is sending the mail. No password or authentication technique is used by SMTP; all you need is a valid user ID. Servers will respond with a 250 code indicating that the user is valid and allowed to send mail. Addressing the message comes next, and this is done with "RCPT TO: <email address>." Again, the server should respond with a 250 response code. To fill out the body of the message, the DATA command is issued which should elicit a 354 response from the server. The 354 command means that the server has accepted our command, but the command will not be completed until we send the <CRLF>.<CRLF> sequence. We are now free to send the body of the message, and the server will not send another response until we send the carriage return line feed combination. Once the <CRLF>.<CRLF> has been sent, the server will send another 250 response. At this point we are finished and can issue the QUIT command. Servers respond to QUIT with a 220 response and then disconnect the line. It is not absolutely necessary to send the QUIT command; we could just close the connection and the server would handle that just fine. (See Table 3.1.)

As we can see, our actions only happen when we receive a response from the server. The likely events we will receive from the server are 220, 250, and 354 responses for "everything is OK." Codes of 400 and 500 are error conditions and

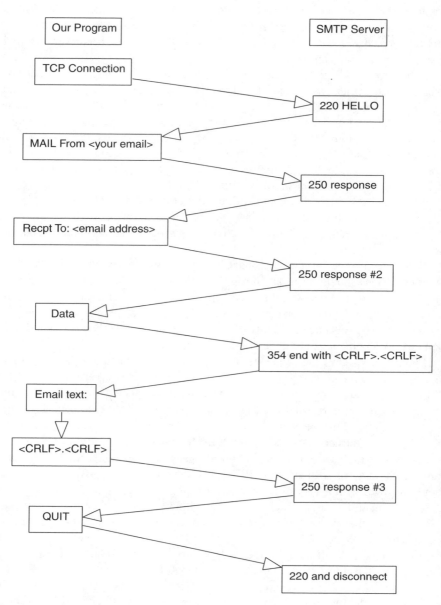

FIGURE 3.13

TABLE 3.1
Event Matrix

State/Event	200 Received	250 Received	354 Received	>400 Received	TCP Error
Waiting For Hello	Waiting For Address/ Send from	Waiting For Hello/ Do Nothing	Waiting For Hello/ Do Nothing	Waiting For Hello/ QUIT	Waiting For Hello/ QUIT
Waiting For Address	Waiting For Address/ Do Nothing	Waiting For Data/ Send Recpt	Waiting For Address/ Do Nothing	Waiting For Address/ QUIT	Waiting for Address/ QUIT
Waiting For Data	Waiting For Data/ Do Nothing	Waiting Send Body/ Send Data	Waiting For Data/ Do Nothing	Waiting for Data/ QUIT	Waiting For Data/ QUIT
Waiting To Send Body	Waiting To Send Body/ Do Nothing	Waiting To Send Body/ Do Nothing	Waiting To Quit/ Send Body	Waiting To Send Body/ QUIT	Waiting To Send Body/ QUIT
Waiting To Quit	Waiting To Quit/ Do Nothing	Waiting To Quit/ QUIT	Waiting To Quit/ QUIT	Waiting To Quit/ QUIT	Waiting To Quit/ QUIT.

we need to handle them differently. Several interactions with the server generate both 250 and 220 response codes, and a state machine will make handling them very easy. Our action taken from these events will be determined by our current state. The control code just became much easier to write.

Our event listing will be 220, 250, 354, >400, and TCP Error. These values will fit nicely into an enumerated type. Five events will make for a fairly simple state machine matrix. We will need states to handle all of the boxes in the right column of Figure 3.13. This will allow us to account for all the possible interactions between our application and the mail server.

Surprisingly, we will need states for only half of the boxes in the right column of Figure 3.13. When we receive a response code, the action we take will allow us to skip over the next box in the diagram as a state. We just go to a state where we are waiting for a response to the last action. The combination of Event Received and Current state will allow us to determine uniquely the next action we need to take. This lets us drive a simple case structure to handle the mail conversation, which is far easier to write than one long chain of SubVIs in which we will have to account for all the possible combinations. The table summarizes all of the states, events, and actions.

We have an action called "Do Nothing." This action literally means "take no action" and is used in scenarios that are not possible, or where there is no relevant action we need to perform. One of the state/event pairs, Waiting For Hello and 354 Received, has a Do Nothing response. This is not a possible response from the server. A response code in the 300 range means that our command was accepted, but we need to do something to complete the action. TCP connections do not require any secondary steps on our part, so this is not likely to happen. We will be using an array for storing the state/event pairs, and something needs to be put into this element of the array. Do Nothing prevents us from getting into trouble.

You can see from the table that there is a correct path through the state machine and, hopefully, we will follow the correct path each time we use the SMTP driver. This will not always be the case, and we have other responses to handle unexpected or undesirable responses. For the first row of the state table, TCP errors are assumed to mean that we cannot connect to the mail server, and we should promptly exit the state machine and SMTP driver. There is very little we can do to establish a connection that is not responding to our connection request. When we receive our 220 reply code from the connection request, we go to the Waiting for Address state and send the information on who is sending the e-mail.

The waiting for Address state has an error condition that will cause us to exit. If the Sending From information is invalid, we will not receive our 250 response code; instead, we will receive a code with a number exceeding 500. This would mean that the user name we supplied is not valid and we may not send mail. Again, there is little we can do from the SMTP driver to correct this problem. We need to exit and generate an error indicating that we could not send the mail.

Developing the state machine to handle the events and determine actions is actually very simple. All we need is an internal type to remember the current state, a case statement to perform the specific actions, and a loop to monitor TCP communications. As LabVIEW is going to remember which number was last input to the current state, we will need the ability to initialize the state machine every time we start up. Not initializing the state machine on startup could cause the state machine to think it is currently in the Wait to Quit state, which would not be suitable for most e-mail applications.

Figure 3.14 shows the state/action pair matrix we will be using. The matrix is a two-dimensional array of clusters. Each cluster contains two enumerated types titled "next state" and "action." When we receive an event, we reference the element in this matrix that corresponds to the event and the current state. This element contains the two needed pieces of information: what do we do and what is the next state of operation.

To use the matrix we will need to internally track the state of the machine. This will be done with an input on the front panel. The matrix and current state will not be wired to connectors, but will basically be used as local variables to the state machine. We do not want to allow users to randomly change the current state or the matrix that is used to drive the machine. Hiding the state from external code prevents programmers from cheating by altering the state variable. This is a defensive programming tactic and eliminates the possibility that someone will change the state at inappropriate times. Cheating is more likely to introduce defects into the code than to correct problems with the state machine. If there is an issue with the state machine, then the state machine should be corrected. Workarounds on state machines are bad programming practices. The real intention of a state machine is to enforce a strict set of rules of behavior on a code section.

Now that we have defined the matrix, we will write the rest of the VI supporting the matrix. Input for Current State will be put on the front panel in addition to a

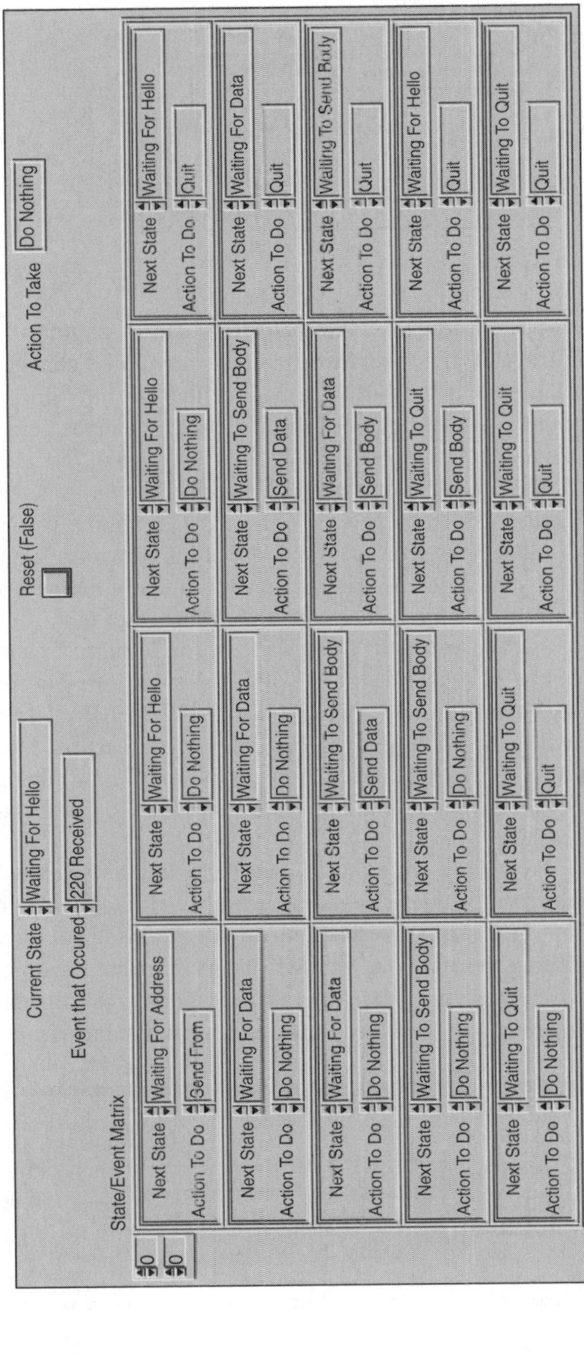

Current State | Waiting For Hello

Event that Occured | 220 Received

Reset (False)

Action To Take | Do Nothing

State/Event Matrix

	Next State	Action To Do	Next State	Action To Do	Next State	Action To Do	Next State	Action To Do
Next State: Waiting For Address	Action To Do: Send From	Next State: Waiting For Hello	Action To Do: Do Nothing	Next State: Waiting For Hello	Action To Do: Do Nothing	Next State: Waiting For Hello	Action To Do: Quit
Next State: Waiting For Data	Action To Do: Do Nothing	Next State: Waiting For Data	Action To Do: Do Nothing	Next State: Waiting To Send Body	Action To Do: Send Data	Next State: Waiting For Data	Action To Do: Quit
Next State: Waiting For Data	Action To Do: Do Nothing	Next State: Waiting To Send Body	Action To Do: Send Data	Next State: Waiting For Data	Action To Do: Send Body	Next State: Waiting To Send Body	Action To Do: Quit
Next State: Waiting To Send Body	Action To Do: Do Nothing	Next State: Waiting To Send Body	Action To Do: Do Nothing	Next State: Waiting To Quit	Action To Do: Send Body	Next State: Waiting For Hello	Action To Do: Quit
Next State: Waiting To Quit	Action To Do: Do Nothing	Next State: Waiting To Quit	Action To Do: Quit	Next State: Waiting To Quit	Action To Do: Send Body	Next State: Waiting To Quit	Action To Do: Quit

FIGURE 3.14

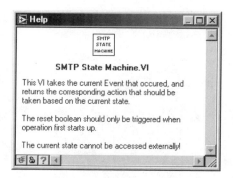

FIGURE 3.15

Boolean titled "Reset." The purpose of the reset Boolean is to inform the state machine that it is starting and the current internal state should be changed back to its default. The Boolean should not be used to reset the machine during normal operation, only at startup. The only output of the state machine is the action to take. There is no need for external agents to know what the new state of the machine will be, the current state of the machine, or the previous state. We will not give access to this information because it is not a good defensive programming practice. What the state machine looks like to external sections of code is shown in Figure 3.15.

The "innards" of the state machine are simple and shown in Figure 3.16. There is a case statement that is driven by the current value of the reset input. If this input is "false," we index the state/event matrix to get the action to perform and the new state for the machine. The new state is written into the local Variable for Current state, and the action is output to the external code. If the reset Boolean is "true," then we set the current state to Waiting for Hello and output an action, Do Nothing. The structure of this VI could not be much simpler; it would be difficult to write code to handle the SMTP conversation in a manner that would be as robust or easy to read as this state machine.

Now that we have the driving force of our SMTP sending VI written, it is time to begin writing the supporting code. The state machine itself is not responsible for parsing messages on the TCP link, or performing any of the actions it dictates. The code that is directly calling the state machine will be responsible for this; we have a slave/master relationship for this code. A division of labor is present; the SMTP VI performs all the interfaces to the server, and gets its commands from the state

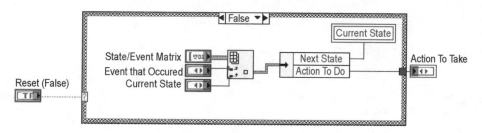

FIGURE 3.16

machine. This makes readability easier because we know exactly where to look for problems. If the SMTP VI did not behave correctly, we can validate that the state machine gave correct instructions. Assuming the state machine gave correct instructions, the problem is with the SMTP VI.

State machines work well for dealing with protocols such as SMTP. SMTP sends reply codes back, and the reply codes may be the same for different actions. The 220 reply code is used for both quitting and starting the mail conversation. If you were not using a state machine to determine what to do when you receive a 220 from the server, "tons" of temporary variables and "spaghetti code" would be needed instead. The matrix looks much easier to work with. Instead of following code and tracking variables, you look at the matrix to determine what the code should be doing.

3.6 QUEUED-STYLE STATE MACHINE

As the name suggests, the queued-style state machine works with an input queue. Prior to entering the state machine, a queue or input buffer is created. As the state machine executes, the state that has executed is removed from the queue during execution of the state machine. New states can be added to or removed from the queue based on what happens during execution. The execution of the queued-style state machine can complete by executing the close state when the queue is empty. We recommend always using a Close state as the last element of the queue. This will enable the program to take care of all communications, VISA sessions, and data handling. There is a way to combine these methods through the use of the Default state in the case statement.

There are two ways to implement the queue. The first method is using the LabVIEW queue functions. The Queue palette can be found in the Synchronization palette in the Advanced palette of the Function palette (are you lost yet?). [Functions>>Advanced>>Synchronization>>Queue]. The VIs contained in this palette allow you to create, destroy, add elements, remove elements, etc. For use with the state machine, the program could create a queue and add the list of elements (states to execute) prior to the state machine executing. Inside the While loop, the program could remove one element (state) and wire the state to the case selector of the case structure. If an error occurs, or there is a need to branch to another section of the state machine, the appropriate elements can be added to the queue. The addition can be either to the existing list, or the list could be flushed if it is desired to not continue with the existing list of states.

The use of the LabVIEW Queue function requires the programmer to either use text labels for the case structure, or to convert the string labels to corresponding numeric or enumerated constants. One alternative is to use an array of enumerated types instead of the Queue function (again, string arrays would work fine). The VI can place all of the states into an array. Each time the While loop executes, a state is removed from the array and executed. This method requires the programmer to remove the array element that has been executed and pass the remaining array through a shift register back to the beginning of the state machine, as shown in Figure 3.11.

3.6.1 WHEN TO USE THE QUEUED-STYLE STATE MACHINE

This style of state machine is very useful when a user interface is used to query the user for a list of states to execute consecutively. The user interface could ask the user to select tests from a list of tests to execute. Based on the selected items, the program can create the list of states (elements) to place in the queue. This queue can then be used to drive the program execution with no further intervention from the user. The execution flexibility of the application is greatly enhanced. If the user decides to perform one task 50 times and a second task once followed by a third task, the VI can take these inputs and create a list of states for the state machine to execute. The user will not have to wait until the first task is complete before selecting a second and third task to execute. The state machine will execute as long as there are states in the buffer. The options available to the user are only limited by the user interface.

3.6.2 EXAMPLE USING LabVIEW QUEUE FUNCTIONS

This first example will use the built-in LabVIEW Queue function. In this example, a user interface VI will prompt the user to select which tests need to be executed. The selected tests will then be built into a list of tests to execute, which will be added to the test queue. Once the test queue is built, the state machine will execute the next test to be performed. After each execution, the test that has been executed will be removed from the queue. This example is not for the faint of heart, but it shows you how to make your code more flexible and efficient.

The first step is creating the user interface. The example user interface here is a subVI that shows its front panel when called. The user is prompted to select which tests to execute. There are checkboxes for the user to select for each test. There are a number of other methods that work as well, such as using a multiple selection listbox. The queue can be built in the user interface VI, or the data can be passed to another VI that builds the queue. We prefer to build the queue in a separate VI in order to keep the tasks separated for future reuse. In this example, an array of clusters is built. The cluster has two components: a Boolean value indicating if the test was selected and an enumerated type constant representing the specific test. There is an array value for each of the options on the user interface.

The array is wired into the parsing VI that converts the clusters to queue entries. The array is wired into a For loop in order to go through each array item. There are two case statements inside the For loop. The first case statement is used to bypass the inner case statement if the test was not selected (a false value). The second case statement is a state machine used in the true case to build the queue. If a test is selected, the VI goes to the state machine and executes the state referenced by the enumerated type constant from the input. Inside the specific cases the appropriate state name (in string format) is added to the output array. In some instances multiple cases may be necessary to complete a given task. In these instances, the cases to execute are all added to the output array. This is why the string value of the enumerated type input is not simply added to the queue. Using the state machine allows a selected input to have different queue inputs. You would be tied to the name

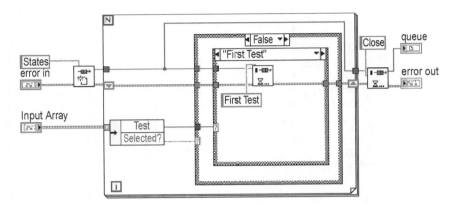

FIGURE 3.17

of the enumerated type if the Format into String function was used. When all of the array items have been sorted, a close state string is added to the end of the array to allow the main program to close the state machine.

The final stage of the VI is to build the queue with the inputs from the output string array. The first step is using the Create Queue function to create a named queue. The queue has a reference ID just like a VISA instrument. The ID is then passed into a For loop with an output array of strings. Inside the For loop, each string is put into the queue using the Insert Queue Element VI. When the VI completes execution, the reference ID is passed back to the main program. The queue-building VI is shown in Figure 3.17.

Now that the queue is built, the actual test needs to be created. The main VI should consist of a state machine. The main structure of the state machine should be a While loop with the case structure inside. Again, each case, except the Close state, should wire a "true" Boolean to the conditional terminal of the While loop. The only trick to this implementation is the control of the case statement. In the beginning of the While loop, the Remove Queue Element VI should be used to get the next state to execute. Once the state executes, the While loop will return to the beginning to take the next state from the queue. This will continue until the Close state is executed and the While loop is stopped. In the Close state, the programmer should use the Destroy Queue VI to close out the operation.

There is one final trick to this implementation: the wiring of the string input to the state machine. There are two ways to accomplish this task. The first is to create the case structure with the string names for each state. One of the states will need to be made the Default state in order for the VI to be executable. Because there are no defined inputs for a string, one of the cases is required to be "default." We would suggest making the default case an Error state as there should not be any undefined states in the state machine. If you do not want to use strings for the state machine, the second option is to convert the strings into enumerated-type constants. The method required to perform this action is described in Section 3.2.4. The enumerated constant can then be used to control the state machine. The main diagram is shown in Figure 3.18.

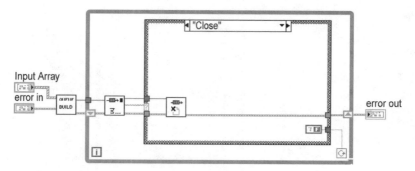

FIGURE 3.18

3.6.3 EXAMPLE USING AN INPUT ARRAY

A second version of the queued-style state machine involves using an array of states to execute instead of the LabVIEW Queue functions. We will use the same example application to illustrate the interchangeability of the methods. The application can use the same user interface. This time, instead of creating an array of strings based on the user inputs, the array of the enumerated types used in the user interface will be built. This array will then be passed to the main state machine. The programmer should make sure to add the Close State constant to the end of the array to prevent an endless loop. As a backup plan, the user should also make the Close state the default state. This will force the Close state to execute if the array is empty. The VI to build the state array is shown in Figure 3.19.

At the beginning of the While loop, the first state is taken off of the array of states by using the Index Array function. This value is then directly wired to the case structure input. The array is also passed to the end of the While loop. At the end of the While loop, the state that was executed is removed. Using the Array Subset function performs the removal. When the state array is wired to this function, with the index value being set to 1, the first element is removed from the array. This is continued until the Close state is executed, or until the array is empty. The diagram of the main VI is shown in Figure 3.20.

3.7 DRAWBACKS TO USING STATE MACHINES

There are very few drawbacks to state machines, and we will go through those instances here. The first issue we have found with state machines is the difficulty following program flow. Due to the nature of state machines, the order of execution can change due to many factors. The code becomes difficult to debug and trace errors. This is especially true with time-critical applications where execution high-lighting is not an option. Documentation is crucial for reading and debugging tests using state machines.

For applications where there are only a couple of tasks that are done sequentially, a state machine can be overkill. Creating an enumerated control for the case statement, setting up Error and Close states, and creating the necessary shift registers

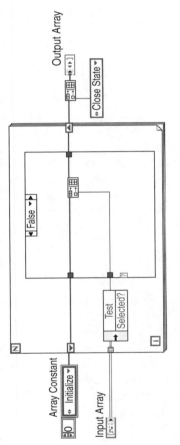

FIGURE 3.19

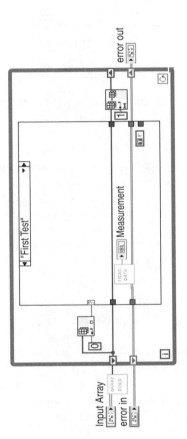

FIGURE 3.20

can be more work than is necessary. This is the case only in very simple sequences where there will not be major changes or additions. If there is a possibility of expanding the functionality of the VI, a state machine should be used. The benefits and issues of using a state machine should be considered during the architecting stage of an application.

3.8 RECOMMENDATIONS AND SUGGESTIONS

As is the case with most programming tasks, there are a number of ways to solve a problem. Although this is true, there are design patterns that can make life easier. This section will outline some of the design tools and methodologies that we have found to help implement state machines.

3.8.1 DOCUMENTATION

The programmer should always spend time documenting all code; however, this is especially true when using state machines. Because the order of the execution changes, thorough documentation can help when debugging. An additional reason to document is for when you attempt to reuse the code. If it has been a while since you wrote the VI, it may take some time to figure out how the code executes and why some inputs and outputs are there. Some of the code written in LabVIEW strives to abstract low-level interactions from the higher levels. Good documentation can help ensure that the programmer does not have to go through the low-level code to know what is required for the inputs and outputs. Chapter 4 also discusses some documenting methods available in LabVIEW.

3.8.2 ENSURE PROPER SETUP

As state machines can change the order of execution, special care should be taken to ensure all equipment is in the proper state, all necessary inputs have been wired, all necessary instruments are open, etc. You should try to make every state a stand-alone piece of code. If you are taking measurements from a spectrum analyzer, and the instrument needs to be on a certain screen, you must make sure to set the instrument to that screen. There is no guarantee that previous states have set the screen unless the order of execution is set. If there is a chance that a prior state will not execute, the necessary precautions must be taken to avoid relying on the prior state to perform the setup.

3.8.3 ERROR, OPEN, AND CLOSE STATES

When creating a state machine, there are three states that should always be created. There should be an Error state to handle any errors that occur in the program execution. If you are not using enumerated types or text labels for the states, you should make the Error state the first state. This way, when states are added or removed, the location of the Error state will always remain the same. An additional benefit to making the Error state the first state is when a Match Pattern function is used to select the state to execute. When no match is found a −1 is returned. If the returned value is incremented, the state machine will go to the Zero state. The Error state can be as simple as checking and modifying the error cluster and proceeding to the Close state or to an Error state that can remove certain states and try to recover remaining portions of the execution. The Close state should take care of closing instruments, writing data, and completing execution of the state machine. This is especially important when performing I/O operations. For example, if a serial port is not closed, the program will return an error until the open ports are taken care of. The Open State should handle instrument initialization and provide a single entry point to the state machine.

3.8.4 STATUS OF SHIFT REGISTERS

Most state machines will have a number of shift registers in order to pass data from one state to another, unless local or global variables are used. National Instruments suggests that local and global variables be used with caution. Depending on the purpose of the state machine, care needs to be taken with regard to the initial values of the shift registers. The first time the state machine runs, any uninitialized shift registers will be empty. The next time the state machine runs, the uninitialized shift registers will contain the value from the previous execution. There are times that this is desirable; however, this can lead to confusing errors that are difficult to track down when the register is expected to be empty. This method of not initializing the shift register is an alternative way to make a global variable. When the VI is called, the last value written to the shift register is the initial value recalled when it is loaded.

As a rule of thumb, global variables should generally be avoided. In state machine programming, it is important to make sure the machine is properly initialized at startup. Initializing shift registers is fairly easy to do, but more importantly, shift register values cannot be changed from other sections of the application. The biggest problem with global variables is their global scope. When working in team development environments, global variables should be more or less forbidden. As we mentioned earlier, a state machine's internal data should be strictly off-limits to other sections of the application. Allowing other sections of the application to have access to a state machine information can reduce its ability to make intelligent decisions. If the value of the shift register is not known at the time the state machine is started, it should be quantifiable during the open state — that's one of the reasons it's there.

3.8.5 TYPECASTING AN INDEX TO AN ENUMERATED TYPE

This was mentioned earlier, but this problem can make it difficult to track errors. When the index is being typecast into an enumerated type, make sure the data types match. When the case structure is referenced by integers, it can be much more difficult to identify which state is which. It is far easier for programmers to identify states with text descriptions than integer numbers. Use type definitions to simplify the task of tracking the names of states. Type definitions allow for programmers to modify the state listing during programming and have the changes occur globally on the state machine.

3.8.6 MAKE SURE YOU HAVE A WAY OUT

In order for the state machine to complete execution, there will need to be a "false" Boolean wired to the conditional terminal of the While loop. The programmer needs to make sure that there is a way for the state machine to exit. It is common to forget to wire out the false in the Close state which leads to strange results. If there is a way to get into an endless loop, it will usually happen. There should also be safeguards in place to ensure any While loops inside the state machine will complete execution. If there is a While loop waiting for a specific response, there should be a way to set a timeout for the While loop. This will ensure that the state machine can be completed in a graceful manner.

It is obvious that the state machine design should include a way to exit the machine, but there should only be one way out, through the Close state. Having any state able to exit the machine is a poor programming practice. Arbitrary exit points will probably introduce defects into the code because proper shutdown activities may not occur. Quality code takes time and effort to develop. Following strict rules such as allowing only one state to exit the machine helps programmers write quality code by enforcing discipline on code structure design.

3.9 PROBLEMS/EXAMPLES

This section gives a set of example applications that can be developed using state machines. The state machines are used to make intelligent decisions based on inputs from users, mathematical calculations, or other programming inputs.

3.9.1 THE BLACKJACK EXAMPLE

To give a fun and practical example of state machines, we will build a VI that simulates the game of Blackjack. Your mission, if you choose to accept it, is to design a VI that will show the dealer's and player's hands (for added challenge, only show the dealer's up card). Allow the player to take a card, stand, or split the cards (if they are a pair). Finally, show the result of the hand. Indicate if the dealer won, the player won, there was a push, or there was a blackjack. Obviously, with an example of this type, there are many possible solutions. We will work through the solution we used to implement this example. The code is included on the CD included with this book.

The first step to our solution was to plan out the application structure. After creating a flowchart of the process, the following states were identified: Initialize, Deal, User Choice, Hit, Split, Dealer Draw, and Result State. The Initialize state is where the totals are set to zero and the cards are shuffled. Additionally, the state sets the display visible attributes for the front panel split pair's controls to "false." The flowchart is shown in Figure 3.21.

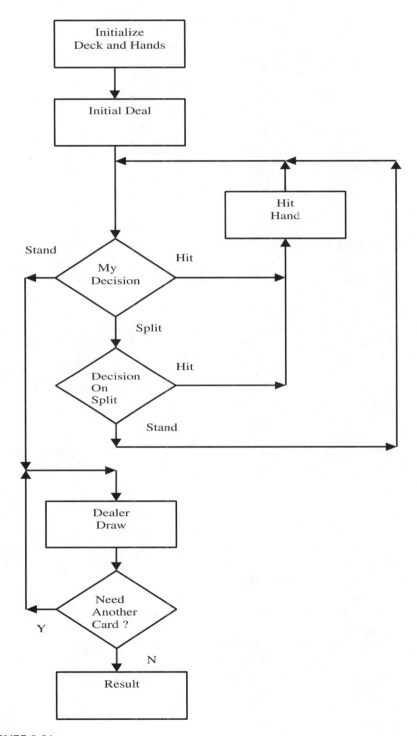

FIGURE 3.21

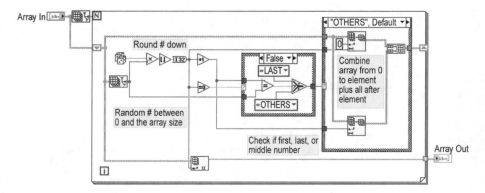

FIGURE 3.22

The shuffling was performed by the following method. A subVI takes an input array of strings (representations of the cards) and picks a card from the array at random to create a new array. The cards are randomly chosen until all of the cards are in the new array. The VI is shown in Figure 3.22.

The next state to define is the Deal Cards state. This state takes the deck of cards (the array passed through shift registers) and passes the deck to the Deal Card VI. This VI takes the first card off the deck and returns three values. The first is the string value of the card for front panel display. The second output is the card value. The final output is the deck of cards after the card that has been used is removed from the array. This state deals two cards to the dealer and to the player. The sum of the player's cards is displayed on the front panel. The dealer's up card value is sent to the front panel; however, the total is not displayed.

The User Choice state is where the player can make the decision to stand, hit, or split. The first step in this state is to evaluate if the user has busted (total over 21) or has blackjack. If the total is blackjack, or the total is over 21 without an ace, the program will go directly to the Result state. If the player has over 21 including an ace, 10 is deducted from the player's total to use the ace as a one. There is additional code to deal with a split hand if it is active.

The Split state has a few functions in it. The first thing the state does is make the split displays visible. The next function is to split the hand into two separate hands. The player can then play the split hand until a bust or stand. At this point, the hand reverts to the original hand.

The Hit state simply calls the Deal Card VI. The card dealt is added to the current total. The state will conclude by returning to the User Choice state. The Dealer Draw state is executed after the player stands on a total. The dealer will draw cards until the total is 17 or greater. The state concludes by going to the Result state. The Result state evaluates the player and dealer totals, assigning a string representing a win, loss, or tie (push). This state exits the state machine. The user must restart the VI to get a new shuffle and deal.

As can be seen by the code diagram of the VI shown in Figure 3.23, the design requirements have been met. There are a number of ways to implement this design; however, this is a "quick and dirty" example that meets the needs. The main lesson

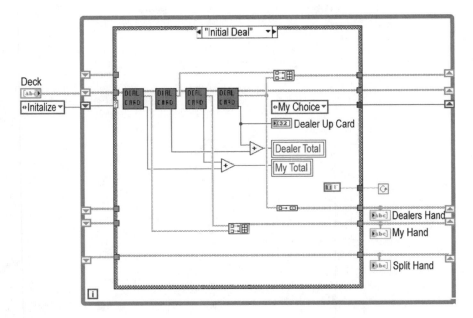

FIGURE 3.23

that should be learned is that by using a state machine, a fairly intricate application can be developed in a minimal amount of space. In addition, changes to the VI should be fairly easy due to the use of enumerated types and shift registers. The programmer has a lot of flexibility.

3.9.2 The Test Sequencer Example

For this example, there is a list of tests that have been created to perform evaluation on a unit under test. The user wants to be able to select any or all of the tests to run on the product. In addition, the user may want to run the tests multiple times to perform overnight or weekend testing. The goal of this example is to create a test sequencer to meet these requirements.

The first step is to identify the structure of the application that we need to create. For this problem, the queued state machine seems to be the best fit. This will allow a list of tests to be generated and run from an initial user interface. With a basic structure identified, we can create a flowchart to aid in the design of the state machine. The test application will first call a User Interface subVI to obtain the user-selected inputs. These inputs will then be converted into a list (array) of states to execute. For this example each test gets its own state. After each state executes, a decision will need to be made. After a test executes, the state machine will have to identify if an error has occurred. If an error was generated in the state that completed execution, the state machine should branch to an error state; otherwise, the state that executed should be removed from the list. In order to exit the testing, an Exit state will need to be placed at the end of the input list of states. In this Exit state, the code will need to identify if the user selected continuous operation. By "continuous

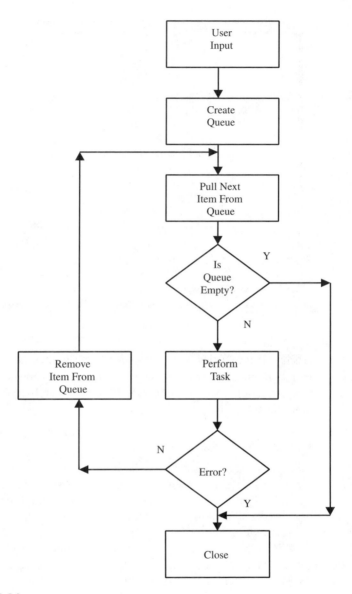

FIGURE 3.24

operation" we mean repeating the tests until a user stops. This option requires the ability to reset the list of states and a Stop button to allow the user to gracefully stop the test execution. The flowchart is shown in Figure 3.24.

The first step is to design the user interface. The user interface for this example will incorporate a multiple select listbox. This has a couple benefits. The first benefit is the ability to easily modify the list of tests available. The list of available tests can be passed to the listbox. The multiple select listbox allows the user to select as many or as few tests as necessary. Finally, the array of selected items in string form is available through the Attribute node. The list of tests can then be used to drive the state machine, or be converted to a list of enumerated constants corresponding to the state machine. In addition to the multiple select listbox, there will need to be a Boolean control on the user interface to allow the user to run the tests continuously, and a Boolean control to complete execution of the subVI. By passing the array of tests into the User Interface VI and passing the array of selected items out, this subVI will be reusable.

The next step is to build the state machine. The first action we usually take is to create the enumerated type definition control. This will allow us to add or remove items in the enumerated control in one location. The next decision that needs to be made is what to do in the event there is no match to an existing state. This could be a result of a state being removed from the state machine, or a mismatch between the string list of tests to execute and the Boolean names for the states. There should be a default case created to account for these situations. The default case could simply be a "pass-through" state, essentially a Do Nothing state. When dealing with strings, it is important to acknowledge that these types of situations can occur, and program accordingly. The code diagram of the Test Sequencer VI is shown in Figure 3.25.

Once the enumerated control is created, the state machine can be built. After performing an error check, the array of states is passed into a While loop through a shift register. The conditional terminal of the While loop is indirectly wired to a Boolean created on the front panel to stop the state machine. This will allow the program to gracefully stop after the current test completes execution. What we mean by "indirectly" is that the Boolean for the stop button is wired to an AND gate. The other input of the AND gate is a Boolean constant that is wired out of each state in the state machine. This allows the Close state or the Stop button to exit execution. One important item to note on the code diagram is the sequence structure that is around the Stop button. This was placed there to ensure the value of the button was not read until the completion of the current state. If the sequence structure was not used, the value of the Stop button would have been read before the completion of the given state. If the user wanted to stop the state machine, and the user pressed the button, the state machine would finish the current test and perform the next test. Only after reentering the state machine would the "false" be wired to the conditional terminal of the While loop.

Inside the While loop, the Index Array function is used to obtain the first state to execute by wiring a zero to the index input. The output of this function is wired to the case structure selector. This will now allow you to add the cases with the Boolean labels.

The Next_State subVI is the most important piece of code in the state machine. This subVI makes the decision of which state to execute next. The first step in the code diagram is to check the current state in the queue. This is the state that has just executed. This value is compared to the error state enumerated constant. If this is a match, the state machine proceeds to the Close state to exit execution. This is the method for this application to exit the state machine after an error if no error handling has been performed. After verifying that the Error state was not the last state to execute, the error cluster is checked for errors. Any errors found here would

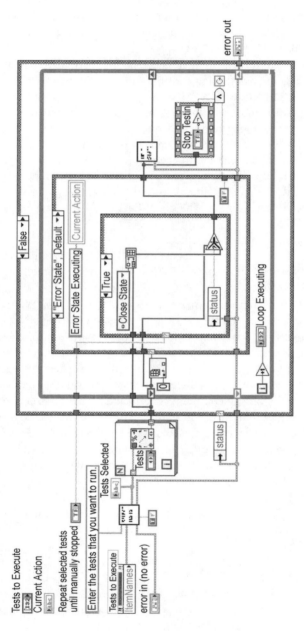

FIGURE 3.25

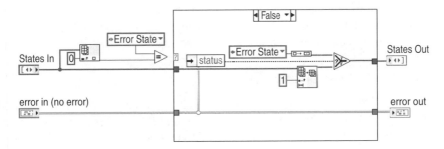

FIGURE 3.26

have been created during the test that last executed. If there is an error, the Error state enumerated constant is wired to the output array. This will allow any error handling to be performed instead of directly exiting the state machine. If no error has occurred, the Array Subset function will remove the top state. Wiring a one to the index of the function performs this action. If there are no more states to execute, an empty array is passed to the shift register. The next iteration of the state machine will force the error state (which was made the default state) to execute. The code diagram for the Next-State VI is shown in Figure 3.26.

The first state in this state machine is the Error state. The Error state in this example will perform a couple of functions. The Error state can have code used to perform testing, or clean up functions in the case of an error. This will allow the user to be able to recover from an error if the testing will still be valid. The second function is resetting of the states if continuous sequencing is selected. The first step is to make this case the default case. This will allow this case to execute if the input array is empty or doesn't match a state in the state machine. If an error occurred, the error cluster will cause the remainder of the state array to be passed to the Next State subVI. If no error occurred, the VI will wire the original array of states to a Build Array function. The other input of this function is an enumerated constant for any state in the state machine except the Error state.

You may be asking yourself why any state would be added to the new queue. The reasoning behind this addition was to allow the sequencer to start back at the beginning. The Error state is only entered when there is an error or when the queue is empty. Because the next state VI uses the Array Subset function to obtain the array of states to be wired to the shift register, the first state in the list is removed. The reason the Error state constant cannot be used is the first check in the Next_State subVI. If the Error state is on the top of the array, the subVI will think that an error has occurred and has been dealt with. The VI will then proceed to the Close state.

The remainder of the test sequencer is relatively straightforward. Each state passes the test queue from the input of the state to the output. The error cluster is used by the test VIs and is then wired to the output of the state. Finally, a "True" Boolean constant is wired to the output of each state. This is to allow a "False" Boolean to be wired out of the Close state. The other states have to be wired to close all of the tunnels. Additional functions can be added to the sequencer such as a front panel indicator to show what state is currently executing, an indicator to show the loop number being executed, and even results for each test displayed in an array on the front panel. The sequencer can be modified to meet the needs of the application. The test sequencer is a simple (relatively speaking) way to perform test executive functionality without a lot of overhead.

3.9.3 THE PC CALCULATOR EXAMPLE

The goal is to create a VI to perform as the four-function calculator that comes on most computer desktops. For this example, the higher-level functions will not be added. Only the add, subtract, multiply, and divide functions will be implemented. The idea is to use the classical-style state machine to provide the same functionality.

Again, the first step is to identify the form and function of the application. There needs to be a user interface designed to allow the user to input the appropriate information. For this example, an input section designed to look like the numeric keypad section of a keyboard is designed. In addition to the input section, there needs to be a string indicator to show the inputs and results of the operations. Finally, a Boolean control can be created to allow a graceful stop for the state machine. The state machine is controlled via the simulated numeric keypad.

Boolean controls will be used for the keys on our simulated keypad. The Boolean controls can be arranged in the keypad formation and enclosed in a cluster. The labels on the keys can be implemented by right-clicking on the control and selecting "Show Boolean Text." The text tool can then be used to change the Boolean text to the key labels. The "True" and "False" values should be changed to the same value. The text labels should be hidden to complete the display. The buttons should be "False" as the default case. Finally, the "Mechanical Action" of the buttons will need to be modified. This can be done by right clicking on the button and selecting the mechanical action selection. There is a possibility of six different types of mechanical actions. The default value for a Boolean control is "Switch when Pressed." The "Latch when Released" selection should be selected for each of the buttons. This will allow the button to return to the "False" state after the selection has been made. The front panel is shown in Figure 3.27.

After the cluster is created, the cluster order needs to be adjusted. Right-clicking on the border of the cluster and selecting "Cluster Order" can modify the cluster order. When this option is selected, a box is shown over each cluster item. The box is made up of two parts: The left side is the current place in the cluster order; the right side is the original order value. Initially, the values for each item are the same. The mouse pointer appears like a finger. By clicking the finger on a control, the value displayed on the top of the window frame is inserted into the left side of the cluster order box. The controls can be changed in order, or changing the value shown on the top window frame can change the value of each in any order. When you are finished modifying the cluster, the "OK" button needs to be pressed. If a mistake has been made or the changes need to be discarded, the "X" button will reset the values of the cluster order.

For our example, the numbers from one to nine will be given the cluster order of zero to eight, respectively. The zero is selected as the ninth input, and the period is the tenth input. The Divide, Add, Multiply, Subtract, and Equal keys are given the 11th to the 15th cluster inputs, respectively. Finally, the "Clear" key is given the

FIGURE 3.27

16th and final cluster position. The order of the buttons is not important as long as the programmer knows the order of the buttons, as the order is related to the position in the state machine.

The code diagram consists of a simple state machine. There is no code outside of the state machine except for the constants wired to the shift registers. Inside the While loop, the cluster of Boolean values from the control is wired to the Cluster to Array function. This function creates an array of Boolean values in the same order as the controls in the cluster. This is the reason the cluster order is important. The Search 1-D Array function is wired to the output of the Cluster to Array function. A "True" Boolean constant is wired to the element input of the search 1-D array function. This will search the array of Boolean values for the first "True" Boolean. This value indicates which key was pressed.

When the Search 1-D Array function is used, a no match results in a –1 being returned. We can use this ability to our advantage. If we increment the output of the Search 1-D Array function, the "no match" case becomes a zero. The output of the Increment function is wired to the case statement selector. In the zero case, when no match is found, the values in the shift registers can be passed through to the output without any other action being taken. This will result in the state machine continually monitoring the input cluster for a keypress, only performing an action when a button is pressed. The code diagram of the state machine is shown in Figure 3.28.

For this state machine, there are four shift registers. The first is used for the display on the front panel. The initial input is an empty string. The resulting value of the display string is sent to the display after the case structure executes. Inside the case structure, the inputs decide how to manipulate the string. There will be more discussion of this function after the remainder of the shift registers are discussed. The second shift register is a floating-point number used to hold the tem-

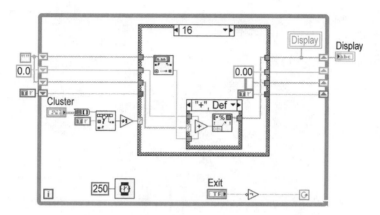

FIGURE 3.28

porary data for the calculations. When one of the operators is pressed, the value in
the display is converted to a number and wired to this shift register. At the beginning
of execution, after computing the function, or after a clear, the intermediate value
shift register is set to 0. When the user presses one of the operators, the third shift
register is used to hold the value of the selected operator. After the equal sign is
pressed, the operator shift register is cleared. The final shift register is used to hold
a Boolean constant. The purpose of this constant is to decide whether to append
new inputs to the existing display, or to start a fresh display. For example, when the
user inputs a number and presses the plus key, the current number remains in the
display until a new button is pushed. When the new button is pushed, the display
starts fresh.

The easiest way to make the discussion clearer is to describe the actions performed in the states. As stated earlier, the zero state does no action. This is the state
when nothing is pressed. States 1–11 are the inputs for the numbers and decimal
point. In these states there is a case statement driven by the value in the final shift
register (Boolean constant). If the value is "True," the value of the input is sent to
the display discarding any previous values in the display. If the value is "False," the
input key value is appended to the data already in the display. In each of these cases
a "False" is wired to the shift register because the only time the value needs to be
"True" is when the display needs to be cleared.

In states 12 through 15, the display string is converted to a floating-point number.
This number is wired to the temporary data shift register. The string value of the
display is also wired back to the display shift register. A "True" is wired to the
Boolean shift register to force the next input to clear the display. Finally, the value
of the operator selection is wired to the operator shift register in order to be used
when the Equal button is pressed. Speaking of the Equal button, this is the 16th
state. This state has a case structure inside. The case structure selector is wired to
the operator shift register. There are four cases, one for each of the operators. The
display string is converted to a floating-point number, and is wired into the case
structure. The previous input is taken from the shift register and is also wired to the
case structure. Inside each case, the appropriate function is performed on the inputs

with the result being converted to a string and wired to the display output. The temporary data shift register and the operator shift register are cleared. The final step in this case is to wire a "True" to the Boolean shift register to clear the display when a new input is selected. The final state is for the Clear button. This state clears all of the shift registers to perform a fresh start.

There are only two other components to this example: the Quit button that is wired to the conditional terminal of the While loop allowing the user to stop the application without using the LabVIEW Stop button, and a delay. The delay is needed to free-up processor time. The user would not be able to input values to the program if there was no delay because the state machine would run continuously. A delay of a quarter second is all that is necessary to ensure that the application does not starve out other processes from using the processor.

BIBLIOGRAPHY

LabVIEW Graphical Programming. Gary W. Johnson, McGraw-Hill, New York, 1997.
G Programming Reference, National Instruments, Austin, TX, 1999.
LabVIEW with Style — A Guide to Better LabVIEW Applications for Experienced LabVIEW
    Users. Gary W. Johnson and Meg F. Kay, Formatted for CDROM included with
    LabVIEW Graphical Programming, second ed., Austin, TX, January 12, 1997.

4 Application Structure

This chapter provides insight into developing well-structured applications, and will be particularly helpful for those applications that are relatively large. Several topics will be discussed that are important to the success of a software project. First, the various issues that must be considered before development can begin will be looked at. Then, the role of structure, or framework, of applications and its importance will be explained. The sections that follow will elaborate on software models, project administration, and the significance of documentation.

The three-tiered approach will then be presented as a framework for well-structured applications, stressing the importance of strict partitioning of levels. This topic will include the main, test, and driver levels of an application. Some of the features discussed in the book to this point have involved the LabVIEW Project. We will now take a look at the project and some of its features. The chapter will conclude with a summary example.

4.1 PLANNING

Complex architectures are not needed when the application being developed is simple. It is relatively easy to throw together a program in LabVIEW for performing specific functions on a small scale. But when the application becomes large in size, several design considerations should be taken into account before coding can begin. The following issues, among others, need to be considered: flexibility, extensibility, maintainability, code reuse, and readability.

Flexibility and extensibility impact the ability of an application to adapt to future needs. The ability to add functionality after the application has been released should be designed into the code. It is almost inevitable that requirements will change after the program is released. The architecture of large applications needs to be designed with the ability to make additions. For example, the end user may demand additional functionality to meet new requirements. If the application was not designed with the capacity to evolve, incremental enhancements can prove to be very difficult. The needs of the user evolve over time, and a well-designed application can easily adapt.

Maintainability of code is necessary for applications so that needed modifications can be made easily. The concept of allowing for change in functionality holds true for the ability to maintain and modify code easily. For example, if a power supply that is being used in the current test setup will not be used in another test rack, you may need to change to a different model. How easily your code can be modified to reflect this change in the test setup is material. The amount of work involved in the alteration depends on how the code is structured.

Code reuse is required for cycle-time reduction on future projects. This attribute is often overlooked because programmers focus on accomplishing the goal of the current project. The time it takes to complete future projects can be reduced if even small pieces of the code can be reused. When software is written in a way that it cannot be reused, efforts are duplicated unnecessarily. Both time and money can be saved when a project is developed with reuse as a design goal.

The ability of the software to provide abstraction is also significant because it improves code readability. Not everyone interacting with the program needs the same level of abstraction. Someone who will use the application, but knows nothing about programming, does not need or wish to see the low-level data manipulation of the program. Operators want an easy user interface that will allow them to use the application for their specific purpose. On the other hand, the person in charge of writing and maintaining the application needs to see all levels of the program. Abstraction allows the programmer to conceal subsections of the application from those who would not benefit from seeing it. Drivers abstract the I/O so it is easier to understand the test level. The test level abstracts the logic so the main level is easier to read.

The concepts presented in this chapter are a good starting point for beginning a project. "Plans are nothing; planning is everything," is a quote by Dwight D. Eisenhower that is applicable to software design. Without adequate planning, large applications are not likely to be successful. Planning provides a roadmap for development and helps minimize the occurrence of unexpected events. You can plan with contingencies depending on the results of the design stages.

Inadequate planning is more likely to result in problems. When designing an application, detailed knowledge of the system — instruments, software requirements, feature sets, etc. — plays a significant role in building a successful application.

4.2 PURPOSE OF STRUCTURE

The topics discussed on application structure may be applied to programming languages other than LabVIEW. Architecture and process are two elements that are important in all languages. The structure of the program or framework that is used is important for future additions, modifications, and maintenance. If the correct process is taken in designing the software system, the application can change as the needs of the user change. These things should be taken into account in the early stages of the development process. Systematically approaching the development of an application means deciding on a process.

The importance of heuristics as discussed by Rechtin and Maier should also be considered. Several rules of thumb that guide the development process will be pointed out as the three-tiered approach is described. These are suggestions and ideas that have been learned through experience.

As is the case in any programming language, the programmer must take the time to understand the nature of the task at hand and what the purpose of the project is. By this, we mean the project requirements should be well defined. There should

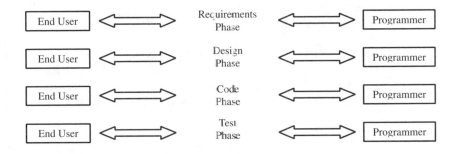

DIAGRAM 4.1

be a clear goal or result for the project. Defining the requirements is one of the first stages in the development of any application. Some people believe that a big portion of the project is completed once the detailed requirements are defined and documented. An example is writing an application to monitor process control. The user requirements might include the number of items to monitor as well as the frequency of the sampling. You must decide what instruments will be used, the type of communications (serial, GPIB, etc.), what specific measurements are needed, where the data will be sent, and who the end users of this application are. These are considered derived requirements. Together, these are some of the more general items that would be included in the enumerated list of requirements.

The requirements are key deliverables for a software project. One of the most common reasons software projects run over budget and beyond due dates involves not having the requirements defined. There is a tendency to add new features into the application late in the development cycle. If the requirements keep changing, it will be difficult to adhere to limited schedules and budgets. You must lock in the requirements to prevent "feature creep."

When the requirements are very loosely defined or not defined at all, the end result has many possibilities. Failing to document the needs of the customer can prove to be very costly. A lot of time and money will be wasted in making changes to fit customer needs. It can be avoided if time is spent earlier in the process. If requirements are not in writing, contract payments may not be made.

The end user of the program plays a key role in development. If the end user of the program is the person writing the code, the requirements do not have to be defined as well because that person will know what is needed. When the code is being written for someone else, they must be consulted at several stages in the process in addition to the early requirements phase. The saying, "You never really understand a person until you consider things from his point of view," holds true here. (See Diagram 4.1 on user involvement.)

4.3 SOFTWARE MODELS

There are many software models that exist, but only the waterfall model and the spiral model will be described in this section. These are two common software models that are widely used in the development process of a software project. They

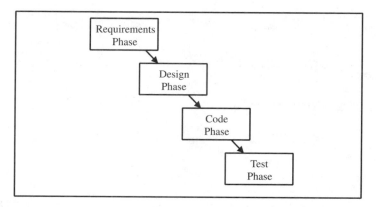

DIAGRAM 4.2

are called "lifecycle models" because they can be applied from the beginning to the end of the application's existence. Both models have several variations, but only the basic models will be presented.

4.3.1 THE WATERFALL MODEL

The waterfall model has been widely used for some time. It is a classic model that consists of distinct phases in the project. In its simplest form, the waterfall model contains the following phases: requirements, design, coding, and testing. The waterfall model is depicted in Diagram 4.2. The modified versions that are in use sometimes include more than one step in each phase.

Documentation plays an important role in the waterfall model. In its purest form, the focus is kept on one phase at a time. Each previous phase is thoroughly documented and approved before proceeding to the next step.

In the requirement phase, the capability and functionality needs are defined in detail. There are many requirements that have to be outlined, such as hardware requirements, software requirements, backward and forward compatibility requirements, etc. There are also user requirements and requirements that are derived.

The design phase consists of deciding on the structure of the application in detail. This is the stage where you decide how you will implement the requirements. It includes developing the design or architecture of the application at a high level, and performing the description of logic to accomplish the objective. This chapter focuses mainly on the design phase of the project.

The coding phase includes the actual implementation and software development. As the name suggests, the actual programming is done in this step. The plans made in the design phase are stepping stones for programming. When working in a team environment where several people are involved, good coordination is necessary. The program can be separated into modules and integrated later.

The testing phase attempts to find and fix all the bugs in the code, and includes integration. The point of this phase is to determine if the specifications and requirements were met as outlined in the earlier stages. The importance of testing is not always emphasized as much as it should be. No matter how much time is spent on testing, finding all of the faults is very difficult. Some faults are hard to find and may eventually slip through the cracks. Generally, test plans will be developed to verify and validate the code's conformance to requirements.

The waterfall model heavily stresses the importance of the requirements phase to eliminate or reduce problems early. The requirements must be explicitly outlined in this model before work can begin on the next phase. Keep in mind that defining detailed requirements will not always translate into a good application structure. However, it does bring to attention the important phases that are involved in application development. This model is aimed at getting the project completed in one pass. Returning to a previous phase to make changes using this model can become costly because one phase is to be completed before the next phase begins.

4.3.2 THE SPIRAL MODEL

The second model is the spiral model, which is essentially an iterative development process. In this model, software is developed in cycles that include the phases described previously. Each iteration either fixes something from the previous one, or adds new features or functionality to the application. The importance that is stressed in this model is that the significant issues are discovered and fixed early in the development process. A goal for a deliverable can be defined for each iteration of the project. The spiral model is depicted in Diagram 4.3.

Each release or version includes going through planning, evaluation of risks, design and code, and software assessment. Planning consists of establishing the goals, constraints, and alternatives that are available. The potential issues and alternatives are analyzed in the second stage. The design and coding stage involves implementation of the design where the application is actually developed and tested. Finally, the application is evaluated during software assessment.

The evaluation of risks related to the project is crucial in the spiral model. You start with the most important risk and continue through one development cycle, working to eliminate that risk. The next cycle begins with the next important issue.

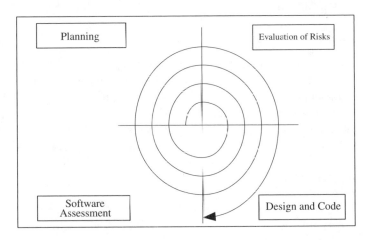

DIAGRAM 4.3

Iterations continue until all issues have been resolved, or the requirements for the finished project have been fulfilled.

The spiral model is based on the concept of incremental development during each iteration. The highest priority items, consisting of risks or features, are addressed and implemented first. Then, the project is reevaluated, and the highest priority item is defined and implemented in the next iteration. The first release of an application can consist of the first loop, and following versions add new features as more iterations are made.

The spiral model is best when the requirements are not fully defined and development work must begin quickly. Iterative models create an early version of the application for demonstration purposes and further refinement. If the risks cannot be identified easily, this model may not work very well.

4.3.3 BLOCK DIAGRAMS

Using a particular model will not guarantee success, but nevertheless following a model provides an orderly roadmap for the development process. Following one model strictly may not be the best course. There should be sufficient flexibility to take the specific circumstances into consideration. Sometimes it is enough to keep in mind that there are different phases in the models and adapt them to the current project.

In either model, block diagrams are useful tools that can assist in the design of the application. When using a top-down design approach, the block diagram, or hierarchy, helps get the structure defined. It also assists in separating tasks for the project and developing timelines for completion. In this way the developer can get the big picture of how the application is laid out and how to progress in the coding phase. The top-down design is more suitable for large applications where work can begin on the user interface and main level first. Work can then continue down to the driver level. In the bottom-up design, the drivers that will be needed for the program are worked on first. At most, a small team should be responsible for the architecture of the project. This facilitates management and prevents pulling the project in various directions.

4.3.4 DESCRIPTION OF LOGIC

A large application must be designed before the coding phase can begin. The requirements determine what the application must do. The design or architecture determines how the application will actually do it. The design phase of a project includes developing the architecture as well as description of logic for the implementation.

The architecture consists of the framework that will be used for the application, taking code reuse, readability, and flexibility into consideration. This can be done by creating flow diagrams or designing the VI hierarchy for LabVIEW applications. The software architecting is followed by definition of the logic that will be used to perform the actual coding.

In the description of logic, the designer must describe the logic of each VI in the hierarchy. The description of logic should capture the intention of the VIs and provide documentation at the same time. This will prevent rework and reduce the time it takes to develop the application. A description should include both the purposes of the VI and how it will accomplish its objective. The purpose should simply be one sentence describing what the VI will do. When several sentences are needed to describe the action of the VI, the code can possibly be broken down into subVIs. This will increase readability and maintainability of the code. For example, "This VI will configure the signal generator for test A," illustrates the intent of the VI. When describing how the objective will be accomplished, include what VIs will be called and why they will be called.

The coding phase will utilize the description of logic for development of the application. This is followed by the test phase where the code and the description of logic are validated. A test plan must be used to verify the description of logic. This is done by developing test cases designed to test different portions of the description of logic.

4.4 PROJECT ADMINISTRATION

A single programmer project can have its process managed by the programmer. Management of the development process may not be a big problem in this situation. However, projects are often worked on by teams composed of several members. When more than one person works on an assignment, the need for a team leader arises. Someone has to drive the project and control the direction of its progress. The whole development process needs management. Decisions must be made on how and whether to continue to the next phase.

One team member must assume the role of project manager. Without someone to focus on the overall goal of the project, the goals of the members can become divergent. When a team works on all phases of the application, the team leader becomes the lead designer and the one who ensures that the process is moving in the right direction. When separate teams are working on each phase of the application, a project manager is needed to work with all the teams involved. The project manager has to make sure that the designers, programmers, and testers work together and are synchronized. Clear roles have to be assigned to the individual members.

Projects have constraints and risks regarding cost, schedule, and performance. The project manager has to practice techniques to control the risks and constraints. Scheduling is a crucial aspect of the administration of a project. Some stages have to be finished before work can begin on the next stage. Goals and milestones have to be achieved in a timely manner. The project manager works with all who are involved and is made aware of problems as they arise. Resources can be shifted where necessary to assist in problem resolution and to meet schedules. Deadlines are strategic issues that must be dealt with in the appropriate manner. In some cases it might be preferable to be over budget and on time than to be late but within budget constraints. In other cases it is better to be late with a high-quality and high-reliability product.

The administrator should have a good understanding of the complete system. If the project manager is involved in the early conception and requirements stages, then this person will have a better grasp of the purpose of the application. Better decisions can be made on the priorities of the task at hand and how to resolve conflicts. Information must be acquired, evaluated, interpreted, and communicated to the group members as necessary.

Software projects are more difficult to manage than other types of projects for several reasons. It is difficult to make estimates on the project size, schedules, scope, and resources needed. Software projects can fail due to inaccurate estimates on any of these aspects. Planning plays a key role in project management.

4.5 DOCUMENTATION

When a software application is being developed, the proper documentation is often overlooked. Many times, the documentation process will begin only after all the coding has been completed. This results in insufficient reports on the procedures followed and the actual code written. When you return to write documentation after completing the project, you tend to leave out design decisions that are important to the development. Then, the record keeping becomes more of a chore and fails to serve its intended purpose.

Good documentation will allow someone not involved in the development to quickly identify and understand the components of the project. It also provides a good design history for reference purposes on future software projects. Accounts should be kept at all of the phases in the development cycle. The requirements documents are significant because they will guide the rest of the phases. The design phase documentation serves as a reference for the coding phase.

Documentation during the coding phase, or Description of Logic, is critical. Major points help understand what the code is supposed to do. Comments that are included with the code help identify the different segments and the purpose of each segment. They aid in the maintenance, modification, and testing of the code. Updating the code becomes easier for someone who was not involved in the development process. The original programmers may be reallocated, transferred, or may even leave the company. Then, problems can arise for those who use the program and have to make modifications.

4.5.1 LabVIEW DOCUMENTATION

LabVIEW has some built-in functions to help in the documentation of code. As with other programming languages, comments can be included in the appropriate places with the code. This allows anyone to look at the diagram and get a better understanding of the code. When modifications have to be made, the comments can help identify the different areas in addition to their functionality.

LabVIEW allows the programmer to enter descriptions for front panel controls and indicators. When Show Help has been activated from the Help menu, simply place the cursor over the control or indicator to display its description. To enter the description, pop up on the control and select Data Operations from the menu. Then select Description and a window appears that will allow you to type in the relevant information. These descriptions will assist anyone who is using the application to identify the purpose of the front panel controls and indicators

Descriptions can also be added for each VI that is developed, using the Show VI Info selection under the Windows menu. You can include relevant details of the VI, inputs, and the outputs. When the Show Context Help is activated from the Help menu, this VI information will appear in the Help window if you place the cursor over the icon. Help files can also be created and linked to LabVIEW in an on-line form. They have to be created in Windows format and compiled before they can be used in LabVIEW.

4.5.2 PRINTING LabVIEW DOCUMENTATION

You can also select Print Documentation from the File menu and LabVIEW will allow you to customize the way you want to print the documentation. The VI Info that was entered will be included. There is a feature that gives you the ability to print documentation to an HTML file. This file can then be published easily on the Web. Options for saving files in RTF format or as plain text files are also available. The user can select this from the Destination drop-down menu in the window after Print Documentation has been selected. The pictures of the code diagram and front panels can be saved as PNG, JPEG or GIF formats.

4.5.3 VI HISTORY

Another way to document LabVIEW applications is to use the Show History selection under the Windows menu. This will allow the programmer to write what changes are made each time the VI is modified. The VI history provides a good reference when trying to track down what changes were made, why they were made, and when they were made. You can force LabVIEW to prompt the user to input comments into the VI History when changes are made. This is a good practice to incorporate in the development process. Select Preferences from the Edit menu, and then select History from the drop-down box. You can then select the appropriate checkbox so that LabVIEW will prompt for comments each time the VI is saved.

Some firms may desire to be ISO 9000 compliant, which requires more effort. The items covered in this chapter are intended to help in the documentation process for those not requiring ISO 9000. The basic documentation will include how to use a VI, will describe the inputs and outputs, and will discuss the necessary configurations for the user. ISO 9000 requires controlled master copies of all documents to ensure that only the newest version is distributed at any time. In addition, a record must be kept of the controlled documents and the location of their storage.

4.6 THE THREE-TIERED STRUCTURE

Once the requirements are defined and the major design decisions are made, the programmer is ready to work on the structure of the application. An application should be divided into three tiers. The first tier is referred to as the "Main Level." The Main Level consists of the user interface and the test executive. The second level is the "Test Level" or the "Logical Level." The Test Level is responsible for performing any logical and decision-making activities. The lowest level is referred to as the "Driver Level." The Driver Level performs all communications to instruments, devices under test, and to other applications.

Before we look at each of these levels in more detail, we shall identify the benefits of using the three-tier approach. First, this strict partitioning of levels and functions maximizes code reuse. Specific functions or code can be immediately identified and reused because VIs in each level have a defined scope. Drivers can be reused when the need to communicate with another application or instrument arises. Test and measurement VIs can be reused when that test has to be performed. The user interface can also be reused with minor modifications for a different application.

The reuse of code is further simplified with the use of a state machine. State machines work well when the three-tiered approach is applied. State machines and the variations that exist are discussed in Chapter 3. Any state within the state machine can be reused by simple copy-and-paste methods.

A second benefit of using the three-tiered approach is that the maintenance time of the code is minimized. Maintenance and modifications are often necessary after the completion of an application. The application design should therefore plan ahead for changes and make them easy to apply. Because distinct layers exist, modifications can be made quickly and efficiently. VIs that need modification can be identified and located easily. The code that has to be changed can be pinpointed along with the interdependencies with little effort. When this is done, the modifications can be made where needed.

Another notable benefit of the strict partitioning and three-tier approach is the abstraction that is gained. Each level provides an abstraction for the layer below it. The Driver Level abstracts the vague commands used in instrument communication. The Driver Level provides an abstraction for the Test Level. The Main Level then provides an abstraction for the subroutines and measurements by supplying an easy-to-understand user interface. The user interface is an abstraction that hides or disguises all the lower levels involved.

The NI Test Executive serves as the Main Level for an application. It supplies the User Interface function that allows you to select the sequence of tests that you want to perform. The Test Executive can be customized to match the specific needs of the situation. The Test Executive also has the structure already defined, reducing the responsibility of the programmer.

Figure 4.1 is a diagram of a VI hierarchy that uses the three-tiered approach and depicts the strict partitioning of different levels. A quick glance at the diagram reveals the three distinct layers in the application. The Main Level, the Mid-Level, and the Driver Level can be distinguished easily in this example. If Test 2, shown in the hierarchy, has to be used in another program, it can easily be cut and pasted into a new application. Maintenance of the code is easy because changes can be made to a specific section. Also, note how each level abstracts the level directly below it.

Now look at Figure 4.2; it displays the VI hierarchy of an application that does not utilize the three-tiered approach. Code reuse is diminished in this case because the tests are no longer stand-alone VIs. Modifications and maintenance become difficult because changes made in one location may affect other things. Changes made in Driver 1 can affect the Main VI, Test 1 VI and the Driver 2 VI. The

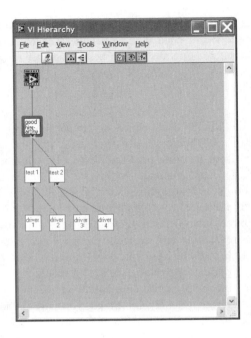

FIGURE 4.1

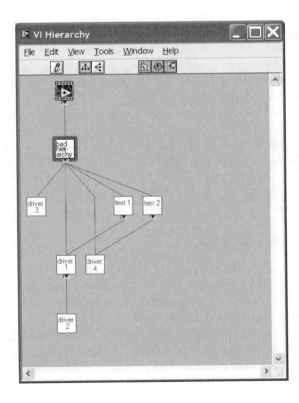

FIGURE 4.2

dependencies are harder to track when a definite structure is not used for the program. Locating a specific section of code will take longer because drivers and tests are mixed. Also note that there is no abstraction below the user interface.

4.7 MAIN LEVEL

Let's first look at the Main Level, which serves as the user interface and test executive. The Main Level should consist of a single VI. Only Test Level VIs are allowed to be called from this first tier. The Test Level will then call the needed drivers for the specific operations. The Main Level should avoid calling drivers because the abstraction benefits are diminished. Reuse is diminished when specific sections cannot be differentiated for copying and pasting methods. Furthermore, the code panel and hierarchy also become difficult to read and maintain. The use of application tiers aids reusability and readability.

The Main Level VI provides user interface functions. It supplies the needed structure for adding application-specific tests, and offers flexibility for changes. National Instruments offers an application called TestStand to be used as a test executive/sequencer. If you are using TestStand, you will already have the extent of partitioning available to gain the benefits of the three-tiered approach. You will be supplying the test and driver level VIs to incorporate into the framework of the executive.

4.7.1 USER INTERFACE

The user interface is part of the main level. The user interface is significant because it is the means by which interaction and control of the program occur. LabVIEW provides various tools for designing an effective front panel. Its graphical nature gives it an edge over other programs when it comes to the user interface. ActiveX and .NET controls can now be used in addition to the basic controls. This section will provide some tips and examples for developing effective interfaces.

As TestStand already has a user interface, you would no longer have the responsibility of creating one. TestStand allows the operator to select the test sequence and control execution of the sequence through the interface. The results of the test sequence are shown in a display that also indicates pass or fail status.

4.7.1.1 User Interface Design

The user interface should be designed with the target operators in mind. The interface should not be designed solely to fulfill the functional requirements, but it should also be user friendly. Unless the programmer is the one using the application, it can be the only interaction the operator has with the program. The Main Level user interface should allow the operator to select settings that are variable. Keep in mind that the user inputs may have to be validated. Unexpected inputs will cause the program to behave in an unexpected manner. Variable inputs by the user may include choosing measurements to perform, inputting cable loss parameters, selecting device addresses, adding file storage tags, selecting processes to monitor, etc. These vari-

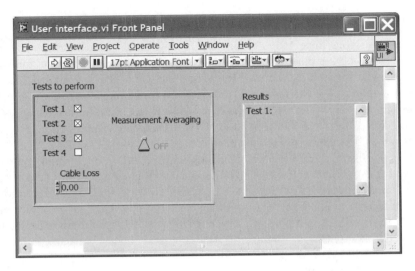

FIGURE 4.3

ables are then passed to the Test Level, ultimately dictating the program flow. Figure 4.3 is an example of a simplified user interface.

Consider using clusters to group-related controls and indicators. Not only does it place the related controls together, but it also reduces the number of wires on the code diagram. When you are trying to manage large amounts of data, the code diagram can get confusing with all of the wires of data being passed to VIs. Clusters can be unbundled as needed, with only one wire representing the group of controls otherwise. Even if you are not using clusters, try to use a frame decoration to group the controls for the user.

The user interface should utilize controls and displays that are appropriate for the situation. If you are entering a cable loss, for example, you should use a digital control with the appropriate precision rather than a knob. This is more practical because the value can be read and changed easily. Using descriptive labels is a good way to differentiate the various front panel controls. Try not to clutter the main user interface with controls or displays that are not needed. The person using the program can easily get confused or lost if too many controls, indicators and decorations are used on the front panel of the user interface. The use of menus will help reduce the clutter, and will give the interface a nice appearance. Use buttons if the function will be used frequently; otherwise, use menus. Dialog controls are also good for user interface functions.

Remember to give the user a way to cancel or abort execution. This is easy to overlook, but is very important to an operator. Users need a way to stop in the middle of execution without having to use the Abort Execution button on the toolbar.

Graphs and charts are useful for displaying data. Sometimes just a glance at a graph can reveal a number of things, but graphs, charts, and other graphics should be used only as needed. Graphing while acquiring data will not only slow the execution of the program, but will take up more memory. Memory concerns will grow as the number of VIs written for the application grows.

4.7.1.2 Property Node Examples

You can use your imagination to develop a professional user interface. The user interface can be simplified by using the Tab control to group similar functions. Another way to simplify the user interface is to allow the user to see only the available options. For example, let's say that you are creating a VI to get user inputs to set up your test system. One portion of the setup requires you to define how to communicate with your unit under test (UUT). There are several options for communications depending on the type of unit.

For our example we are communicating with a UUT through serial communications. There are several options for communicating serially. You could use the computer serial port, a terminal server or even a GPIB to RS232 controller box. If you are using a terminal server to communicate with your UUT you would want to know the IP address and port you are connecting to. You would not want to see a GPIB address or COM port number. By using property nodes, you can control what the user sees based on the chosen input.

Figure 4.4 shows the user interface for the above example that utilizes property nodes. There are 3 tabs for entering information. On the UUT Communications tab there is a control to select the method of communication used. Based on this selection the appropriate options are displayed. In this case, the Terminal Server option was selected. To the right of the selector control are the inputs for IP address and port number.

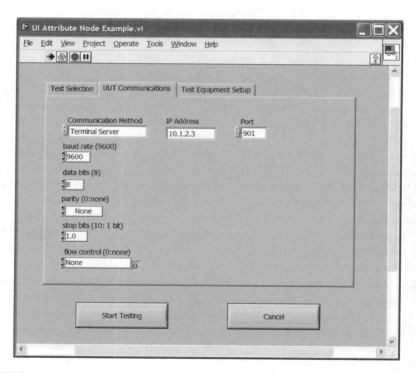

FIGURE 4.4

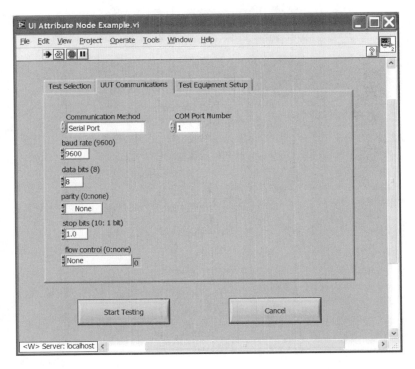

FIGURE 4.5

Figure 4.5 shows what appears when the serial port option is selected. Figure 4.6 captures the code diagram of the example VI. Property nodes are used to make the controls visible or to hide them. Property nodes can be created by popping up on the control from the code diagram. Select Create from the menu, then select Property Node. The available options for the selected control will be available. The Visible property was used in the example. Some common properties that can be manipulated include Visible, Disabled, Key Focus, Blinking, Position, Bounds, Caption, and Caption Visible.

Using Property nodes once again, Figure 4.7 is an example of a menu and submenus structure that is simple to implement. The front panel shown has the main menu that appears on the user interface panel. A Single Selection Listbox is used from the List & Rings palette. Figure 4.8 shows the submenu that appears when the first item, Select Tests, is selected from the listbox. A Multiple Selection Listbox becomes visible and allows the operator to select the tests that have to be executed. When all the needed settings have been completed, the user can hit the Start Testing button to begin execution of the tests.

Figure 4.9 is an illustration of the code diagram for this example. The case structure is driven by the main menu selection. The structure is placed inside a main While loop that will repeat until the Start Testing button is pressed. The Visible Property node is used to make the submenus appear when a selection is made on the main menu. Note that you must first set the Visible Property node for all of the submenus to "false" before the While loop starts.

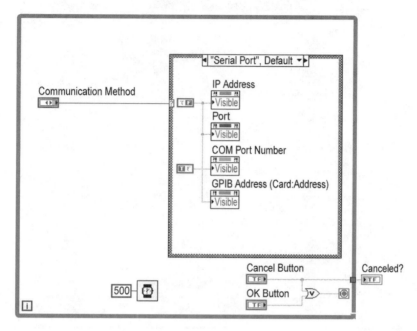

FIGURE 4.6

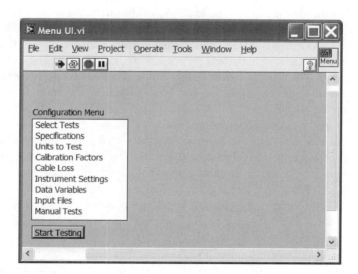

FIGURE 4.7

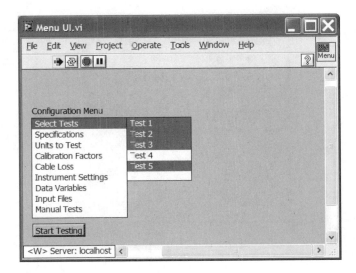

FIGURE 4.8

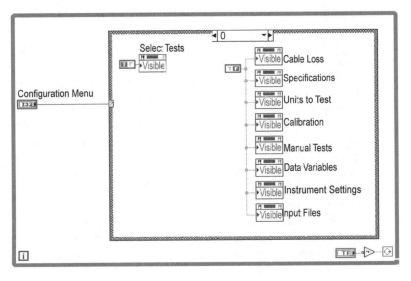

FIGURE 4.9

4.7.1.3 Customizing Menus

LabVIEW run-time menus can be customized to suit specific needs by using the Menu Editor. The menu contents can be modified by selecting the Run-Time Menu item from the Edit menu. The drop-down box allows the programmer to select either the default, minimal, or custom menus to appear during VI execution. The default selection displays the menus that are normally available while the program is not executing. The minimal selection is a subset of the default that appears during run-time. The custom selection requires the programmer to create a new menu structure

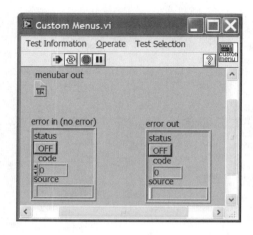

FIGURE 4.10

and save it as a real-time menu (*.rtm) file. Once a real-time menu file is created, it can be used for multiple VIs. A shortcut key combination can be specified for each menu item that is created. The on-line help explains how to customize menus, including how to add User Items.

Figure 4.10 is an example of how custom menus appear when a VI is executing. Three main menus are displayed on the front panel: Test Information, Operate, and Test Selection. The Operate menu is an application item that is normally available if the menu is not customized. The Menu Editor allows you to utilize application items in addition to user items.

Figure 4.11 illustrates how the custom menus may be used programmatically once they have been created. The Current VI's menu bar returns a refnum for the current VI. This VI is available in the Menu subpalette of the Application Control palette. This refnum is passed to the Get Menu Selection VI, which is available in the same subpalette. The Get Menu Selection VI returns the menu item that was selected, as well as the path of the selection in the menu structure. In this diagram, the Get Menu Selection VI is used to monitor selections that are made through the custom menus. The menu selection is then wired to a case structure that takes the appropriate action depending on the selection that is made. The Begin Testing case

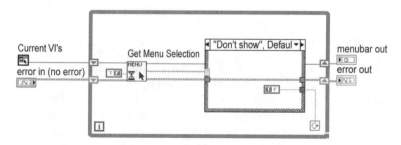

FIGURE 4.11

that is shown corresponds to a menu item in the Test Information menu from the previous figure. When Begin Testing is selected the While loop terminates and the VI completes execution. By utilizing other VIs in the Menu subpalette, a programmer can dynamically insert, delete, or block menu items.

4.7.2 EXCEPTION-HANDLING AT THE MAIN LEVEL

Error handling is one element of the project that is often overlooked or not well implemented. Planning for the possibility of something going wrong is difficult, but necessary. A well-designed program will take into account that errors can and do occur. Building exception handling into a program has several benefits. It is a way to notify the operator something has gone wrong that needs attention. It is also very useful for troubleshooting and debugging purposes, as well as for finding out where and why the problem occurred.

There are different ways to control the error situations that can arise. One way is to let the program attempt to correct the problem and continue execution. For errors that cannot be corrected, the application may complete tests not dependent on the failed subsection. Another possibility would be to halt execution of the program and notify the user via a dialogue box, e-mail, or even a pager.

Error handling is an important task that should be managed in the Main Level. This forces all errors to be dealt with in one central place, allowing them to be managed better. The Main Level controls program flow and execution. The Main Level should also determine the course of action when faults occur. When errors are handled in several locations, or as they occur, program control may have to be passed to lower levels and may be difficult to troubleshoot. Also, when errors are handled in more than one location, the code for the handling may have to be repeated.

When a state machine is used, this significant task is made easy because one state is assigned specifically for error handling. When errors are generated, the state machine jumps to the Error state to determine the course of action. Based on the severity of the fault that has occurred, the Error state in the Main Level will decide what will be done. If the error is minor, other states that might be affected will be parsed and the remaining will be executed. If it is a major fault, the program will perform all closing duties and terminate execution in the normal manner while notifying the user of the error. Chapter 6 discusses exception handling in more detail.

Handling execution based on pass or fail criteria should also be considered. If you are using TestStand the user can specify the course of action when a test fails. You can continue to the next test, stop execution of the whole sequence, or repeat the same test again. Dependencies can be created for the individual tests. A dependency, once created, will execute a test based on the result of another test. The result can be defined as either pass or fail.

4.8 SECOND LEVEL — TEST LEVEL

The Test Level is called by the Main Level. The VIs in this level should be written on a stand-alone basis to allow reuse. Each Test Level VI should perform one test or action only. The code should be broken up so that each test that needs to be performed can be written as a separate VI. When multiple tests are combined in one VI, they are not easily reused because either the extra tests that are not needed would have to be removed, or the extra tests must be executed unnecessarily. These second tier VIs are basically test and measurement subroutines, but can also include configuration and dialog VIs.

Writing each test exclusively in its own VI facilitates reuse in cases where the measurement subroutine has to be executed more than one time. An example of this is making temperature measurements at multiple pressure levels. When a temperature is measured, it will vary with the pressure conditions. A VI that performs a temperature measurement can be written and called as many times as needed to test at the different pressures. Note that the efficiency of the VI is maximized when the pressure is set outside of the temperature measurement VI, and a call is made to it as many times as needed.

The measurement subroutine VIs should perform the initialization of the instruments and any configuration needed to make the measurement. This may include setting RF levels, selecting the necessary instrument fields, or placing the device under test in the appropriate state. These initialization steps must be taken within the VI because the previous condition of the devices may not be known. This is especially true when using a state machine because the program jumps from one state to another; the order of execution is not necessarily predetermined.

When a state machine is being used, only one test or measurement VI should be placed in each state. The benefit of this is that when a particular test needs to be performed, the program executes only the associated state. Placing more than one test in one state causes the additional tests to be executed even if they are not needed. This results in an application that takes more time to run. It also results in loss of flexibility of the state machine. An example of a single test in each state of the state machine is shown in Figure 4.12. The state shown will be executed whenever the particular test has to be performed; the purpose of this state is clearly defined. This method reduces clutter and makes the code diagram self-explanatory.

Flowcharts can assist in the implementation of subroutines and Test Level VIs. They help in defining the execution flow of the VI and the specific decisions that it must make. Flowcharts are especially helpful in LabVIEW because it is a dataflow-based programming language, similar to a flowchart. Once the flowchart is formed,

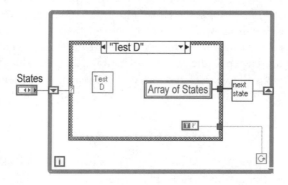

FIGURE 4.12

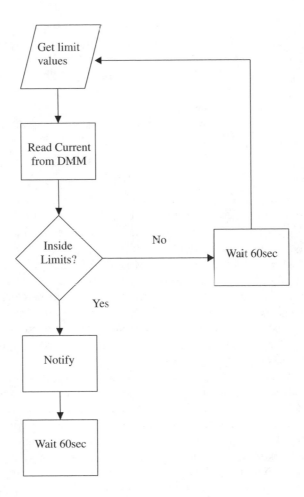

FIGURE 4.13

it is relatively easy to code it in LabVIEW. Figure 4.13 is an example of a simple flowchart. It is checking to see if the current draw from a source is within the limits. This is easily coded in LabVIEW because of the similarities between the two. If you compare the flowchart with the actual implementation in Figure 4.14, the similarities and ease of conversion become apparent. The VI reads the current every 60 seconds to check if the value is within the specified limits. When the value is outside the limits, it will terminate execution and the front panel indicator will notify the user. The flowchart and the LabVIEW VI perform the same function.

4.9 BOTTOM LEVEL — DRIVERS

The Driver Level is an abstraction that makes the Test Level easier to understand. It conceals little-known or unclear GPIB, serial or VXI commands from the user. This level performs any communications necessary to instruments and devices being used. Drivers can be classified into measurement, configuration and status categories.

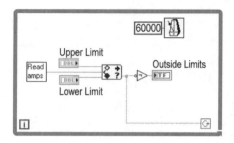

FIGURE 4.14

An efficient way to write drivers is to write each measurement command to one VI each, group configuration commands logically into VIs, and write each status command to one VI. As an example of this driver architecture, examine the HP8920A driver set by National Instruments.

Simply put, measurement drivers are used to perform a measurement. One VI should be used to perform one measurement to maximize the reuse of the VI. By writing measurement drivers in this manner, the driver can be called in the same application for different cases, or in a different application where the same measurement needs to be performed. If more than one measurement is grouped into a single VI, either one of the measurements must be stripped out for reuse or other measurements will have to be performed unnecessarily.

Configuration drivers set up the instrument to make a measurement or place it in a known state at the start of an application. Configuration commands can be grouped logically in VIs. When a measurement has to be performed, usually more than one configuration command is needed to prepare the instrument. Sometimes many parameters have to be configured for a single test. Writing one configuration command to a VI would create difficulty in maintenance because of the number of VIs that will result. Grouping the configuration commands according to the type of measurement will minimize the number of VIs on the Driver Level. In addition, memory space can also be used more efficiently by following this style.

Drivers can range from very simple to very complicated, depending on the instruments being used. A driver for a power supply might need only a few parameters and commands, but an instrument like a communication analyzer might have upwards of 100 different commands. In this case you must group configuration commands to reduce the number of VI drivers that need to be written.

Status drivers simply check the state of an instrument by reading status registers. These are usually written as needed, one to a VI. An example of a status driver is a register that holds the current state of a particular measurement. One bit will be set if the measurement is under progress, and cleared when it is finished. You must ensure that the measurement is finished before reading the value so the status register is checked. Remember that drivers can be created for other types of communication needs as well.

If you need to use TCP, DDE, ActiveX, or PPC, you can use a similar logic when developing the lower layer of the application. When VIs are created to perform a specific action, configuration, or status inquiry, they can be reused easily.

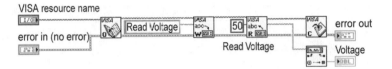

FIGURE 4.15

Figure 4.15 demonstrates a simple driver that can be used to read the voltage from a voltmeter. It has been written so that it can be called whenever the voltage needs to be read. The instrument handles are opened and closed inside the driver.

4.10 STYLE TIPS

We have seen numerous applications in the past few years that do not incorporate good practices that could increase the efficiency, readability, maintainability, and reuse of the code that has been developed. Most of this chapter covers topics to assist the reader in developing successful applications by revealing the programming style that has been effective for the authors. This section was intended to provide more tips on programming, this time by uncovering inefficient programming styles and common pitfalls that can be avoided.

4.10.1 SEQUENCE STRUCTURES

The first inefficient programming style is a result of the overuse of stacked sequence structures in the code diagram. Stacked sequence structures were described in Chapter 1 in detail. The main purpose of the sequence structure is to control the execution order of the code. Code that must execute first is placed in the first frame, and then pieces of the code are placed in the appropriate frame in the order that execution is desired. If your code is data-dependent, however, sequence structures are not needed. The execution order is forced by dataflow; VIs will execute when the data they need becomes available to them.

Overuse of sequence structures is the consequence of not utilizing the structures as they were intended. We have seen VIs that contained sequence structures with 50 or more frames. When the architecture or design phase of the application is omitted, an application with so many frames can be an outcome. It signals that perhaps the VI is performing too many actions and that subVIs are not being used sufficiently. When too many actions are performed, the code that is developed is no longer modular. Lack of modularity hampers the ability to reuse code. Consider using the three-tiered structure approach to your application if your sequence structures have too many frames. The use of subVIs for tests and subroutines will reduce the need for many frames while increasing the ability to reuse code. The frames in the sequence structure can be easily modularized into VIs. At the same time, the code will become more readable and maintainable. By developing the description of logic for the VIs during the design phase, you can determine what each VI should do as part of the whole application.

Figure 4.16 displays a stacked sequence with only four frames, but notice how the wires are already beginning to degrade readability of the code. The sequence

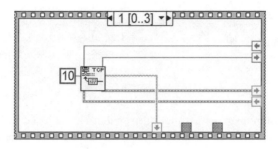

FIGURE 4.16

locals are not easy to follow as the data is being passed from one frame to the next. Code reuse is also becoming difficult in this example. Now imagine what the code would look like if there were 20 or 30 frames in the sequence structure.

If your VIs are data-dependent, you do not have to use sequence structures. For example, execution order can be forced through VIs that utilize error I/O with error clusters. The need for excessive sequence locals may indicate that execution order can be forced simply through dataflow. When many locals are used, problems arise in remembering which local is passing what data. It also degrades readability because of the wiring that is needed to support them. You must; however, be aware of any race conditions in your code.

4.10.2 NESTED STRUCTURES

Nested case structures, sequence structures, For loops, and While loops are sometimes necessary in the code diagram to accomplish a task. However, creating too many levels of nested structures can lead to inefficient code that lacks modularity and readability. The arguments presented previously on the use of stacked sequence structures apply to the use of nested structures.

Try to avoid nesting your structures more than three levels deep. When too many levels of nesting are used, the code becomes difficult to read. Data wires being passed into and out of the structures are not easy to follow and understand. When case structures are being used, you must look at each case to determine how the data is being handled. This, along with the use of sequence locals or shift registers, For loops, and sequence structures, adds to the readability problem.

Figure 4.17 shows the code diagram of a VI that utilized nested case structures four levels deep. Although the case structures are nested only four levels, it is difficult for anyone looking at the code to determine how the final result was actually obtained. You have to look at all the possible true and false combinations to figure out how the data is being manipulated. Imagine if this VI had more than four levels, or if there were more than just the two true and false cases used in each nest. The readability would be degraded further, while code reuse would be impossible.

Utilizing too many levels may also be a signal that your VI is performing too many actions. When too many actions are being performed, the resulting code has no modularity. This hinders the capability to reuse your code. The use of subVIs can reduce the need for excessive nesting of structures, as well as improve code reuse.

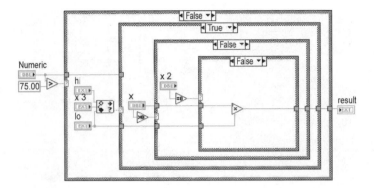

FIGURE 4.17

4.10.3 DRIVERS

Another bad programming style that we have seen is that drivers are sometimes underused. When communication with an external device or program is being performed, the I/O operation is executed in the Test Level, or even in the Main Level, instead of utilizing a driver. The concept of drivers is not fully understood by some LabVIEW programmers. A question that was posed at one user group meeting was, "Why do I need drivers when I can simply look up the command syntax and perform the I/O operation where it is needed?"

There are definite advantages that can be gained by creating and using drivers. The abstraction that drivers provide is a notable benefit for the application. The actual communication and command syntax is hidden from those who do not need or wish to see this code. This also improves the readability of the code when these obscure operations are not mixed with the Main and Test Level VIs.

The use of drivers also facilitates reuse of code. When drivers are not used, the actual code that performs the communication is difficult to reuse because it is part of another VI. Cutting and pasting part of the code is not as easy as inserting a new VI. However, when drivers are written to perform specific actions, they can be reused easily in any application by inserting the driver VI. Drivers must be developed in a way that will simplify its reuse. A thorough discussion on drivers and driver development is presented in Chapter 5.

Figure 4.18 demonstrates some of the reasons why drivers should be used. The VI shown is performing both instrument communications and other activities, using the results obtained. Compare this diagram to the driver shown in Figure 4.15 earlier. Notice that the instrument communications could have been placed in a separate VI, exactly as was done in the driver in Figure 4.15. Abstraction, readability, and reuse could have been improved through the use of a driver.

4.10.4 POLLING LOOPS

Polling loops are often used to monitor the occurrence of particular events. Other parts of the code are dependent on the execution of this event. When the event takes

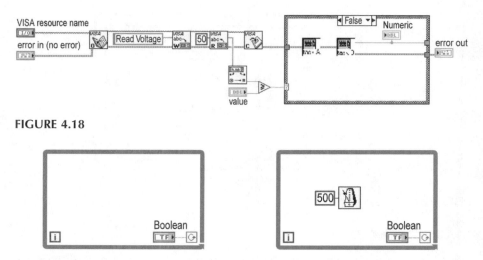

FIGURE 4.18

FIGURE 4.19   **FIGURE 4.20**

place, the dependent code executes in the appropriate manner. Using polling loops to monitor an event may not be the best way to accomplish this goal, however.

Tight polling loops can use all of the available CPU resources and degrade the performance of a system noticeably. If you are working on a Windows platform, you can use the System Monitor to view the kernel processor usage while you are performing activities on the computer. We can try a simple exercise to demonstrate this point. Open a new VI and copy the simple VI diagram shown in Figure 4.19. Set the Boolean to "true," run the VI, and monitor the processor usage; almost 100% of the processor will be used for the simple polling loop shown. What happens if we introduce a simple delay in the same polling loop? Use the Wait until Next ms Multiple in the loop with a 500-millisecond delay as shown in Figure 4.20, and monitor the processor usage again. The resources being used are significantly lower when a delay is introduced. Polling loops will certainly reduce the efficiency of your application.

If you are using polling loops, try to use delays where tight polling loops are not necessary. When loops are used for the user interface, the operator will not perceive a delay of 250ms. If you are using polling loops to synchronize different parts of your code, consider using the Synchronization VIs that are available in the Synchronization palette. These include Notification, Queue, Semaphore, Rendezvous, and Occurrences VIs.

4.10.5 ARRAY HANDLING

The manner in which arrays are handled can affect the performance of an application considerably. Special care should be taken when performing array operations with For loops. A scalar multiplication of an array is a good example for demonstrating the methods available to perform this action. Figure 4.21 illustrates one way to perform the multiplication. The array is passed into the For loop, where the element

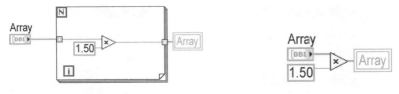

FIGURE 4.21   FIGURE 4.22

is multiplied by a constant of 1.5, and then passed out. The correct result is acquired; however, the method chosen to perform the multiplication is very inefficient. The same result could have been acquired without using the For loop. Figure 4.22 shows that the array could simply have been multiplied by the constant without the For loop. The first method is inefficient because it requires the array to be broken down into its elements, then each element of the array must be multiplied by the constant separately, and, finally, the array must be rebuilt with the results.

Whenever possible, you should try to avoid passing arrays into loops to perform necessary operations. Passing large arrays will result in longer execution times for applications, as well as the use of more memory during execution. Both the speed and performance of your application will be affected. The Show Buffer Allocations function can help you find places where arrays are being allocated. The function can be found under the Profile folder under the Tools menu. This function is described in more detail in Chapter 2.

4.11 THE LABVIEW PROJECT

Building, maintaining and compiling source code for an application can be a difficult task at times. The location of source files must be tracked, a list of supplemental files must be maintained, and build specifications need to be created for the application. You must also be able to handle the complex task of multiple developers working on the same project. All of these tasks can now be accomplished using the LabVIEW Project. We will discuss some of the project features and terminology. Some functions were also discussed in Chapter 2. The Shared Variable is discussed in depth in Chapter 7.

4.11.1 PROJECT OVERVIEW

A project is a grouping of LabVIEW VIs, controls and build specifications. Other files relating to the application, such as documentation and support files, can be stored in a project as well. All the information related to the project is stored in a project file. The project file has a .lvproj extension. The project file includes references to the files contained in the project; build information, deployment information, and configuration information. The project file is an XML file that can be viewed by opening it with a text editing application. Figure 4.23 shows a project file that has been opened in WordPad.

Fortunately you do not have to edit a project in a text file. In LabVIEW, you can open up a project in the Project Explorer. The Project Explorer window is a

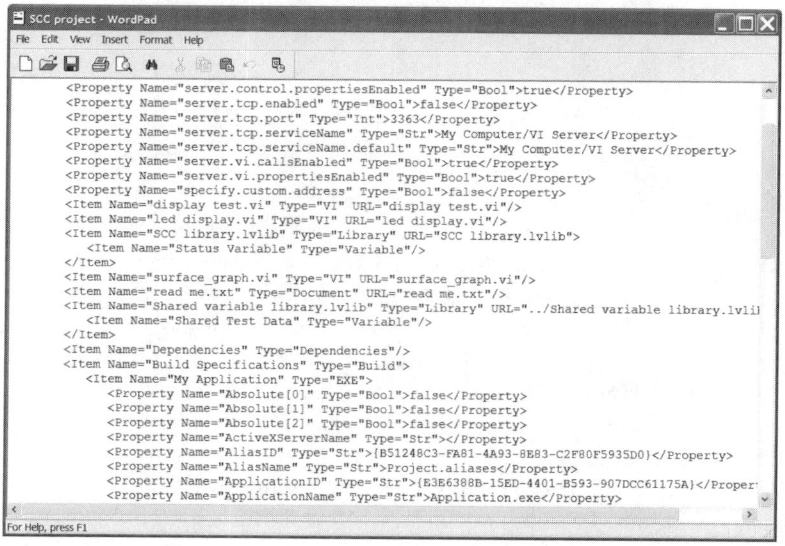

```
<Property Name="server.control.propertiesEnabled" Type="Bool">true</Property>
<Property Name="server.tcp.enabled" Type="Bool">false</Property>
<Property Name="server.tcp.port" Type="Int">3363</Property>
<Property Name="server.tcp.serviceName" Type="Str">My Computer/VI Server</Property>
<Property Name="server.tcp.serviceName.default" Type="Str">My Computer/VI Server</Property>
<Property Name="server.vi.callsEnabled" Type="Bool">true</Property>
<Property Name="server.vi.propertiesEnabled" Type="Bool">true</Property>
<Property Name="specify.custom.address" Type="Bool">false</Property>
<Item Name="display test.vi" Type="VI" URL="display test.vi"/>
<Item Name="led display.vi" Type="VI" URL="led display.vi"/>
<Item Name="SCC library.lvlib" Type="Library" URL="SCC library.lvlib">
    <Item Name="Status Variable" Type="Variable"/>
</Item>
<Item Name="surface_graph.vi" Type="VI" URL="surface_graph.vi"/>
<Item Name="read me.txt" Type="Document" URL="read me.txt"/>
<Item Name="Shared variable library.lvlib" Type="Library" URL="../Shared variable library.lvlib]
    <Item Name="Shared Test Data" Type="Variable"/>
</Item>
<Item Name="Dependencies" Type="Dependencies"/>
<Item Name="Build Specifications" Type="Build">
    <Item Name="My Application" Type="EXE">
        <Property Name="Absolute[0]" Type="Bool">false</Property>
        <Property Name="Absolute[1]" Type="Bool">false</Property>
        <Property Name="Absolute[2]" Type="Bool">false</Property>
        <Property Name="ActiveXServerName" Type="Str"></Property>
        <Property Name="AliasID" Type="Str">{B51248C3-FA81-4A93-8E83-C2F80F5935D0}</Property>
        <Property Name="AliasName" Type="Str">Project.aliases</Property>
        <Property Name="ApplicationID" Type="Str">{E3E6388B-15ED-4401-B593-907DCC61175A}</Proper]
        <Property Name="ApplicationName" Type="Str">Application.exe</Property>
```

FIGURE 4.23

separate window that is loaded when a project is opened. The Project Explorer window is similar to the Windows Explorer. The window contains a graphical listing of all the items stored in the project. Figure 4.24 shows the Project Explorer window for the project file in the previous illustration. You can execute all project related functions through the Project Explorer window. A discussion of some of these functions will follow.

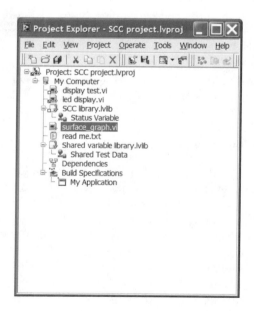

FIGURE 4.24

4.11.2 PROJECT FILE OPERATIONS

To start using a project you will need to create a new project. This can be done by selecting New Project under the File or Project menu. There is also an option to create an empty project in the New dialog box (comes up when you select New from the File menu). If any VIs are currently open, you will see a dialog box asking you if you want to add the currently open files to the new project. Once you address this dialog a Project Explorer window will pop up. By default an empty project will display the name of the new project with the My Computer listed below. My Computer is the target of the project. By default the local computer is the target. Additional targets can be added by right clicking on the project name, which is called the project root, and selecting Targets and Devices from the New category. Here you will have the option to add an existing target or device, an existing target or device on a remote subnet, or a new target or device. This is the only way you can distribute an application to an FPGA, Real Time or PDA target.

Each target will start with two items; Dependencies and Build Specifications. The Dependencies category includes items that are required by the VIs in a target. The Build Specifications category includes build configurations for source distributions, executable files (EXE), installers, shared libraries (DLL), source distributions and zip files. Only source distributions are available without the application builder or LabVIEW Professional Development System installed.

Once your project is opened you are able to start adding files. To add a file to your project you can right click on the target (My Computer) where the item will be added. You will have a lot of options including adding a new or existing VI, folder, Variable and library. You can add non-LabVIEW files such as documents or spreadsheets to the project as well. As with all operations on the project, you can access the operations through the shortcut menu as well as through the main menus. In this instance you can add files through the File menu. There is also the ability to use the New dialog box to add a file to the project. To insert the new item into the project, you need to click on the Add to Project checkbox. The New dialog box is shown in Figure 4.25.

To remove a file from a project you can right-click on the file and select remove. Removing a file from the project does not remove the item from the file system. The reference to the file is simply removed from the project file. The file itself was never physically a part of the project. This is different than the LabVIEW LLB where the VIs are actually part of the LLB file. Because of this method of using links to a file for a project, there needs to be some care taken with the files in the file system. If you move a VI that is part of a project to a new location through the Windows Explorer, the project will not initially find the file. When the project is opened it will try to find the file in the original location, and then start searching elsewhere for the file. You do have the option to browse for the file as well. There is a risk of the project finding a version of the file that you do not want used. Once the project finds the file, the location is updated in the project file. You will have to save the project; otherwise the location change will be lost. As the information is stored in a file, you could manually update the information as well.

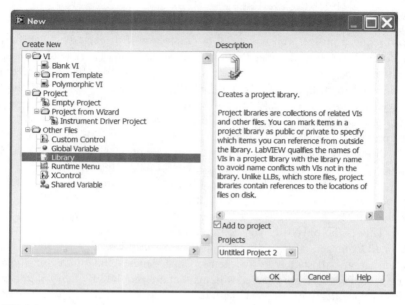

FIGURE 4.25

To add a Shared Variable you use the same procedure as with other files. A Shared Variable is similar to a global variable in that you obtain and pass data to other parts of your program or other VIs locally or over the network without having to connect the data by wires. How to setup a Shared Variable is described in Chapter 2. An in-depth look at the Shared Variable is offered in Chapter 7.

The Project Explorer supports drag and drop functionality. You can add a new file to a project by selecting the VI icon from the corner of a front panel or block diagram and dragging the icon to the Project Explorer. This is equivalent to the methods for adding files already discussed. The process works in the other direction as well. If you are editing a VI you can click on the VI or variable name and drag it to the VI you are editing. The item you had selected will be inserted in your VI. On Windows and Mac operating systems you can select an item or folder and drag it to a target in the Project Explorer.

Finally, you will probably want to use a file that you have inserted into the project. You can open the item by double-clicking on the item name in the Project Explorer. If the item is not a LabVIEW item, the appropriate application will launch in order to open the file.

4.11.3 PROJECT LIBRARY

A LabVIEW project library is an object that contains links to several types of objects including VIs, type definitions, variables and other files. A project library is access through the Project Explorer or the project library window. A library maintains the links to the objects it contains in a library file. A library file has a .lvlib extension. A project library should not be confused with a LLB file. A LLB file is a collection of VIs that are physically a part of the LLB. A project library owns files that are stored in individual files on disk. The files do not move from their original location.

Managing a large application can be difficult. You have to keep track of the hundreds of files that are in the application. The files needed may also include external code, project documentation and support files. By using a project library you have the ability to organize all of the files needed for an application in a single hierarchy. This has the benefit of being able to set permissions on groups of files in a single action. This also will make distributing the files easier. In order to distribute a project library you can either distribute the actual library file along with the corresponding files that the library owns. You can also create a zip file that contains the entire library. The zip file function is part of the Application Builder application.

One problem that pretty much all LabVIEW developers have had at one time or another is the issue of linking to a VI that has the same name as the original VI, but is not the correct VI. It could be a VI that was already in memory that gets saved in a calling VI, or when LabVIEW cannot find a VI and finds a VI with the same name in the wrong location. The project library eliminates this issue. By using XML namespaces, a project library can guarantee that a VI is correct. The VI is stored using the filename and the project library filename. This filename information is known as a qualified name.

To help illustrate the concept of the namespace, an example will be presented. Let's say you are creating a VI to communicate with two pieces of equipment through serial communications. Let us also assume one piece of equipment is connected through the serial port and one is connected through a terminal server. Previously you may have created a library of VIs for communicating through the serial port and a library of VIs for communicating through the terminal server. Now in your test VI you want to insert a VI created for checking the equipment status for both pieces of equipment. You never envisioned using both serial communications and the terminal server, so the status VI you created for both libraries has the same name. As the namespace for each of the VIs is different you could insert both VIs in your test program. The names that LabVIEW would use for these VIs would be in the following (URI) format: project library name:VI name.

To create a new library you would right-click on your target in the Project Explorer and select Library from the New menu item. This will create an untitled project library. You can either start adding files or save the library first. To save the library you can right-click on the library and select Save. The Save function will save the library in the location of your choosing with a .lvlib extension. If you have an existing folder in a project, you can right-click on the folder and select Convert to Library.

The library settings can be configured through the Project Library Properties dialog. This dialog can be launched by right-clicking on a library in the Project Manager and selecting Properties. The dialog is shown in Figure 4.26. The dialog has 3 sections. A General Settings section gives you access to the library protection level, the default palette and default icon. You can also set a version number for your library. The documentation section is where you can add a VI description as well as link to a help file. The Item Settings section gives lets you set the access scope for the entire library or for each individual file.

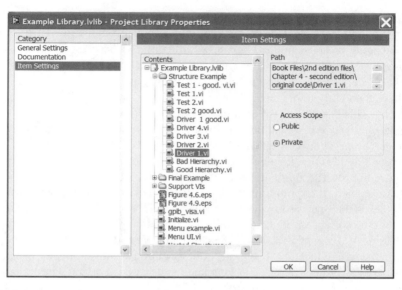

FIGURE 4.26

The access scope lets you set the Project Library or files as either public or private. A public VI is a standard VI that can be opened directly or called by other VIs. A private VI can only be called by VIs that are located in the same project library. If a folder in a project library is marked as private, all VIs contained in that folder will also be private. By using private VIs you can make sure the VI can only be run where they were intended. Figure 4.27 shows a Project Manager view of a library that contains some private VIs. The Private VIs have a key next to their icon.

Setting the protection level in the Project Library Properties window can protect a library or VI. You can set the protection level to Unlocked, Locked or Password-protected. When a project library is locked, items cannot be added or removed. Items that are set to private cannot be viewed. If the library is set to password-protected, the library can only be edited when the password is entered. This also holds true for the viewing of private files. They can only be viewed after the password has been entered.

To add files to a project library you can right-click on the library and select the item you want to add from the menu. You can also drag and drop an item in the project to the project library. A VI that is added to a library will need to be saved in order to properly link to the library. If you add a VI to a library that calls subVIs, you will need to add the subVIs to the project as well. The subVIs are not automatically loaded when the calling VI is loaded. You should also be aware that a VI can be linked to only one project library. If you want to link a VI to a new library you will need to break the connection by selecting Disconnect VI from Library from the File menu.

4.11.4 PROJECT FILE ORGANIZATION

The larger the application, the more difficult it can be to manage all the files in the Project Explorer. You may want to take the time to organize the project so that

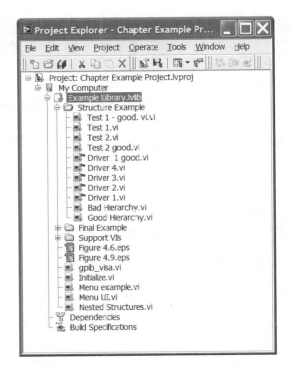

FIGURE 4.27

related items are in common folders. This can make maintenance easier by being able to work on items that are logically grouped together.

To add a new folder to a project or library you can right click on the project item and select Folder from the New option. If you want to add a directory from your files to the project you can right click on the item and select Add Folder. LabVIEW will create a folder in the project that has the same name as the folder on your computer. The files and subfolders that were contained in the original folder will then be added to the project. As a reminder, the actual files on your computer do not change location, the project maintains links to the files on your machine.

The link between the Project Explorer and your machine is not an active link. If you change files in the folder or the folder location, the changes will not be reflected in the Project Explorer. If you want the changes to be available in the project as well, you will need to manually update the project.

National Instruments recommends that you create a separate directory for each project. You should include your project and library files in the directory. If all of your project files are stored in a common directory it will be easier to locate and manage files on your computer.

4.11.5 BUILD SPECIFICATIONS

While storing and editing your application in a Project file has many benefits, ultimately you will need to get the code you have written to an end user or to a test system. There are several ways to distribution your LabVIEW code. You could build a source distribution, an executable, a shared library or a zip file. These options are configured and executed through Build Specifications. A Build Specification can only be created through the Project Explorer.

Once you have a project started you may decide you need to make a copy of the code in another directory in order to make some experimental changes. You might also want to give a copy of the code you have been working on to a coworker to tryout or debug. Regardless of why you want to make a copy of your code, you need to be able to pull all the needed files for a project together. This is where the Source Distribution comes in. The Source Distribution is similar to the Development or Application Distribution operations in earlier versions of LabVIEW. You can save all VIs to a new location. You can choose to include vi.lib, user.lib and instr.lib files. You can customize VI properties, apply passwords and remove diagrams.

The first step in creating a Source Distribution is to open a Project. Under project you can right click on the Build Specifications and select Source Distribution. A dialog box will open with three categories of inputs. The first is the Distribution Settings. Here you will select your Packaging Option. You have the choice of distributing the files to a single location, a single location with the same hierarchy that is under My Computer in the Project Explorer and you have a custom option. If you select custom, you will be able to select where all of the files are installed. You can create multiple locations for files. One benefit of distribution of the files in this method is the application will keep track of the file locations, so you will not get a broken arrow after the files are moved.

The second option for Source Distribution is the Source File Settings. Here you will have options for every item in the Project Hierarchy based on the type of item. If you select a VI, you will have the option to include the file, which destination it is going to and password options. You will also have the option to customize the VIs settings, including removing the code diagram. Once you have configured your file settings you can preview your distribution. The preview will show you where the files will be distributed.

Once you are done you can select build to create the specification. Your new build specification will appear in the Project Explorer window. You can right click on the Source Distribution to Build (distribute), remove the distribution or edit the settings.

Source Distributions are great for making copies of your code and sharing code with other developers, but you might be creating an application for a customer. Even if you are the end user, you may not want to purchase a full LabVIEW license for each test machine you are using. These examples are both excellent reasons to build an application. A LabVIEW application is an executable program with an .EXE extension (.APP for Mac users). What if your end user is programming in another language? This would require building a shared library (DLL) so that the end user can reuse your functions. You can accomplish both of these tasks by using Build Specifications.

The application or shared library requires only that the LabVIEW runtime engine be installed on the target computer. The runtime engine includes the files necessary to run your application. The runtime engine can be installed with your application or can be downloaded from the National Instruments website. To be able to build an application or shared library you need to have either the Professional Development version of LabVIEW or an add-on package. The creation of an executable or shared library is similar to the procedure for creating the Source Distribution. An explanation on how to build an application using the Project Explorer is given in Chapter 2.

The final option is for the Build Specifications in the ZIP file. Here you can add all or some of the Project files to a compressed Zip file at a specific location. You can insert comments into the build specification. The addition of source files is simply a side-by-side window that allows you to add or remove files in a Project to the destination Zip file.

4.11.6 SOURCE CODE MANAGEMENT

Working on large projects often requires using source code management software (SCM). SCM applications allow you to maintain versions of your code as well as be able to prevent other users from modifying your code while you are working on it. The project works with several SCM applications for the management of your project files. Chapter 2 discusses SCM management through the project in detail.

4.12 SUMMARY

Developing an application requires good planning and a design process that should be followed. Following a formal process helps avoid costly mistakes and revisions to the code at the end. One software model will not be suitable for everyone. However, following the requirements, design, code, and test phases will aid in developing applications. The structure of the application, which is decided on during the design phase, is an essential piece of the process. It will determine many crucial aspects of the program. The three-tiered approach, as described, embodies the desired characteristics of most applications: the ability to make future modifications, the ability to add features, ease of maintenance, the ability to reuse code, and layers of abstraction. When the strict partitioning of levels is used in conjunction with the state machine, all the characteristics are further enhanced.

A summary example will help in applying the topics and ideas presented in this chapter. Suppose that Company A is involved in the sale and production of Widget A. Let's follow the steps that would be required to develop an application that would be used in testing the widgets to determine if they meet specifications.

This first step involves defining the requirements for the test program. The goal of this application must be defined before beginning to code. After discussing the program with the appropriate people you enumerate the following requirements:

1. Parameters H, W, and D are to be tested using instruments H, W, and D, respectively.
2. The operator will be a factory technician and will need the flexibility to select individual tests as needed.
3. The program should alert the operator when one of the widgets is out of specification limits.

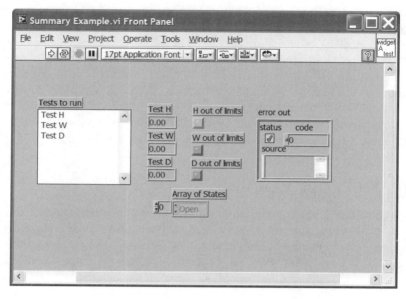

FIGURE 4.28

4. There should be provisions for the addition of tests; measuring Parameter Z using Instrument Z is in the foreseeable future.
5. Company A is planning to produce Widget B next year, which will be tested for H, W, and several other parameters.

The next step is to decide on the structure of the program. We will utilize the three-tiered approach and take advantage of the benefits it provides. The user interface of this application should be simple but flexible enough to provide the operator the level of control needed. Figure 4.28 shows what the User Interface looks like for this application.

The Main Level will use the state machine. It will also abstract the lower levels for the operator. The following states will be needed for this application: Open (to open all communication handles to instruments), Error, Initialize (to put all instruments into a known configuration), Test H, Test W, Test D, and Close (to close all communication channels). Figure 4.29 shows this state machine.

Each test is contained in its own VI, and each state consists of a single test. The Test Level is composed of Test H, Test W, and Test D. Each test VI calls the necessary drivers to perform the measurement. If the operator selects a single test to perform, the other tests will not be executed unnecessarily. An array of states will be built using the selections made by the operator. The first state that will be executed is the Open, and the last state is the Close.

The application takes into account that the needs may evolve over time. If additional tests have to be added, that can be done quickly. Suppose we were asked to add Test M to the current program. All we have to do is follow a few steps to get the needed functionality. First we have to modify the state machine to include the

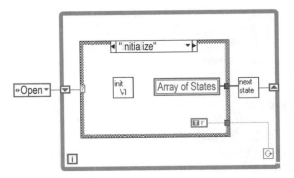

FIGURE 4.29

extra state for this Test M. Then we have to modify how the array of states is built to include the new state if the test has been selected by the operator. Next, we can modify the user interface to include the new test in the list for selection. We would also have to add a display for the measured value, and an LED that would indicate when a widget fails Test M.

Reuse is also made simple by the strict partitioning of levels. When the company begins to produce Widget B, Tests H and W can be reused. Tests H and W are stand-alone test VIs and call the appropriate drivers for performing the tests. If we decide to write another application to test Widget B, all we have to do is place the test VIs in the new application. If the new application were to use the state machine also, then we can copy and paste entire states.

The VI hierarchy for this example is shown in Figure 4.30. The strict partitioning of levels is illustrated by the distinct layers in the hierarchy. At the top is the Main

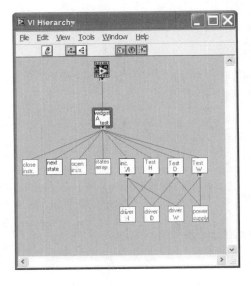

FIGURE 4.30

Level, which controls program execution. The second layer depicts the Test Level VIs. The bottom layer consists of the drivers used to test the widgets. The middle layer VI icons are in blue, and the Driver Level icons are in red. This was done purposely to distinguish the layer that the VI belongs to.

BIBLIOGRAPHY

G Programming Reference, National Instruments, Austin, TX, 1999.
The Art of Systems Architecting. Eberhardt Rechtin and Mark W. Maier, CRC Press, Boca Raton, FL 1997.

5 Drivers

This chapter discusses LabVIEW drivers. A driver is the bottom level in the three-tiered approach to software development; however, it is possibly the most important. If drivers are used and written properly, the user will benefit through readability, code reuse, and application speed.

LabVIEW drivers are designed to allow a programmer to direct an instrument, process, or control. The main purpose of a driver is to abstract the underlying low-level code. This allows someone to instruct an instrument to perform a task without having to know the actual instrument command or how the instrument communicates. The end user writing a test VI does not have to know the syntax to talk to an instrument, but only has to be able to wire the proper inputs to the instrument driver.

The following sections will discuss some of the common communication methods that LabVIEW supports for accessing instruments and controls. After the discussion of communication standards, we will go on to discuss classifications, inputs and outputs, error detection, development suggestions, and, finally, code reuse.

The standard LabVIEW driver will be discussed first. This standard driver is the basis for most current LabVIEW applications. In an effort to improve application performance and flexibility, the Interchangeable Virtual Instrument (IVI) driver was created. IVI drivers will be described in depth later in this chapter.

5.1 COMMUNICATION STANDARDS

There are many ways in which communications are performed every day. Communication is a method of sharing information. People can share information with each other by talking, writing messages, sign language, etc.... Just as people have many different ways to communicate with each other, software applications have many ways to communicate with outside entities. Programs can talk to each other, to instruments, or to other computers. The following communication standards are just some of the methods LabVIEW uses to communicate with the outside world.

5.1.1 GPIB

The General Purpose Interface Bus (GPIB) is a standard method of communication between computer/controller and test equipment. The GPIB consists of 16 signal lines and 8 ground return lines. The 16 signal lines are made up of 8 data lines, 5 control lines, and 3 handshake lines. The GPIB interface was adopted as a standard (IEEE 488). The maximum GPIB data transfer rate is about 1Mbyte/sec. A later version of the standard with added features was defined in 1987. This standard is the ANSI/IEEE 488.2. This enhancement to the standard defines how the controller should manage the bus. The new standard includes definitions of standard messages for all compliant instruments, a method for reporting errors and other status information, and the protocols used to discover and configure GPIB 488.2 instruments connected to the bus.

HS488 is a standard that NI has created. This standard is an extension of the IEEE 488 standard and increases the GPIB data transfer rate. By using HS488 controllers and compatible instruments, the data transfer rate can be increased up to 8 Mbytes/sec. The biggest benefit of the higher data transfer rate is the use of instruments that return large data sets. Instruments such as oscilloscopes and spectrum analyzers send large amounts of data to the control computer. The HS488 standard allows you to increase your test throughput.

There are two types of GPIB commands. There are device-dependent messages and interface messages. Device-dependent messages contain programming instructions, data measurements, and device status. Interface messages execute the following operations: initializing the bus, configuring remote or local settings on the device, and addressing devices.

Many current instrument manufacturers have standardized remote commands. This allows the user of an instrument to learn how to program an instrument in a shorter period of time and makes instruments more interchangeable. In order to try to make programming instruments easier, the SCPI (Standard Commands for Programmable Instrumentation) command set was developed. SCPI commands are standardized for basic functions that almost all instruments support. There are a number of instruments on the market that are not SCPI compliant. These instruments have their own command sets, and formats. This can make writing automation software difficult. One example is the T-BERD PCM analyzer. This instrument is not SCPI compliant. If you wanted to reset the instrument, you would have to search through the reference manual for the command. In this instance, to reset the instrument, you would have to write "FIRST POWER UP" to the instrument. Not only is the command not obvious, but it would require the developer to spend time hunting down commands. Figure 5.1 illustrates the GPIB driver.

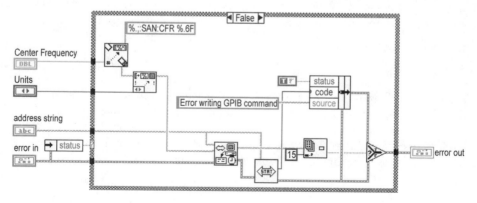

FIGURE 5.1

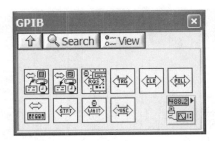

FIGURE 5.2

In the Instrument I/O section of the Functions palette there is a subpalette that contains GPIB drivers. The GPIB palette contains the traditional GPIB 488 commands. On the GPIB palette there is a subpalette, GPIB 488.2, which contains GPIB 488.2 commands. The VIs from these subpalettes can be used in conjunction with a GPIB 488.2 instrument. If the instrument you are using is not GPIB 488.2 compliant, you can only use the VIs in the traditional GPIB palette.

The primary VIs in the GPIB palette are GPIB Read and GPIB Write. These two VIs are the basis for any program using GPIB instruments. There are also VIs used to wait for a service request from the instrument (Wait for GPIB RQS), obtain the status of the GPIB bus (GPIB Status), and initialize a specific GPIB bus (GPIB Initialization). Among the remaining GPIB VIs, there is a GPIB Miscellaneous VI. This VI allows you to execute a low-level GPIB command. The GPIB palette is shown in Figure 5.2.

The GPIB 488.2 palette contains additional functions. The GPIB functions are broken into five categories: single device functions, multiple device functions, low-level I/O functions, bus management functions, and general functions. The single device functions are VIs that communicate with a specific instrument or device. Some of the functions include Device Clear, Read Status, and Trigger. The multiple device functions communicate with several devices at the same time. The VIs define which devices to communicate with through an array of addresses that are input. This category of VIs includes VIs to clear a list of devices, enable remote, trigger a list of VIs, and VIs to perform serial or parallel polls of the devices.

Low-level I/O VIs allow you to have more control over communications. The VIs in this category include functions to read or write bytes from a device, send GPIB command bytes, and configure a device in preparation to receive bytes. The Bus Management functions are VIs used to either read the status of the bus or to perform functions over the entire GPIB. The VIs in this category include VIs to find all listeners on the GPIB, to reset the system, to determine the state of the SRQ line, and to wait until an SRQ is asserted. Finally, the general functions are used to make an address or to set the timeout period of the GPIB devices. The GPIB 488.2 palette is shown in Figure 5.3.

5.1.2 SERIAL COMMUNICATIONS

Serial port communications are in wide use today. One of the advantages of serial communication versus other standards like GPIB is availability: every computer has

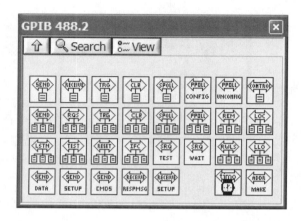

FIGURE 5.3

a serial port. Another benefit to serial communications versus GPIB is the ability to control instruments at a greater distance. The serial standard allows for a longer cable length.

The most common serial standard is RS-232C. This protocol requires transmit, receive, and ground connections. There are other lines available for handshaking functions, but they are not necessary for all applications. Macintosh serial ports use RS-422A protocols. This protocol uses an additional pair of data lines. Due to the additional data lines, the standard is capable of transmitting longer distances and faster speeds reliably. There are other serial protocols available, but these are the most widely used at this time.

The serial port VIs are in the Instrument I/O section of the Functions palette. This subpalette consists of VIs used to read data from the serial port, write data to the serial port, initialize the serial port, return the number of bytes available at the serial port, and to set a serial port break. The Serial Port Initialize VI allows you to configure the serial port's settings. In order to have successful communications between a serial port and a device, the settings of the port should match the device settings. The settings available are buffer size, port number, baud rate, number of data bits, number of stop bits, data parity and a flow control cluster. This flow control cluster bundles together a number of parameters, including a number of handshaking settings. The Serial VI palette is shown in Figure 5.4.

A programmer using the serial standard must ensure that the serial write does not overflow the buffer. Another issue is making sure all of the data is read from the serial port. There are a number of LabVIEW built-in functions designed to configure the buffer size and to query the number of bytes available at the serial port. Figure 5.5 shows a VI

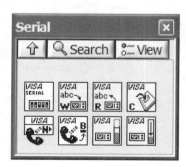

FIGURE 5.4

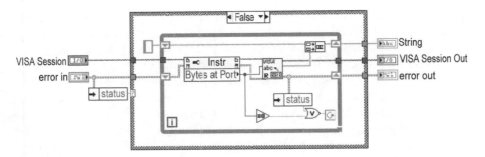

FIGURE 5.5

written to read information from the serial port. This VI performs the read until all of the desired data has been read.

There was a change in LabVIEW serial port communications starting in Lab-VIEW 7. Whereas the older version of serial port VIs still work and can be found in C:\Program Files\National Instruments\LabVIEW 8.0\vi.lib\Instr\serial.llb, the main serial port VIs in the Serial palette were converted to VISA. In addition to the VISA serial VIs, the legacy serial port VIs were also built on VISA communications. This change requires that VISA be installed with LabVIEW even if your application is only doing serial communications.

There are additional serial port standards, which would require a separate discussion. These standards are the Universal Serial Bus (USB) and Firewire (IEEE 1394). USB allows you to plug devices into a common port, and gives you the ability to "hot swap" instruments. There are a number of hardware devices available that are USB capable. In addition, National Instruments builds devices to take advantage of this technology, including a GPIB-to-USB Controller. This external box connects to the PC through the USB port and allows the user to connect up to 14 GPIB instruments without having to have a GPIB port on the PC. This is especially useful when using a laptop computer without I/O slots; the controller can plug into the USB port. You can also find boxes available for adding serial ports to your computer through a USB converter. The new serial ports look just like a standard serial port to your computer. USB supports transfer speeds of 1.5 to 12Mbps.

Firewire is actually the Macintosh implementation of the IEEE 1394 standard. Firewire is a name brand like Coke or Kleenex. Firewire allows hot-swapping of devices and can daisy-chain up to 16 devices. The main benefit to Firewire is speed. The 1394a Firewire standard boasts speeds of 100, 200, and 400 Mbits/Sec. Revisions to the IEEE 1394 standard (1394b) allow for speeds of 800Mbps, 1.6 Gbps and 3.2 Gbps. Currently, there are only 800Mbps devices commercially available. The 1394b standard also increases the allowable cable length from 4.5m to 100m.

5.1.3 VXI

VME Extensions for Instrumentation (VXI) is a standard designed to support instrument implementation on a card. VME is a popular bus architecture capable of data rates of 40MB/s. VXI combines the speed of the VMEbus with the easy-to-use command set of a GPIB instrument. The goal of VXI instrumentation is to produce a small, cost-reduced hardware system with standardized configuration and programming. The VXI Plug&Play standards promote multi-vendor interchangeability by standardizing the instrument commands for all VXI instruments. By implementing instruments on cards, the size necessary to implement a test station can be greatly reduced. The ability to implement a number of instruments in a small frame allow the test developer to create a test site in places that were not practical before, freeing up resources for other applications. The VXI standard also gives the user the flexibility of custom solutions. Cards can be made and utilized to implement solutions that are not available off the shelf. The VXI VIs are contained in a subpalette of the Instrument I/O palette.

NI advises against using VXI for new applications. The availability of VXI code is only to maintain compatibility with existing applications. The push is to use VISA functions for instrument communications going forward.

5.1.4 LXI

Due to users' desire to be able to setup existing and new devices without having to use special cables or controllers as well as to be able to reduce cost (specialized VXI modules increase the cost due to lower volume), a new standard is being developed. The LXI standard uses LAN (Ethernet) as the system backbone. This has several benefits including the reduction in cost (no card cages or interface cards), speed, and availability (every computer has a LAN port and many newer instruments already have LAN support as well). This standard is still in the definition phase, but has the support of the major instrument manufacturing companies. I think we will be hearing more from LXI in the future.

5.1.5 VISA DEFINITION

Virtual Instrument Software Architecture (VISA) is a standard Application Programming Interface (API) for instrument I/O communication. VISA is a means for talking to GPIB, VXI, or serial instruments. VISA is not LabVIEW specific, but is a standard available to many languages. When a LabVIEW instrument driver uses VISA Write, the appropriate driver for the type of communication being used is called. This allows the same API to control a number of instruments of different types. A VI written to perform a write to an instrument will not need to be changed if the user switches from a GPIB to a serial device. Only the resource name must be modified where Instrument Open is used.

Another benefit of using VISA is platform independence. Different platforms have different definitions for items, like the size of an integer variable. The programmer will not have to worry about this type of issue; VISA will perform the necessary conversions. Figure 5.6 is a side-by-side comparison of GPIB and a VISA driver.

As is seen in Figure 5.6, the main work in a VISA application is in the initialization. GPIB communications require the address string to be passed everywhere a driver is called. If there were a change in the instrument, like using a serial instrument instead of a GPIB instrument, a large application would require consid-

GPIB DRIVER EXAMPLE VI

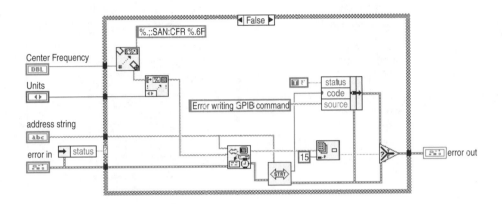

VISA DRIVER EXAMPLE VI

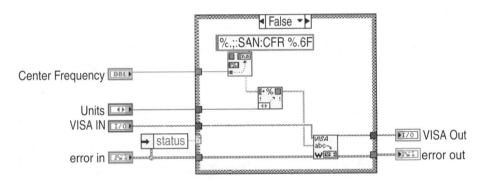

FIGURE 5.6

erable changes. All the drivers would have to be changed. An application using VISA would require changing only the input to the VISA Open VI. The resulting instrument reference would still be valid for the VISA drivers, requiring no change. VISA drivers offer flexibility.

The VISA driver VIs are located in the Instrument I/O section of the Functions palette. The VISA subpalette contains a wide range of program functions. In the main palette are the standard VISA driver VIs. These VIs allow you to open a communication session, read and write data, assert a trigger, and close communications. In addition to the standard VISA VIs, there are a number of advanced VISA functions. These are contained in the VISA Advanced subpalette and three subpalettes on the Advanced palette.

The first subpalette on the VISA Advanced palette is the Bus/Interface subpalette. It contains VIs used to deal with interface-specific needs. There are VIs to set the serial buffer size, flush the serial buffer, and send a serial break. The VISA GPIB

FIGURE 5.7

Control REN (Remote Enable) VI allows you to control the REN interface line based on the specified mode. The VISA VXI CMD or Query VI allows you to send a command or query, or receive a response to a previously sent query based on the mode input.

The next subpalette is the Event Handling palette. The VIs in this palette act on specified events. Examples of events are triggers, VXI signals, or service requests. Finally, the Register Access subpalette allows you to read, write, and move specified-length words of data from a specified address. The Low-Level Register Access subpalette allows you to peek and poke specified bit length values from specified register addresses. The VISA palette is shown in Figure 5.7.

5.1.6   DDE

Dynamic Data Exchange (DDE) is a method of communication between Windows applications. This communication standard is no longer supported in the current versions of LabVIEW. The following text discusses support in earlier versions of LabVIEW.

In DDE communications, there is a server and a client application. The DDE client is the program that is requesting data or sending a command to the DDE server. Assuming both applications are open, the client first establishes communication with the server. Connections are called "conversations." The client can then request the server to send or modify any named data. The client can also send commands or data to the server. A client can either request data or request to be advised of data changes for monitoring purposes. Like the other forms of communication, when all tasks have been completed, the client must close communication with the server.

LabVIEW can act as the server or the client. One example of LabVIEW acting as a client would be a VI that obtains data from an Excel spreadsheet or writes the data to the spreadsheet. If LabVIEW is acting as a server, another Windows program could open and run a VI, taking the data obtained to perform a task.

The DDE VIs are in the Communications palette. There are VIs for opening and closing conversations, and performing advise functions, requests, and executions. In addition to the DDE function drivers, there is a subpalette contained in the DDE palette. This subpalette contains the DDE server functions. These functions are used to register and unregister DDE service and items. There are also VIs used to set and check items.

5.1.7 OLE

OLE, like DDE, is no longer supported in the current versions of LabVIEW. The following text discusses support in earlier versions of LabVIEW.

Object Linking and Embedding (OLE), or automation, is the ability to place objects from other software programs into another application. This ability allows both the expansion of the program's abilities and the ability to manipulate data in another application. An example of this would be taking a movie clip (AVI file) and embedding it in a Word file. Even though Word has no idea what a movie clip is, it can display it in the word processing environment. OLE is a method by which objects can be transferred between applications.

OLE works with objects using a standard known as the Component Object Model (COM). The COM standard defines common ways to access application objects to determine if an object is in use, is error reporting, or if there is object exchange between applications, and a way to identify objects to associate them with specific applications. OLE is a superset of the ActiveX standard and uses the same VIs. There is an in-depth discussion of ActiveX with examples in Chapter 8.

5.1.8 TCP/IP

There are three main protocols for communication across networks: Transmission Control Protocol (TCP), Internet Protocol (IP), and User Datagram Protocol (UDP). TCP is built on top of IP. TCP breaks the data into packets for the IP layer to send. TCP also performs data checking to ensure the data arrives at its destination in a singular, complete form. TCP/IP data consists of 20 bytes of IP information, followed by 20 bytes of TCP information, followed by the data being sent. The TCP/IP protocol can be used on all platforms of LabVIEW and BridgeVIEW.

Every computer on an IP network has a unique Internet address. This address is a 32-bit integer, usually represented in the IP dotted-decimal notation. The address is separated into 8-bit integers separated by decimal points. The Domain Name Service (DNS) system is a database of IP addresses associated with unique names. For instance, a user looking up the National Instruments Web site (www.ni.com) will be routed to the appropriate IP address that corresponds to the name. This process is known as "hostname resolution."

There are a number of standards using TCP/IP that can be implemented using LabVIEW. Telnet, SMTP, and POP3 are a few applications built using the TCP/IP protocol. Telnet can be used for providing two-way communications between a local and remote host. POP3 and SMTP are used to implement mail applications.

FIGURE 5.8

With TCP/IP, the configuration of your computer depends on the system you are working on. With Windows, UNIX, and Macintosh Version 7.5 and later, TCP/IP is built in. For earlier versions of Macintosh Operating systems, the MacTCP driver needs to be installed.

The TCP palette is located in the Communication section of the Function palette. The VIs in the TCP palette allow you to open and close connections. Once the connection is opened, you can read and write data through the VIs in the TCP palette. There are also VIs to create a listener reference and wait on listener. The IP to string function allows you to convert an IP address to a string. There is an input to this function to specify if the address is using dot notation. A function to convert a string to an IP address is also available. The VIs in this palette are shown in Figure 5.8.

5.1.9 DataSocket

DataSocket is a programming technology that facilitates data exchange between applications and computers. Data can easily be transferred between applications over an Internet connection. DataSocket is built using TCP/IP and ActiveX/COM technologies. The DataSocket server can reside on the local machine or on another machine on the network. You can read data using DataSocket http, ftp, and local files. DataSocket can also read in live data through a dstp (DataSocket transfer protocol) connection. You also have the ability to control your LabVIEW application through a Web interface by using CGI functions with DataSocket. The Shared Variable in LabVIEW 8 is replacing the DataSocket functionality. Support of DataSocket will remain, but new applications should start transitioning to the new variable. The Shared Variable is discussed in detail in other chapters.

The DataSocket VIs are in a subpalette of the Communication section of the Function palette. The DataSocket VIs work in the same way VISA or other standard LabVIEW VIs operate. There are VIs for opening and closing connections. The Open function will open communication based on the URL input and the access mode input. The URL input must be one of the above-mentioned protocols. The output of the Open function is a DataSocket reference. This reference is used in the same manner as a typical connection refnum. The remaining VIs use this reference to perform actions on the desired information. You can then read or write a string, Boolean, integer, or a double value. If you want to read or write arrays of these data types, the necessary VIs are available in the DataSocket Write and the DataSocket

FIGURE 5.9

Read subpalettes. The Advanced subpalette gives you the ability to read or write variants. In addition to the variant functions, there are also low-level functions for performing DataSocket communication. These functions include VIs to connect and update data. Finally, there is a VI to control the DataSocket server programmatically. You should also be able to access the DataSocket server from your Start menu under the National Instruments DataSocket name. The DataSocket function palette is shown in Figure 5.9.

If you want to perform live data updates, you first need to determine if the DataSocket server is running on the local machine. The typical format for a local write data to a DataSocket server is dstp://localhost/test. This assumes that "test" is the label for the data you are writing to the server. If you are using a local server, the DataSocket server will need to be launched through the function in the DataSocket Advanced subpalette. Then, you will need to open a DataSocket connection with Write Attribute selected. You can then write the data you want to share to the DataSocket server. If you are running the DataSocket server on another computer, the machine address will need to be in the DSTP address.

To read the data from the server, you will again need to determine if the server is local or on a remote machine. Once you have the server name resolved, and have a connection open to the server with the read attribute, you can use the Read DataSocket VIs to read the data in. You will need to use the Update data VI if you want to read new data after it has been written to the server.

To read and write static data, the process is the same. The only difference is the URL used to connect to the DataSocket. Examples of a VI used to generate live data to the DataSocket server, and a VI to read the data from the DataSocket server, are shown in Figure 5.10. This example includes additional attributes. This allows items like time and date stamps to accompany the data that is being transferred. The DataSocket server is launched on the same PC as the Data Write VI. There are additional examples in the LabVIEW on-line reference.

5.1.10 Traditional DAQ

Data acquisition (DAQ), in simple terms, is the action of obtaining data from an instrument or device. In most cases, DAQ is performed using plug-in boards to collect data. These plug-in boards are made by a number of manufacturers, including

DataSocket Write VI

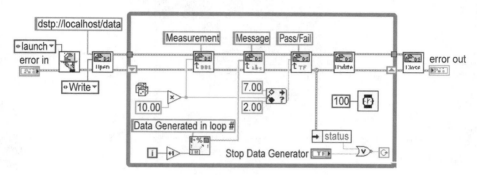

DataSocket Read VI

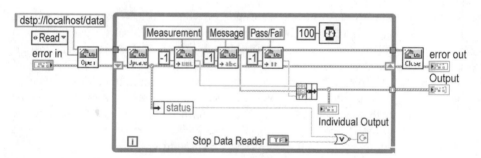

FIGURE 5.10

National Instruments. These DAQ boards perform a variety of tasks, including analog measurements, digital measurements, and timing I/O. One convenience is the ability to obtain boards for PC, Macintosh, and Sun workstations.

One of the benefits of using National Instruments boards is the availability of NI-DAQ drivers for the boards. Although other manufacturers' boards are compatible with LabVIEW, the DAQ library will most likely not be compatible with the board. Most board manufacturers do provide their own drivers for their equipment; some even have drivers written in LabVIEW. Even if the code is not written in LabVIEW, DLLs can be implemented by using the Call Library function. Code Interface Nodes (CINs) can be used to implement drivers written in C source code.

The Data Acquisition subpalette is a part of the Functions palette. The Data Acquisition palette is made up of six subpalettes: the Analog Input VIs, Analog Output VIs, Digital I/O VIs, Counter VIs, Calibration and Configuration VIs, and Signal Conditioning VIs. The Data Acquisition subpalette is shown in Figure 5.11. Each of the subpalettes is comprised of a number of VIs of varying complexity and

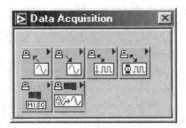

FIGURE 5.11

functionality. There are four levels of DAQ VIs. They are Easy VIs, Intermediate VIs, Utility VIs, and Advanced VIs. As a rule, the Utility VIs are stored in their own subpalette. The Advanced DAQ VIs are also stored in their own subpalette. The main difference between the Easy VIs and the Intermediate VIs is the ability of the Easy VIs to run as stand-alone functions. These VIs call the higher-level VIs to perform the task. The Easy VIs allow you to pass in the device number and channel numbers. The VIs will also perform error-handling functions to alert you if an error has been encountered.

Let's look at the Analog Input subpalette. The palette consists of the four types of VIs described above. The Easy VIs include functions to acquire one or multiple waveforms from an analog input. There are also functions for acquiring samples at the designated channels. The Intermediate VIs allow you to configure the hardware and associated settings, start an acquisition, read the buffered data, make single scan acquisitions, and clear the analog input task. The Analog Input palette contains two subpalettes. The first subpalette contains the Utility VIs. These VIs include functions to initiate a single scan, a waveform scan, or a continuous scan. The second palette contains the Advanced functions. The Advanced functions palette contains VIs to perform configurations, read the buffer, set parameters, and control analog input tasks. We could devote a number of chapters on DAQ functions, but the DAQ functions are described in great detail in the NI Data Acquisition Basics manual. We will not attempt to cover material that is concisely covered already.

5.1.11 NI-DAQmx

Whereas the traditional DAQ (Legacy) VIs have been used successfully to automate data acquisition applications for more than a decade, support of new capabilities and functional improvements resulted in the creation of DAQmx. DAQmx is a superset of the DAQ Legacy VIs. You can still do all the functions that you used to be able to do, but now you have additional features such as multithreaded execution, additional driver functionality and configuration applications like DAQ assistant and express VIs. DAQmx is available for Windows and Linux operating systems. NI-DAQmx Base was created to provide a subset of DAQmx functionality for MAC OSX, RTX and Pocket PC operating systems.

The first benefit of DAQmx is the support for newer devices. New devices for data acquisition are continually created with added functionality over older model devices. In order to support the new functionality new drivers will need to be created.

The new drivers will be created for DAQmx only. Although the support for traditional NI-DAQ will continue, no new drivers will be created. In order to continue to develop applications it is advised to use DAQmx in order to make sure the code will continue to be supported.

In most cases DAQmx improves application speed. This is due to a couple of new features. First, the legacy drivers ran only in a single thread. DAQmx now supports multithreaded execution speeding applications that can do two or more acquisition tasks in parallel. The second improvement in speed is by application design. Now with more control over operations such as reserving resources and configuration so that the user application can be designed to perform these operations only when needed to reduce expensive overhead.

Finally, DAQmx tools can make application development easier. The ability to use the express VI for configuring an acquisition task can shorten the amount of time needed to get a test running. This can be valuable when the test program will change often, and having to recode a VI each time a new test is needed could be tedious. The DAQ Assistant is an application that can make coding easier by walking you though each step of building an acquisition task step by step. An example will now be shown for a simple read of information from an analog voltage input from a DAQ card.

The first step for the example is to set up the hardware. The hardware setup is done in the measurement automation explorer (MAX). Under the Devices and Interfaces folder all available hardware should be shown. Any installed DAQ cards should show up here. If you are developing the application at your desk before installing it on your acquisition system, your device will not be here. You can add a device with the drivers for your hardware, or you can create a simulated device in order to be able to develop and test your application. You can do this by right clicking on the folder and selecting Create New. Here you can select the appropriate drivers. Once you have the device selected you can configure settings such as initial settings and connector type. For this example we have created a simulated DAQ card (PCI-6220) at DEV1. The MAX window is shown in Figure 5.12.

Now on the code diagram the DAQ Assistant will be selected from the input palette under the Express Palette. The initial DAQ Assistant interface that automatically launches is shown in Figure 5.13. For this example we are going to select Voltage under the Analog Input heading. After selecting the type of acquisition the input screen gives you the options to set up what hardware channels to use. The available information is based on what has been already set up in MAX. Here I am able to select specific analog input lines on DEV1 (the simulated 6620 card).

Now that you have completed the initial setup the DAQ assistant opens a window for configuring the channel parameters. This interface is shown in Figure 5.14. Here you can add, remove and test the channels to make sure you have all the settings needed. There is another tab on the bottom of the window. This tab will show the connection diagram. Here you can select each channel and see what wires are connected based on the defined DEV1 connector in MAX. This window is shown in Figure 5.15. Once configuration is complete the resulting DAQ block will be on the code diagram. The block is the standard Express VI block with the blue border. Now you can wire controls to any inputs you want to be able to change. The output

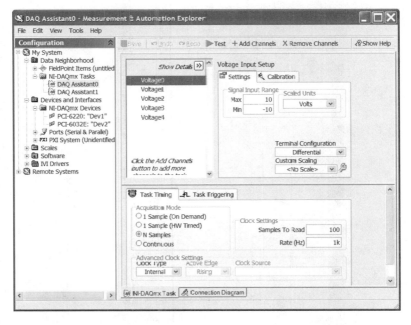

FIGURE 5.12

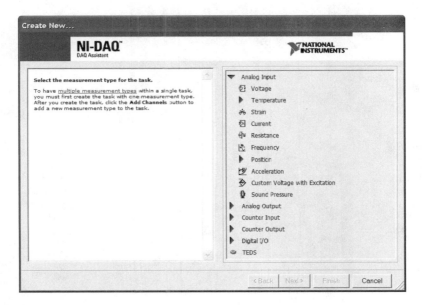

FIGURE 5.13

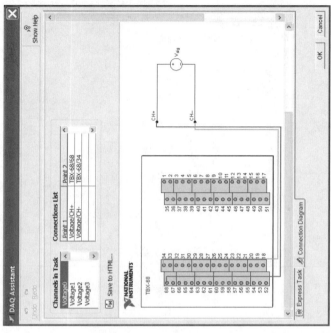

FIGURE 5.15

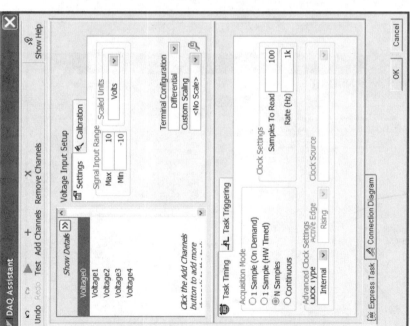

FIGURE 5.14

waveform can be connected to the DAQ Assistant block output. Running the VI produces a graph of the input channels.

Some people do not like to use the express VIs because the code is hidden from the user. You have two options. You can wire the application using the DAQmx subVIs the same way you would use the Legacy DAQ VIs or you can configure the application with the DAQ assistant above and choose to generate DAQmx code by right clicking on the icon. Figure 5.16 shows a simple single channel analog input acquisition. Notice the DAQmx VIs are polymorphic. A polymorphic VI is a VI that can perform different functions based on an input value. In this case there is a VI for creating a channel. In this instance analog input is selected, so the VI will create an analog input voltage channel. If the programmer changed the text below to read analog output voltage (AO Voltage) an analog output voltage channel would be created. This setting can be changed by right clicking on the text selector at the bottom of the VI as is shown in Figure 5.17. This is another way to make programming easier. The same code can be used for different applications by changing the functionality of the polymorphic VIs without having to insert a new VI.

5.1.12 FILE I/O

File input and output is a type of driver that people do not often think of. The ability to read data from a file and write data to a file in many ways is similar to reading data from and writing data to a GPIB instrument. You require a means to identify the file you want to communicate with. Instead of a GPIB address you have a file path. You also need to be able to transfer data from one place to another. Instead of passing data between the computer and the GPIB instrument, you are passing data between the LabVIEW program and a file. The File I/O functions are very similar to instrument or communication drivers.

The File I/O VIs can be found in the File I/O section of the Function palette. This subpalette contains a number of file functions as well as subpalettes containing VIs pertaining to binary files, file constants, configuration files, and advanced file functions. The standard file I/O functions include VIs for opening/creating a file, reading data from a file, writing data to a file, and closing a file. In addition to these functions, there are VIs for writing and reading data from a spreadsheet file, writing or reading characters from a file, and reading lines from a file. The File I/O palette is shown in Figure 5.18.

There are two remaining functions that are included with the standard file I/O functions. The first VI allows you to build a file path. This VI creates a new file path by appending the file name or relative path from the string input to the base path. The default value of the base path is an empty path. The result is the combined file path. If there is a problem in one or both of the inputs, the VI will return "not-a-path." The second function takes a file path and breaks it apart. The last section of the path is wired out as a string filename. The remainder of the path is wired out as a path. The VI will output an empty string and "not-a-path" if there is an invalid input. The binary file VIs allow you to read and write 1- or 2-D arrays of data to a byte stream file. The byte stream file can be in a signed word format or a single precision format. The configuration file palette contains VIs used to read and modify

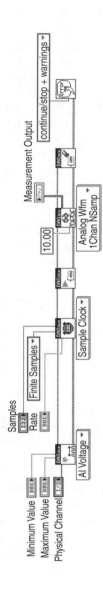

FIGURE 5.16

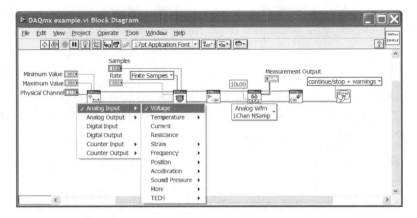

FIGURE 5.17

FIGURE 5.18

FIGURE 5.19

information in the configuration files. The File Constants palette contains VIs that allow you to access the current directories, paths, or VI library directories. In addition to these functions, there are constants that can be used to create inputs to the file I/O VIs.

The Configuration Files Palette contains VIs used to read from and write to INI formatted files. These VIs can be very useful when requesting configuration information from a user to set up the code execution. Using these VIs you can save the settings that the user had entered to be loaded the next time the application is run so that the user does not have to reenter all the information. There could even be options to save configurations in the event there may be multiple setups needed.

The Advanced palette contains VIs that perform a number of file-related tasks. The Advanced palette is shown in Figure 5.19. The File Dialog function displays the file dialog box for the user to select a file. The output is the path of the file selected. The Open File VI allows you to specify a datalog type. There is a function used to find the offset of the end of file (EOF). The seek function allows you to begin a file in a position other than the beginning of the file. There are VIs used to set access rights for a specified file, as well as to find out information on the file, directory, or volume.

There is a set of five VIs in the Advanced palette that performs actions on directories. There is a VI that allows you to move a file or directory. There are also VIs that allow you to copy a file or directory, as well as delete a file or directory. The New Directory function allows you to create a directory at the specified path. The List Directory function lists all of the file names and directory names that are found in the directory path.

The final set of functions in the Advanced palette are VIs used to convert between strings and paths. The functions can perform the functions on a single string or an array of strings. There is also a VI that converts a refnum to a path. These VIs are useful when converting string paths created by the user in a user interface to a file path to perform file functions.

We will now give a quick example of how to read and write data when dealing with datalog files. The first step is to create the data type used for storing the data. For this example we will be recording three distinct values per datalog value. The first is the index of the data. This is simply the value of the For loop index used to create the data. The second item in the data cluster is the data. The data for this example is simply random numbers generated between 0 and 10. The final data type used for the cluster is a date and time stamp. This value is written as a string. To summarize, our data type consists of an integer, a real number, and a string.

The first step is to create the code to perform the data generation. The For loop executes 100 iterations. Inside the For loop, the loop index, the test data, and the time and date string are bundled into a cluster. This cluster is wired to the output of the For loop, where auto indexing is enabled. When all the data has been collected, the New File VI is used. The File Path contains the name and location of the file you are writing the data to and will be needed when you want to retrieve the data. The file path is the only required input. There are a number of other inputs to the VI that can be wired, or left as default. To write and read datalog files, you will need to wire a copy of the data format to the datalog type. Wiring the actual data to the input, or wiring a constant with the same data type, can do this. The other inputs are permissions, group, deny mode, and overwrite. The overwrite input for our example will be given a "true" value. This allows the program to overwrite an existing file with the same name as specified in the file path input. If the input were "false," the program would error out when trying to create a new file that already exists.

Once the file is created, the next step is to write the data out. The Write File VI is used to send the collected data to the datalog file. The inputs of the Write File VI include convert eol (end of line), header, refnum, positive mode, positive offset, error in, and the data. The only required inputs are the refnum and data inputs. The data from the For loop is wired to the data input. The final step of this subVI is to close the file using the Close File VI.

The next step is to create a VI to read the data back from the file. In this VI, the Open File function is used to create a connection to the file. The File Path input

Create Logfile VI

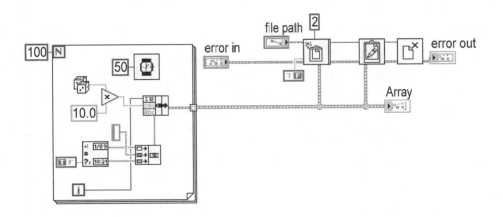

Read Logfile VI

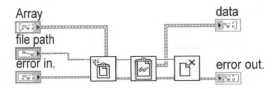

FIGURE 5.20

is used to point the VI to the datalog file. In addition to the file path, the data type is wired to the Datalog Type input. This data type needs to match the data type of the cluster we wrote to the file. This allows you to read the information back in the appropriate format. In addition to the datalog type and file path, you can set the open mode and deny mode for the file. This allows you to determine the file permissions. Once the file is opened, you need to use the Read File function. This VI is used to acquire the data from the file, and write the data to an indicator. Again, the final step is to close the file. The code diagram for the Datalog Write VI and the Datalog Read VI is shown in Figure 5.20.

5.1.13 CODE INTERFACE NODE AND CALL LIBRARY FUNCTION

LabVIEW has the ability to execute code written in C as well as to execute functions saved in a DLL. There are two methods for calling outside code. The programmer can call code written in a text-based language like C using a Code Interface Node (CIN). The programmer also has the ability to call a function in a DLL or shared library through the use of the Call Library function. A short description of each will follow.

The CIN is similar in some respects to a subVI. The CIN is an object on the block diagram of a VI. The programmer can enter inputs required to execute a function, and wire the outputs of the CIN to the remainder of the program. The main difference is a subVI is code written in the G language to perform a function, whereas the CIN executes text-based code to perform the function. The CIN is linked to compiled source code. When the execution of a block diagram comes to the CIN, LabVIEW calls the executable code, returning the final outputs to the VI.

There are a number of reasons for using the Code Interface Node. One benefit is the ability to use existing code in your LabVIEW program. If a function is already written in C, you have the ability to integrate the code into your LabVIEW program to reduce development time. Another benefit to using a CIN is to expand the functionality of LabVIEW. Certain system functions that do not have corresponding LabVIEW functions can be implemented using code written in C. This can help a programmer to perform low-level programming with LabVIEW's graphic-based interface. A final consideration for using CINs is speed. Whereas LabVIEW is fast enough for most programming tasks, certain time-critical operations such as data acquisition and manipulation can be done more efficiently in a programming language like C. The use of the CIN allows the programmer to use the right tool for the right job.

The ability to use prewritten code is a key to reducing development time. Functions to perform many Windows functions have already been written. These functions are typically written in C, and are stored in Dynamic Link Libraries (DLLs). LabVIEW can call these Windows functions in two ways. The first way is through the use of a Code Interface Node. An easier method for calling DLL functions is through the use of the Call Library function. The main difference between calling C code in a CIN and using the Call Library function to call a DLL is the integration of the source code. When using a DLL, the code remains in its library; it is not copied into the executable files of the application. The other obvious difference is the fact that DLLs are Windows-specific, whereas the Code Interface Node can be used across platforms.

For more information on the Code Interface Node, the Code Interface Reference Manual can be found on National Instruments' Web site. The PDF file covers how to integrate a CIN on any platform. For information on using DLLs, there is an application note on the NI Web page. Application Note 087, "Writing Win32 Dynamic Link Libraries (DLLs) and calling them from LabVIEW," discusses the methods for using DLLs.

5.2 DRIVER CLASSIFICATIONS

There are three main functions a driver performs. The three types correspond to the three main purposes of a driver: configure an instrument, take a measurement, or check the status. These three main types of drivers will be discussed below. When creating driver VIs, National Instruments recommends a standard format the drivers should follow. Driver libraries should contain the following functions: Initialize, Configure, Action/Status, Data, Utility, and Close.

5.2.1 CONFIGURATION DRIVERS

The first type of driver is a Configure VI. These VIs should open or close communications with the instrument, initialize the instrument, or configure the instrument for the desired use. The Initialize driver first performs the initial communications. This should include opening a VISA session if VISA is being used. The Initialize driver can also perform instrument setup and initial configurations. This can allow the instrument to begin in a known or standard state. The Configuration Instrument drivers send the necessary commands to the instrument to place the instrument into the state required to make the desired measurements. There may be a number of configuration VIs for a particular instrument, logically grouped by function or related purpose. The Close driver closes the instrument communication, the VISA handle, and any other required items to complete the testing process. It is important to close the instrument communications, especially when doing serial and TCP communications. When a serial port is open, no other applications can use the port. If the port is not closed, the port is inaccessible until LabVIEW is closed. With TCP, when you connect to another machine, the port on that machine will stay open unless you close the session or the session time out.

5.2.2 MEASUREMENT DRIVERS

Measurement drivers are used to take measurements or read specific data from the instrument. The user should be aware that a data driver does not always require reading data from an instrument. The data driver could also be used to provide data to an instrument, like sending a waveform to a signal generator. It is important to note that only one measurement should be taken per driver. This is done to promote reusability as well as to ensure the application speed is not compromised by taking unneeded measurements.

5.2.3 STATUS DRIVERS

The action/status drivers are used to start or stop a specified process, check errors, and general instrument-related information. One example would be a VI written to start and stop a Bit Error Rate (BER) test or a waveform capture from a spectrum analyzer. Another example is checking a status register to find out if a test that has been initiated is completed so the result can be read from the instrument. The VI would not change any of the instrument configurations, only the initiation or termination tasks are performed. As checking the status of an instrument can require the instrument to be reset, a set of utility drivers should also be designed. The utility drivers are used to perform tasks such as reset, self-test, etc.

5.3 INPUTS/OUTPUTS

An important aspect of a driver is the interface with the calling VIs. There are a number of standard inputs and outputs for drivers. The Error In and Error Out clusters are the most important I/Os in a driver. These clusters have three components. For the Error In cluster, the first control is a status Boolean control; a "true" indicates there is an error. The second is a numeric control to display an error code. The final control is a source string. This string can indicate where an error occurred. There are two primary reasons for using the Error In and Out clusters. The first reason is obviously error handling. If an error has already occurred in a program, the Error In cluster will pass this information to the driver, preventing the execution of the intended task. The error cluster can also pass error information out of the driver if an error occurred while the driver was executing. A discussion of error handling is described in the following section.

The second reason for using the Error In and Out clusters is flow control. The wiring of the Error Out of one VI to the Error In of another forces the order of execution because of data dependency. For example, an instrument needs to be configured prior to taking a measurement. Wiring the Error Out of the configuration driver to the Error In of the Measurement driver forces the order of execution.

The other required inputs are the instrument communication handles. Depending on the communication VIs being used, a number of different inputs could be used. We suggest using VISA standards in your drivers. This will allow the same driver format regardless of what type of communication is used to address your instrument or device. The standard method for wiring the connector pane has the VISA session in and out in the top left and right positions, respectively. The Error In and Out are in the bottom left and right positions, respectively. This consistency of location makes connections easier to wire and find.

For readability and ease of use, the programmer should use as few inputs and outputs to a driver VI as possible. The use of clusters should be avoided unless the information is packaged in a form that other subVIs would use like the error cluster. If the cluster is not passed on, the main program will need to bundle and unbundle the items. This can obscure the intention of the code and complicate the code diagram. Additionally, the complex data type will have an effect on performance.

5.4 ERROR HANDLING

Error handling is one of the most important considerations when a programming task is begun. For this reason there is an entire chapter in this book dedicated to error handling. This section will just highlight some of the driver-specific error-handling issues.

The main error handling that should be performed in the driver is the detection of errors that are passed in. If an error is passed into a driver, the driver should not execute any tasks. The driver should consist of a case statement controlled by the status field of the error cluster. The driver code would then execute only if no error passed in. When an error is passed into a driver, the instrument communication VIs will not execute if an error cluster is passed to them. Error processing should only occur in the upper levels of the program, as prescribed by the three-tiered design architecture. The benefit of not processing errors in the driver is the ability of the driver to be reused. If error processing is performed in the driver, the results of the processing may not be applicable to a new program using this driver. Doing error

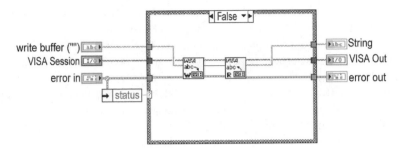

FIGURE 5.21

processing in the driver would cut down on code reuse. An example of the use of this "bypass" is shown in Figure 5.21.

The next issue with error processing in drivers is the implementation of timeouts. A driver should have a way out. If a driver is written to read the status of a register through the use of a While loop to read the data from the device, there should be a way to exit after a specified time if the desired response does not occur. This can result in setting an error if the program will not function without the desired value.

In writing applications that read data from a device, you should add code to ensure that any errors that occur during the data acquisition are handled in an appropriate manner. For example, assume you are reading data from a serial instrument. In this example you are reading the information from the serial port until the desired data is read. To perform this task, the read operation is in a While loop that is executing until the desired input is read. When the desired input is received, a "false" Boolean is wired to the conditional terminal of the While loop. If an error would occur, the desired input would never be received, resulting in the While loop continuing to execute until you stop the application. You should check the Boolean value of the error cluster in each iteration of the While loop to check for an error. The result of this error check can be combined with the result of the data check to determine whether to execute another iteration of the While loop. The Boolean from the error cluster and the data check can be combined through the Boolean logic functions to control the conditional terminal of the While loop. An example showing all three of the above-mentioned techniques is shown in Figure 5.22.

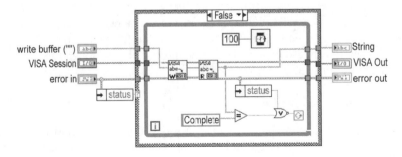

FIGURE 5.22

One type of error detection that should be mentioned is the ability to set error traps in the driver code for debugging purposes. During the development stages of a driver, "traps" can be put in place to trap and isolate errors. This can lead to faster error detection for the purpose of debugging the driver being developed. These traps can be either disabled or removed when the driver development has been completed. Some instances of error traps can be simply collecting the data being read in from a serial port, and saving the data to be reviewed by the developer. As some errors will only occur when running at full speed, recording the data for later analysis could be of great benefit. The recording of this same data would be considered unnecessary in the final driver version, hence the need for an error trap. Once the driver has been fully debugged, the trap can be eliminated. Data logging, discussed in the error-handling chapter, is a similar tool that allows you to save and view data after the VI has been executed.

When measurements are being made in a loop, or setup is being performed in a state machine, care needs to be taken with error handling. There should always be a shift register passing the error cluster to each iteration. When this is forgotten, errors become difficult to track because the error cluster gets cleared with the next iteration of the While or For loop.

5.5 NI SPY

It is difficult at times to debug drivers. Commands are sent to the instrument by the program, but are the parameters correct, how long do the calls take, is there a problem with the instrument, etc.? The developer performing the application debugging needs a way to monitor and verify that the program is doing what was intended. One tool provided by National Instruments can aid in code verification. The NI Spy utility is an application that monitors, records, and displays API calls made by National Instruments applications. The NI Spy can be used to locate and analyze any erroneous API calls that your application makes, and to verify that the instrument communication is correct.

5.5.1 NI Spy Introduction

The NI Spy program is similar to a GPIB analyzer. The NI Spy displays function call names, parameters, and GPIB status as the developer's program executes calls. The NI Spy allows access to information like the contents of data buffers, process and thread IDs, and time stamps for the start and finish times of the function calls. The spy program can also create a log of the information, although this can produce a significant performance loss.

5.5.2 Configuring NI Spy

The first step is to open the NI Spy program. If you go to the Start menu of your computer and then to the Programs folder, there should be a folder labeled National Instruments and there should be an icon for the NI Spy. When this icon is selected, the window shown in Figure 5.23 comes up. In the title bar, the name "NI Spy"

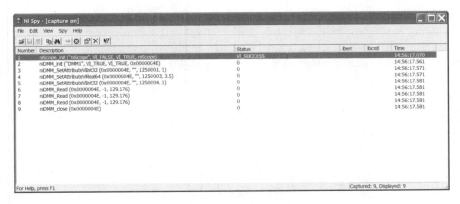

FIGURE 5.23

should appear followed by the program's status. In parentheses, the title bar will indicate whether capture is on or off. By default, Capture is off when you open the NI Spy application. Figure 5.23 shows the NI Spy window with Capture on.

Before starting the NI Spy program, the first step should be to configure the options for the application. By selecting the Spy menu, the following options are available to you: Start Capture, Options and Calculate Duration.

To modify the NI Spy capture options, select Options from the Spy menu. The NI Spy options can be modified only when Capture is off. NI Spy, by default, displays 1000 calls in the Capture window, displays Small Buffers and does not enable file logging. The Call History Depth option identifies how many API calls the NI Spy will display. If more than the selected number of API calls are made, the Capture window will show the most recent calls, discarding the calls at the beginning. If the NI Spy program is unable to display all of the API calls due to low system memory, a message box will appear giving the user the option to stop the capture or free up system resources before continuing.

The Buffer Limit per Parameter selection allows you to choose between Small or Large Buffer mode. The Small Buffer mode displays up to 64 bytes of data, whereas the Large Buffer mode displays up to 64K bytes of data. For either of these modes, if there is more data than the allowed buffer, the middle data will be removed. For example, in the Full Buffer mode, the first 32K bytes and the last 32K bytes of data will be displayed. A row of dashes between the two halves of the buffer is inserted to indicate that part of the data has been omitted.

The File Logging selection in the NI Spy options allows the program to record all calls to a log file. File logging is useful when debugging an application that causes the system to crash. If file logging in the Fail-Safe Logging mode, you can view the API calls that were captured prior to the system crash by opening the saved log file. In order to use this function, a file name must be provided to store the logged API calls. There are two modes of file logging available. The first is Fail-Safe Logging. Fail-Safe Logging is a method of guaranteeing that the log file will not be corrupted if the system crashes. The logging is accomplished by opening the log file, writing the data, and closing the log file after each API call. It should be obvious that this method of logging the data is slow. If performance and time are an issue, Fast Logging is available. This method of logging opens the file at the start. The data from each call is written to the log file when the call is captured. The file is not closed until the capture is stopped or logging is disabled. The Fast Logging method of file logging is much faster than Fail-Safe Logging, but if your system crashes, data will be lost.

If you have more than one National Instruments driver installed on your computer, you can specify which APIs you want to spy on at any time. The View Selections tab shows the API choices that are available to monitor. You also have a choice of what columns to display. Types of National Instruments drivers are GPIB-488.2, VISA, and IVI-type drivers. By default, all installed APIs are enabled. There will be a check next to the API types selected for capture. You can omit any driver on the list by clicking on the name; the check will be removed.

Finally, there is an Error tab for specifying what to do if an error occurs. You have the option of ignoring errors, stopping if any error occurs or stopping only on specific error parameters. The ability to pinpoint the specific breakpoint is a big plus when trying to isolate an elusive fault.

5.5.3 RUNNING NI SPY

There are three ways to start capturing API calls. The first is to select Start Capture from the Spy menu. The second method is to click on the arrow button on the toolbar. Finally, the user can press F8 to turn Capture on. Once you turn Capture on, you can run your application. When you want to view the captured information you can return to NI Spy to view the captured calls. To turn Capture off, click on the red "X" button on the toolbar.

You can view the API calls in the main NI Spy window as NI Spy captures them. The captured API calls are displayed in the order in which they are received. There is one line of information displayed for each captured call. The information includes the number of the call, a C-style function prototype, and the start time for the call.

By using the Properties dialog box you can see detailed call information for every captured API call. To see the properties of a specific call, double-click on the call in the Capture window, right-click on the call and select properties, or select Properties from the View menu. The Properties dialog box includes one to five pages of detailed information on the captured call. All API captured calls have a General tab, most captured calls have Input and Output tabs, some captured calls have a buffer page, and some IVI captures can have an Interchange Warning tab. The General section displays the process and threads IDs, the Windows handles, and the start and stop time statistics. The Input page displays the API call's input parameter types and values. The Output section displays the parameters that were returned after the call completion. The buffer page is present only for calls that involve the transfer of a buffer of data; this page displays the contents of the data buffer. Finally, the Interchange Warning section displays warnings about the specific call with respect to instrument interchangeability. This option is available for IVI drivers.

To search through the list of captured calls to find a specific string in the API function names, parameter values, or any other string, select Find from the Edit menu. Enter the text that you want to search for in the Find What box. Click the Find Next button to find the next captured call containing the specified string. The Match Errors Only selection can be used to limit the search to captured calls that have an error. If no search string is specified, the search locates the next captured call that failed. The Match Case selection specifies whether the search is case sensitive.

5.6 DRIVER GUIDELINES

Aside from the general driver information, there are a number of implementations that can add robustness and reusability to a driver. This section will give an overview of some of the functionality that should be added to a driver to accomplish the desired results.

One guideline that should be followed is the method of making only one measurement per driver. Since the programmer will want different measurements at different times, the programmer should keep one measurement to a driver. This allows the code to be reused easily. The user of the driver will not have to take a number of measurements in order to receive one desired value. Making multiple measurements when only one measurement is desired limits performance.

When developing a driver, the programmer should try to combine configuration settings into logical groups. If configuring an RF generator requires setting four different parameters every time, the configuration of those parameters should be in a common driver. This would allow the user to set the generator with the appropriate settings through the access of one driver.

When you are linking the controls and indicators to the connector panel of the icon, you should choose a connector configuration that will provide extra connectors. When all of the inputs and outputs have been wired, extra connectors allow for expansion without disconnecting all existing connections. When a driver is already called in a program, and if the programmer adds a new input or output, the user will not have to rewire all of the existing connections. When there are extra connectors, the existing connections do not change, allowing the current wiring to remain unchanged.

5.7 REUSE AND DEVELOPMENT REDUCTION

The biggest benefit of developing quality drivers is the ability to reuse the drivers. Even when the programmer does not expect to use a specific driver again in the future, things change quickly. There is no better feeling in software development than, when developing an application, you realize that the underlying code has already been written. If a driver has been properly written, applications that are completely different could still use the same driver. The ability to reuse code is the biggest factor in cycle-time reduction. By not having to rewrite drivers, which includes time to learn the equipment, coding, and debugging, the user can dramatically reduce the time required to develop an application. Making drivers generic enough to reuse can require more time and effort up front, but the benefits that can be realized are substantial.

There are many drivers for numerous instruments and manufacturers that have already been written. The first place you can look for an instrument driver is on the installation CD that came with your LabVIEW application. The second disk is a disk of instrument drivers. In addition to these drivers, many of the drivers are available on the National Instruments Web page. Not only is this resource a comprehensive list of drivers, but also they are the most recent versions. The National Instruments ftp site is ftp.ni.com. Your login is "anonymous" and your password is your Internet address.

Many drivers available on the National Instruments Web page have been submitted to NI and accepted for distribution. There are standards to which NI requires all drivers submitted to adhere. Many of the standards have already been discussed, and these standards can be found in the application note, AN106. As the drivers have already been designed to the required standards, they should be easily inserted into your application with no modification. This allows the programmer to concentrate on developing the application without concern about the underlying communications. This can lead to significant development time reduction.

For unusual or difficult-to-find instrument drivers, there are some other resources available. The LabVIEW Info Group is a place you can try. The Info Group is a large knowledge base that you can utilize. For subscription requests you can send an e-mail to info-labview-on@labview.nhmfl.gov. To post a message to the Info Group, send an e-mail to info-labview@labview.nhmfl.gov. There are also some other user groups such as LAVA (LabVIEW Advanced Virtual Architects). There are discussion groups as well as code examples available at the LAVA Website. The LAVA address is lavausergroup.org.

5.8 DRIVER EXAMPLE

To tie together some of the driver techniques and guidelines, we will present an example set of drivers. This set of drivers will communicate with Microsoft Word using ActiveX. This example will create only a couple of relevant drivers for illustration purposes. If you want more information on ActiveX, Chapters 7 and 8 will give a detailed description and numerous examples.

The fist step is to define the task we want to accomplish. We will want to open Word, create a new file, set the margins, set the page size, set the page orientation, write text to the file, save the file, and close Word. The first step is to identify the driver types needed. You will need configuration drivers and measurement drivers. As configuration drivers perform instrument communication and configuration, the VIs needed to open Word, close Word, and configure the settings will be contained in these drivers. The action of reading or writing data to an instrument or application requires measurement VIs. The write text to file will fall into this classification.

A driver to open an automation reference to Word will need to be created. This action will be combined with the creation of a new file. This allows the user to open Word with a new document in the initial step. The next driver to be created will configure the page setup parameters. Most times when you are modifying a one-

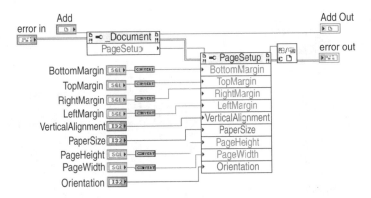

FIGURE 5.24

page setup parameter, you will want to modify additional page setup parameters. This is a good place to combine the configuration settings into one subVI to facilitate ease of programming. Not only will the programmer be able to see all of the input parameters that can be changed in one location, but the driver can also ensure order of execution. Some of the page setup parameters need to be modified after other parameters have been set. For example, you need to modify the page style prior to setting the orientation. The orientation setting will be reset after modifying the page style. If you are placing individual VIs to set these parameters, you could forget or be unaware of certain data dependencies, causing parameters to not be set in the desired manor. The code diagram for the Page Setup Configuration VI is shown in Figure 5.24. In addition to the data dependencies there are issues with data conversions. For example, when writing a value to a margin input, you would attempt to write data in inches. However, to get a margin value of one inch, a 72 needs to be wired to the input of the property node. Inside the driver, there is a function to convert an inch input to the required automation input. This allows you to abstract this information from the person using the driver.

The Write Text VI takes a string input and inserts it into the file at the specified index. If making multiple write statements, you could wire the end value from the previous write to the start value of the current Write VI. This allows you to do incremental data storage in the file. You would only want to have this VI write the text to the file. Any additional functions added to this VI would limit your ability to reuse the VI. For example, if you wanted to perform a spell check on the document, you would have to perform this spell check each time text is written to the file. You may want to check the spelling only after all of the text has been written to the file. If the spell check function is in its own VI, you can invoke this function when you need it. There is also the possibility you do not want to perform a spell check at all. Measurement VIs should be in their own VIs unless you are sure you will always want to do the multiple tasks together. An example using these VIs is shown in Figure 5.25. In the example, Word is opened; a new file is created (testfile); some of the page setup parameters are modified; two strings are written to the file, separated by a time delay; and the file is closed. More information on controlling Microsoft Word using ActiveX is included in Chapter 8.

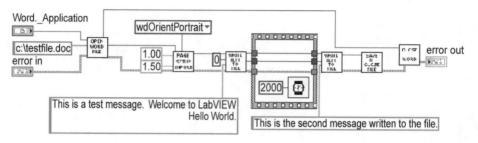

FIGURE 5.25

5.9   INSTRUMENT I/O ASSISTANT

The instrument IO Assistant is a tool that can be used to set up communications with an instrument or device. This tool is available in the Instrument IO palette. When the tool is selected it places an Express VI on the code diagram. The configuration window will come up automatically (depending on your LabVIEW settings) . Here you can select what connected device you want to communicate with and configure any needed settings.

Now that you have configured your instrument you can add steps to the code. If you click on add step you will have the choice of Query and Parse, Write, and Read and Parse. In this example we are communicating with a switch. The code will have the choice of opening or closing the switch and which channel to operate on. For this example, Write has been selected. When you select Write, you can configure several settings. Here two variables are added, a string to input open or close and a number for channel. This input is shown in Figure 5.26.

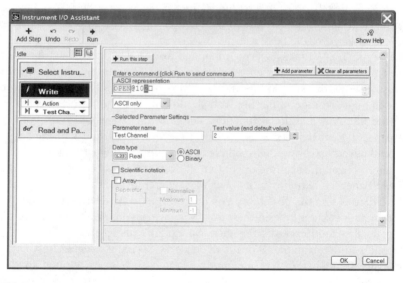

FIGURE 5.26

You can continue to add read and write actions until you have completed the needed actions. You can run the code directly from this interface or you can close the window, which will update the express VI on the code diagram. You can connect the inputs and outputs as needed and run the code from here.

5.10 IVI DRIVERS

IVI drivers were developed to allow hardware-independent test programs. In 1997, a number of manufacturing companies approached National Instruments to develop generic drivers that would be interchangeable. The IVI Foundation was a direct result of this effort. The organization, made up of representatives from National Instruments and a number of the instrument manufacturing companies including Hewlett Packard, Tektronix, Rohde & Schwarz, and Anritsu, has developed a set of standards and requirements for "generic" drivers. The IVI Foundation is an evolving group that is open to end users and interested parties. Anyone who is interested in joining can find more information on the IVI Foundation Web site (www.ivifoundation.org).

The goal of the IVI Foundation was to build upon the standards set by the VXI Plug&Play Systems Alliance. The VXI Plug&Play standards promote multi-vendor interoperability by standardizing the instrument commands for all VXI instruments. IVI instruments go one step further by trying to standardize an instrument type regardless of format. A power supply would have the same API regardless of the standard (GPIB, Serial, VXI, other) or the manufacturer.

IVI drivers are not language specific. By using DLLs to convert the commands from a uniform API to the required instrument code, there is a wide range of programming languages that can be used. LabVIEW and LabWindows/CVI are both capable of using IVI drivers; however, the DLLs can be written using only LabWindows. Due to the use of DLLs, IVI drivers are not platform independent. If you do not want to write your own drivers, or are not using LabWindows/CVI, a library of IVI drivers is available from National Instruments.

5.10.1 CLASSES OF IVI DRIVERS

The initial rollout of the IVI standards encompassed five classes of IVI drivers. The current IVI standard includes eight classes of IVI drivers. The eight classes are the DC Power Supply, Oscilloscope, Digital Multi-Meter (DMM), Arbitrary Waveform/Function Generator, Power Meter, RF Signal Generator, Spectrum Analyzer and Switch. New classes may be defined as the technology advances.

Let's look at the DMM class as an example. The IVI driver for the DMM class (IviDmm) is designed to operate a typical DMM, as well as support advanced functions found in the more complex instruments. The IVI class description divides the DMM into two sections: fundamental capabilities and extensions. The fundamental capabilities cover functions like reading a measurement or setting a range. An extended capability would be like setting auto-range, making multiple measurements, or other advanced features not available on all DMMs. For the DMM, there are fourteen groups defined (13 extension groups). Groups refer to the defined classification of commands. Examples of extension groups are IviDmmMultiPoint and IviDmmDeviceInfo. The IviDmmMultiPoint extension group supports the base DMM functions and also the ability to accept multiple triggers and acquire multiple samples per trigger. The IviDmmDeviceInfo extension group supports the base DMM functions and also the ability to return additional information concerning the instrument's state, such as accuracy and aperture time. Documentation on all the IVI classes and their groups is available on the IVI Foundation website.

5.10.2 INTERCHANGEABILITY

This section will discuss how IVI drivers allow for instrument interchangeability. One problem that has been seen in production testing for a long time is the lack of instrument interchangeability. This problem can arise for a number of reasons. An instrument that needs to be taken out for calibration or maintenance is one example. Other possible scenarios are when an instrument needs to be replaced and is no longer available; if the test system developer wants to use an instrument from another manufacturer; if the test software is going to be used by a group in another area with their own set of instruments. These issues are problems because the test software would have to be altered to replace an instrument with one from another manufacturer, or a newer model with new functions and commands. These problems force test system developers to stay with the same system instead of improving or cost reducing. The ability to change instruments would allow greater flexibility.

The first benefit of IVI drivers is the ability to interchange instruments. A power supply from a different manufacturer can replace the existing power supply without changing the test software. This will allow the development of a generic test station; users would be able to change instruments based on availability and cost.

5.10.3 SIMULATION

This section will discuss how an IVI driver can be used in simulation mode to allow debugging and input checking without the instrument being connected to the computer. When a programmer is developing software, the ability to incrementally debug the code is a technique that helps reduce development time. This would be an implementation of the spiral software development model. There is a full discussion of software development models (spiral and waterfall) in Chapter 4. By using IVI drivers in simulation mode, the test code can be debugged without the instrument being connected to the computer. The driver will return an instrument handle to allow a program using VISA to run without the instrument physically present. The user can also use the driver in simulation mode to choose the measurement that will be returned to the test program. This will allow the designer to test the program's response to common and unusual measurements returned by the instrument. The measurement returned can be set to random number generation within a range.

When using instrument-specific drivers, another feature is realized. The developer can perform range and status checking while developing the software. The driver will verify that the inputs sent to the instrument are within the specifications of the instrument. These are options that can be turned on or off. Turning on the range-checking feature helps the developer debug the test software. Turning off range-checking allows for faster execution time when the program is run in the final environment.

5.10.4 STATE MANAGEMENT

An IVI driver can speed up application execution when state caching is used. One problem encountered when programming a test application, particularly when utilizing state machine architecture, is the lack of knowledge of the instrument's current state. The user will not know what state the instrument is in at a given time, requiring the programmer to set all necessary configurations, even if the instrument is already configured properly. This can add substantial time to a test application.

The solution is to use state caching. This can be performed when using LabVIEW or LabWindows/CVI. When using state caching, the last setting for each function on an instrument is stored. When the driver goes to change the setting of a function, the driver checks to see what the last known state of that function was. If the setting is the same, the driver will not execute the command. The driver also tracks changes in settings when different screens are displayed.

5.10.5 IVI DRIVER INSTALLATION

When the IVI driver CD is inserted into the drive, the IVI Driver Library Installation interface starts. In the interface you have the options of viewing the release notes, installing the IVI driver library, installing instrument drivers, and browsing the CD. To install the IVI software you will need to click the IVI Driver Library Installation selection. This will begin the standard installation interface. After making the typical selections, a selection screen will appear. The installer will prompt you to select the instrument drivers to install. This is the initial place to obtain and install the IVI instrument drivers. There are a number of items on this installer screen. On the left of the screen is a selection for the IVI class. On the right side is a listbox containing the specific drivers. In order to install the drivers you need to use in your development, you must first select the desired IVI class. This will list the available IVI drivers in the specific driver input. In the specific driver input is the list of available drivers with a checkbox selection on the left of the individual drivers. To select the needed driver, you need to select the appropriate checkbox.

In addition to the IVI class input and the specific drivers input, there are three additional options on the IVI driver installation screen. There is a button to select all instrument drivers, a button to deselect all instrument drivers, and a control to replace the existing drivers. This control can be set to either replace the instrument drivers currently installed with the IVI drivers, or to leave the existing instrument drivers. This is an important selection if you have made modifications to the current standard drivers; it will prevent the IVI installation from overwriting your changes.

The IVI installation will set up three categories of software. The installation categories are instrument drivers, utilities, and driver software. The instrument driver installation includes the IVI class drivers, the IVI class simulation drivers, and the IVI-specific drivers. The utility installation includes NI Spy, the Virtual Bench software, and the Measurement and Automation Explorer. The driver software includes the IVI engine, NI VISA, NI DAQ, and the CVI run-time engine. When the installation is complete, the computer will need to be restarted.

5.10.6 IVI CONFIGURATION

The first step, after installing the IVI software, is to run the IVI Configuration Utility. The IVI Configuration Utility can be started by double-clicking the Measurement & Automation Explorer (MAX) icon on the Windows desktop, or by selecting the utility from the National Instruments folder in the Programs folder in the Start menu. The IVI settings are available under the IVI Drivers folder on the left window. Figure 5.27 shows the IVI configuration parameters in a MAX window.

There are three categories of IVI configuration items in the IVI folder. The main sections are Logical Names, Driver Sessions and Advanced. The logical name is what is used by LabVIEW to select the appropriate IVI driver similar to calling COM 1 to connect to the first serial port on a computer. To add a new logical name you can simply right click on the logical names folder and select add. Here you can set the logical name and what driver session it is linked to.

The Driver Sessions folder contains the loaded IVI drivers. If you have the equipment connected, and the software was installed properly the name should show up in the list. Don't worry if it is not there; you can download the drivers for the instrument you are looking for from NI's website. When you install the IVI drivers the instrument name will show up here.

When you click on the IVI driver name a configuration window will show up in the right MAX window. There are five tabs to configure: General, Hardware,

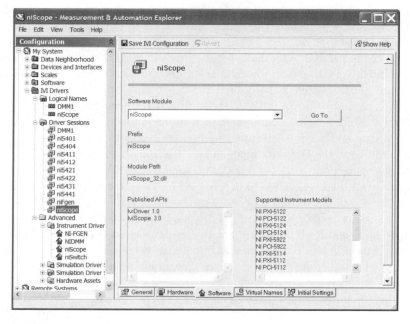

FIGURE 5.27

Software, Virtual Names and Initial Settings. Here you have the ability to select whether to simulate with a specific driver, choose the software module to use (shows what DLL contains the source code and the supported instrument models), virtual names for hardware options such as CHANNEL1, hardware links to actual installed equipment if available and initial instrument settings.

The Advanced folder gives you access to the driver software modules and hardware. In the Hardware Assets sub-menu you can define equipment in your test system and link to the resource descriptor. Here you will have a name that can be used in your code that is not tied to a specific instrument. You can define a DMM that is connected to a GPIB address. If you replace the DMM with one from another manufacturer, you would not have to change your code. You would still reference the DMM name. Be aware that the names used in the IVI configurations are case sensitive.

5.10.7 How to Use IVI Drivers

IVI class drivers are used in the same manner as standard instrument drivers. The IVI class drivers can be found in the Instrument I/O subpalette of the Functions palette. Each type of IVI class driver has its own subpalette. Each subpalette contains an Initialize and Close VI. There are also groups of VIs to perform instrument configuration, instrument functions, and utility functions that are necessary for the specific class driver. The developer can use these class drivers like typical VISA drivers. The programmer would put an Initialize VI on the diagram first. The main input to the Initialize VI is the logical name. The logical name is what tells the LabVIEW program what instrument and drivers to reference. As you will recall, in the setup of the IVI configuration items, the logical name references a particular virtual instrument. These logical names can be altered as needed using the Configuration utility. It is recommended that you set the name initially after installation and do not change it often. Applications that have been developed use this name, and may not work once the logical name has been altered. The virtual instrument refers to a specific driver in the Instrument Drivers folder, and a device. The specific driver then specifies the DLL containing the code module used to communicate with the device. The VIs associated with the instrument driver are placed in the Instrument Drivers palette during the installation.

The Initialize VI also has inputs to do an ID query and reset the instrument. The outputs of the VI are the Instrument Handle and the Error Out. The Instrument Handle can be passed throughout the VI and subVIs, just like a standard VISA instrument handle. Once the instrument is initialized, the functions required to perform the necessary programming task can be accomplished in two ways: the user can utilize the function VIs from the class driver subpalettes or make use of the LabVIEW Property node. When doing IVI driver configurations, the LabVIEW Property node is used in the same manner as ActiveX controls. As with all applications using communications, the final step is calling the Close IVI Class Driver VI. The following diagram shows an IVI example written with standard VIs and with the Property node. The VIs perform exactly the same function. Figure 5.28 illustrates the IVI example with and without the Property node.

DMM Measurement Example

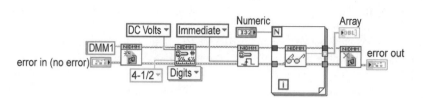

DMM Measurement Example Using Property Node

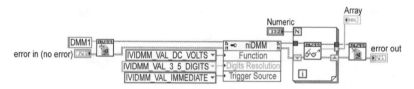

FIGURE 5.28

5.10.8 Soft Panels

The Soft Panels (IVI Virtual Bench) VIs were designed to simulate the front panel of an instrument. The main use for the soft panels is manual instrument control. The user can use the Front Panel VI to manually control the instrument. The similarity of the soft panel VI to the actual instrument interface allows the user to be familiar with some of the function immediately. The key is when the specific type of instrument changes. If a Hewlett Packard oscilloscope is replaced with a Tektronix oscilloscope, the user should still be able to control the instrument with no noticeable change. As the interface is the same, there are no new knobs or menus to learn. The IVI configuration files do all the work. An example of the niScope soft panel is shown in Figure 5.29.

5.10.9 IVI Driver Example

The information above can become confusing. Every name seems to include either virtual, instrument, or driver. In addition, each type of file references one or more of the other IVI file types. In order to alleviate some of the confusion, an example will be provided to help clarify things. It should be noted that there are a few examples that come with the IVI library. The examples are contained in the following path: labview\examples\instr\iviClass.llb.

For this example we will be simulating an oscilloscope using the IVI drivers. The first step is to create a logical name. In the IVI Configuration we need to go to the Logical Name folder and select Create New. You can enter a name and description. For this example we will call the instrument "niScope." You will be able to select a driver to associate with the name. We will use an existing driver. The selections available in the pull down menu depend on the drivers you have loaded. The driver we selected was the standard niScope driver.

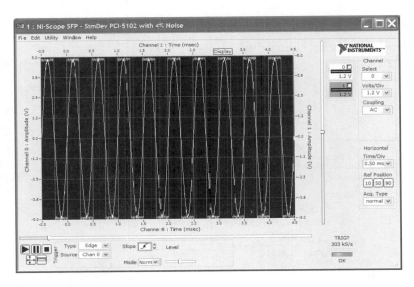

FIGURE 5.29

If you go to the Logical Name folder, you will see the summary page. Here you can modify the driver associated with the name. This is where you would make the modification if an instrument were exchanged for another in your equipment rack. If the program were written properly, with only class drivers being used in the code, no modifications would be necessary to the application. If we were going to use a real oscilloscope, the device could be changed here.

In the Driver Sessions folder you can see the driver properties. The General tab allows the programmer to select options such as Range Checking and Query Instrument Status. In the software tab, the programmer has the choice of using the class driver or the specific driver for the output data simulation. We will be using the class driver for this example. The programmer also has the opportunity either to modify or change the simulation virtual instrument and its settings. If you click on the initial settings tab you can configure the starting state of the instrument. You can go through the list of properties and change default values to match your instrument defaults. We will leave the current settings. The Virtual Names tab allows the programmer to associate a virtual channel name with the specific channel string.

Now that the IVI configurations are set up, the application can be written. As has been mentioned before, only IVI class drivers should be used in the application to reduce the amount of modifications if a new instrument is used. This example is based on the niScope example that comes with the IVI library (Acq Wfm Edge Triggered). The first coding step is to put the Initialize VI for the IVI oscilloscope class on the code diagram (niScope Initialize.vi). For the logical name input, a string constant or control with the text "niScope" should be wired to the first terminal. An alternate method is to create a resource name control Here you can select the appropriate name from a pull down list. The inputs for ID query and reset device are both defaulted as "true." And, as always, the Error In cluster should be wired to the final terminal of the input connector.

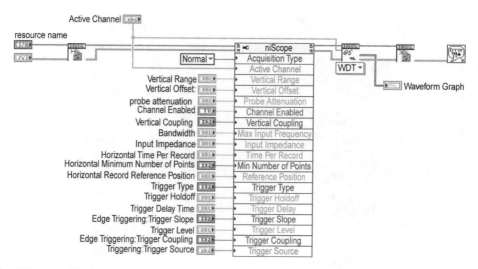

FIGURE 5.30

Once the instrument is initialized, the inputs for the vertical and horizontal parameters need to be configured. In addition to the scope parameters, the triggering inputs need to be set. For this example, we will be using the Property node to configure the necessary parameters instead of the class drivers. The first code diagram is shown in the following figure. The first trial of this VI used one Property node to set all of the vertical, horizontal, and triggering parameters. Selecting the item from either the ActiveX subpalette or the Application Control subpalette of the Function palette created the Property node. By using the positioning tool, you are able to increase the number of inputs by pulling the bottom corner down. The same task could also be accomplished by right-clicking on an input and selecting Add Element. A control was created for each input that was necessary with the appropriate default values being set. Figure 5.30 displays the Scope Example Code Diagram.

After setting the chart and triggering parameters, set the class VI to determine the actual number of data points to acquire. The output of this VI is then wired to the NIScope Read WDT VI. The data from the Read Waveform VI is then bundled together and wired to the waveform graph on the front panel. Finally, the Instrument Close VI is added to the code diagram.

Now it is time to debug our driver. The best method for debugging an application like this is to use the NI Spy utility to monitor the instrument communications. The NI Spy was discussed earlier in this chapter. There are a couple of IVI-specific items that need to be mentioned. When you go to the Spy menu of the NI Spy utility you will notice the installed IVI drivers available in the monitor list. For this example we will want to turn off the NI-VISA and the NI-488.2 monitoring options. They are being turned off to aid in interchangeability checking. If those items are turned off, any items with conflicts will be listed in blue. This will aid in spotting conflicts without having to go through all of the items captured. Once Capture has been turned on, we are ready to test our application.

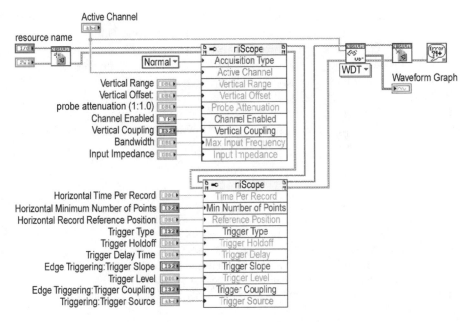

FIGURE 5.31

When we press the Start button, the program starts executing. As you should
now be aware, there is an error in the application. A message box comes up stating
that an error occurred at the tenth argument of the Property node. The listed possible
reasons are that a Null is required for the channel name when setting an attribute
that is not channel-based. The message box also lists the bad attribute. Clicking
Continue will close the message box and complete the execution. As the horizontal
parameters and the triggering parameters are not channel-based, they cannot be on
the same property node as the vertical parameters. Figure 5.31 shows the modified
code diagram that corrects this programming error.

Before we move on with the testing, let's take a look at the API captures from
the NI Spy utility. The NI Spy display is shown in Figure 5.32. The tenth entry in

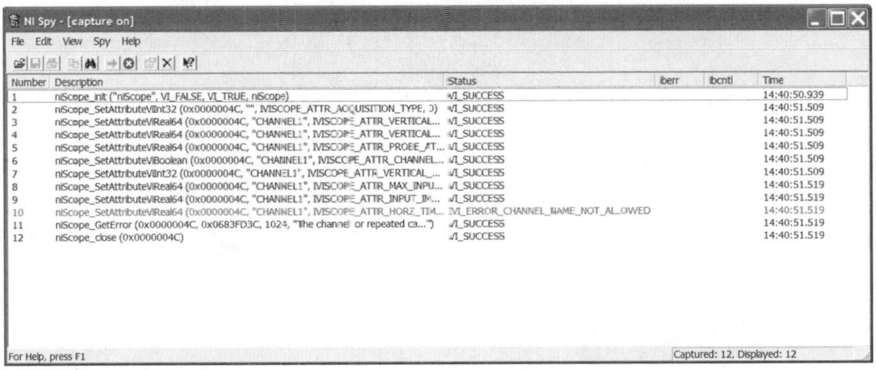

FIGURE 5.32

the list is the attempt to write the horizontal time per record parameter to the scope. Because an error occurred in this step, the line is in a red font. Double-clicking on the line opens up the Properties box. In the Properties box you can see what the inputs and outputs of the communications were. If you click on the Output tab, you will see the instrument returned the following error statement: (IVI_ERROR_ CHANNEL_NAME_NOT_ALLOWED). This is followed by a description in the text box below the error statement. Now that we have modified our VI, we can attempt to run the application again. This time the application runs without error.

BIBLIOGRAPHY

LabVIEW Graphical Programming — Practical Applications in Instrumentation and Control. Gary W. Johnson, McGraw-Hill, New York, 1997.

G Programming Reference, National Instruments, Austin, TX, 1999.

Data Acquisition Basics Manual, National Instruments, Austin, TX, 1999.

Code Interface Reference Manual, National Instruments, Austin, TX, 1999.

IVI-1: Charter Document, IVI Foundation, San Diego, CA, November 1999.

IVI-4.2: IviDmm Class Specification, IVI Foundation, San Diego, CA, January 2004.

Developing COM/ActiveX Components with Visual Basic 6, Dan Appleman, SAMS, 1998.

Using TCP/IP. John Ray, QUE Corporation, Indianapolis, 1999.

Application Note No. AN006, Developing a LabVIEW Instrument Driver, National Instruments, Austin, TX.

Application Note No. AN087, Writing Win32 Dynamic Link Libraries (DLLs) and Calling Them from LabVIEW, National Instruments, Austin, TX.

Application Note No. AN111, LabVIEW Instrument Driver Standards, National Instruments, Austin, TX.

Application Note No. AN120, Using IVI Drivers to Simulate Your Instrumentation Hardware in LabVIEW and LabWindows/CVI, National Instruments, Austin, TX.

Application Note No. AN121, Using IVI Drivers to Build Hardware-Independent Test Systems with LabVIEW and LabWindows/CVI, National Instruments, Austin, TX.

Application Note No. AN122, Improving Test Performance through Instrument Driver State Management, National Instruments, Austin, TX.

6 Exception Handling

Code is often written without considering the potential that an error might occur. When events occur that an application is not expecting, problems arise. Then, during the debugging phase, an attempt is made to go back to the code and implement some error traps and correction. However, this is usually not sufficient. Exception handling must be taken into account during the early stages of application development. The implementation of an error handler leads to more robust code.

This chapter discusses errors and the topic of exception handling in LabVIEW. First, exception handling will be defined along with its role in applications. This explanation will also clarify the importance of exception handling. Next, the different types of errors that can occur will be discussed. This will be followed by a description of the available LabVIEW tools for exception handling. as well as some of the debugging tools. Finally, several different ways to deal with errors in applications will be demonstrated.

6.1 EXCEPTION HANDLING DEFINED

Exceptions are unintended or undesired events that occur during program execution. An exception can be any event that normally should not take place. This does not mean that the occurrence of the exception is unexpected, but simply should not happen under normal circumstances. An error results when something you did not want to happen, does. Therefore, it makes sense to make alternate paths of execution when exceptions take place. When exceptions or errors occur, they must be dealt with in an appropriate manner.

Suppose that you have written a program in which you divide two variables, Integer x by Integer y. The resulting quotient is then used for some other purpose. On some occasion, y may be set to zero. Some programs do not trap errors such as dividing by zero and allow the CPU to throw an exception. When the CPU throws an exception for a program, the program will be terminated by the operating system — in most cases this is an undesirable result. In LabVIEW, the result of this division is undefined. LabVIEW returns the result Inf, or infinity, on a floating point division by zero. If you were to use the integer quotient and remainder, a division by zero results in a quotient of zero. In both cases, the application does not throw an exception and close. However, this is an example of an unexpected and unintended outcome. Infinity can be converted successfully into a word integer in LabVIEW. If the value is converted for other uses, several other errors can result. This is an example of a simple error that has to be managed using exception handling.

Exception handling is needed to manage the problems or errors that occur. It is a mechanism that allows a program to detect and possibly recover from errors during execution. Exception handling leads to more robust code by planning ahead for potential problems. Depending on the purpose of an application, the ability of an application to respond to unexpected events can be critical. Typical programs that are used by one person, at their desk may not need a lot in terms of robust performance. An application that is automating an assembly line that is producing hundreds of thousands of production units for revenue would benefit significantly from robust and stable code. The implementation of an error handler increases the reliability of the code. It is difficult to prepare for all the possible errors that might occur, but preparing for the most probable errors can be done without much effort.

You can write your code to try to catch as many errors as possible, but that requires more code to implement. After a certain point you will have more code involved in catching errors than you do for performing the task that you originally set out to do. The exception handling code itself may sometimes contain errors. You also create a problem of what to do when the error is caught.

Error detection and error correction are two different activities, but are both part of exception handling. Error detection consists of writing code for the purpose of finding errors. Error correction is the process of managing and dealing with the occurrence of specific errors. First you have to catch the error when it occurs; then you have to determine what action to take.

Performing error detection is useful for debugging code during the testing or integration phase. Placing error checks in the code will help find where the faults lie during the testing phase. The same detection mechanisms can play a dual role. The detection mechanism can transfer control to the error handler once the handler has been developed. This will be beneficial if you are using an iterative development model, where specific features can be added in each cycle.

Exception handling is performed a little differently in each programming language. Java uses classes of exceptions for which the handler code can be written. For example, an exception is represented by an instance of the class "Throwable" or one of its subclasses. This object is used to carry information from the point at which an exception occurs to the handler that catches it. Programmers can also define their own exception classes for their applications.

C++ uses defined keywords for exception handling: Try, Catch, and Throw. The Try and Catch keywords identify blocks of code. Try statements force the application to remember their current location in the call stack and perform a test to detect an error. When an exception occurs, execution will branch directly to the catch block. After the catch block has executed, the call stack will be "rolled back" to the point where the program entered the Try block.

LabVIEW provides some tools for error detection. But just like other programming languages, implementation of exception handling code is left to the programmer. The following sections will guide you in creating error handling code for your application. Chapter 10 covers topics relating to Object-Oriented Programming, including definitions for object, class, and subclass, but exception handling in Java and C++ are beyond the scope of this book.

6.2   TYPES OF ERRORS

Errors that occur in LabVIEW programs can be categorized into either I/O-related or logic-related. I/O errors are those that result when a program is trying to perform operations with external instruments, files, or other applications. A logical error is the result of a bug in the code of the program. The previous example of dividing an integer value by zero is a logical error. These types of errors can be very tricky to find and correct. Both I/O- and logic-related errors are discussed in the following sections.

6.2.1   I/O ERRORS

Input/Output encompasses a wide range of activities and VIs within LabVIEW. Whether you are using communication VIs (TCP, UDP, .NET, USB, Bluetooth, etc.), data acquisition, instrument I/O, or file I/O, there is a probability that you will encounter related errors at some point.

I/O errors can be the consequence of several things. The first circumstance that can cause this type of error is improper initialization or configuration of a device or communication channel. For example, when performing serial communication, the baud rate must match between the external device and the controller. If this initialization is performed incorrectly an error will result. For some devices a command must be sent to put them into remote mode, which will allow communication with the controller. When reading or writing to a file, the file must be opened first. Similarly, when writing to a database, a connection has to be established before records can be inserted. Initialization can also include putting an instrument or device into a known state. Sometimes this can be done by simply sending a reset command, after which the device will enter a default state.

A second cause of I/O errors is simply sending the wrong commands or data to the instrument or application. When invalid data is sent, a write error will result. Some devices simply ignore the data whereas others return an acknowledgment. This can play a role in what type of correction and handling you perform. When data is being communicated to an external device, you have to ensure both the correct data and the correct format are being sent. You must adjust the information you are sending to suit what the device is expecting to receive. Typographical errors can also be classified in this section.

Another I/O-related error takes place when there is a problem with the instrument or application being used. When dealing with applications or files, this can occur for several different reasons. The file may not be in the specified path or directory. Alternatively, you may not have the needed file permissions to read or write to the file. Instrument I/O errors of this nature usually occur if the instrument is not powered-on or not functioning properly. A similar problem happens when the instrument locks up or freezes. Power cycling may return it to a known state and make it operational again. These types of errors can also be a result of incorrectly configuring the external device. Instruments can return unusual results when they are not configured appropriately.

Missing hardware or software options can be a source of I/O errors. You may also need to check if you have the correct interface drivers installed. Interface incompatibility and component incompatibility should be investigated.

The last and most common issue is a network or communication bus is interrupted. Internet Protocol (IP) based communication will periodically find a message has not gone through. IP itself does not guarantee message delivery. IP also does not guarantee delivery in order of transmission — its possible for IP packets to arrive in an order other than what they were transmitted in. Applications using any Ethernet communications needs to have the ability to sort packets when they arrive out of order, or simply do not arrive.

6.2.2 LOGICAL ERRORS

Logical errors happen when there are faults in the code itself. The code diagram in Figure 6.1 illustrates an innocent mistake that can occur. In the While loop, the programmer intends the loop to stop executing when the temperature reaches 75.0 degrees or higher. However, the loop, as it stands currently, will stop when the temperature is lower than 75.0. This is an example of an easy mistake that can cause an error in applications. These types of problems can be difficult to find and are also very time consuming. Debugging tools are invaluable when looking for the source of faults.

Errors can sometimes occur when the inputs specified by the user are not validated. If the user does not provide reasonable inputs expected by the program, an error can occur. The application must validate the data to ensure it is within the acceptable range. For example, the user may have to specify which unit number, between one and ten, to perform a sequence of tests on. The program has to verify that only the acceptable range is entered before beginning execution. Unit zero may not exist for test purposes, therefore the code must check for the appropriate inputs. Be aware of numeric precision errors and conversion errors that can also be difficult to track down.

LabVIEW allows the programmer to set acceptable ranges for Numeric, Boolean, and List & Ring controls. This can be done by popping up on the control and selecting Data Range from the menu. The programmer also has the option of coercing the input value so that it is within the valid range on front panel controls only. The two options when setting a data range are to ignore or to coerce the value when a range is specified. This option is available in the drop-down box. The coercion option

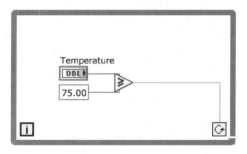

FIGURE 6.1

reduces the need to write code for performing the same task, but is a practice we generally do not recommend. If a value is entered out of range from the user interface it is preferable to notify the user the value is not valid and attempt to get the user to correct the input.

6.3 BUILT-IN ERROR HANDLING

LabVIEW notifies the user of some run-time errors for instrument and file I/O operations through dialog boxes. LabVIEW does not deal with the errors and, in general, leaves exception handling to the programmer. However, LabVIEW does provide some tools to aid the programmer in exception handling. The first tool that will be discussed is the error cluster. The error cluster is used in transporting information from the detection mechanism to the handler. After the error cluster, a brief description of VISA error handling will be presented. Next, the error-handling VIs will be considered. There are three error-handling VIs in particular: the Simple Error Handler VI, the General Error Handler VI, and the Find First Error VI. Section 6.4 will then discuss the implementation of exception handling code.

6.3.1 ERROR CLUSTER

The error cluster is a detection mechanism provided for programmers. The cluster consists of a status, code and source. Each of these provides information about the occurrence of an error. The status is a Boolean that returns "true" if an error condition is present. The code is a signed 32-bit signed integer that distinguishes the error. The source is simply a string that gives information on where the error originated. The error cluster as a whole provides basic details about the error that can be used for exception handling purposes. In LabVIEW 7 and 8, the coerce function of a control does not work when the VI is used as a subVI.

Figure 6.2 shows the Error In and Error Out clusters as they appear on the front panel. The Error In and Error Out clusters can be accessed through the Array & Cluster subpalette in the Controls palette. The error clusters are based on National Instruments' concept of error I/O. VIs that utilize this concept have both an Error In control and Error Out indicators, which are usually located on the bottom of the front panel. The cluster information is passed successively through VIs in an application, consistent with data flow programming.

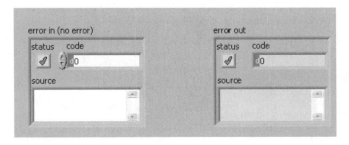

FIGURE 6.2

The error clusters can serve a dual purpose in your application. By using error I/O, the order of execution of VIs can be forced. This eliminates the need for sequence structures to control the order of execution. Simply pass error clusters through VIs for detection and order.

When the cluster is passed in to a VI, the VI should check if an error condition is present. By default, a new, blank VI doesn't have an error cluster input or code to check the status. If there is no existing error, execution will continue. The cluster picks up information on whether an error has occurred during the VI's execution and passes this information to the next VI, which performs the same check. In the simplest case, when an error does occur in any VI, the VIs that follow and use the cluster should not execute. When the program completes, the error is displayed on the front panel.

The error I/O concept and the error clusters are easy to use and incorporate in applications. Many of the LabVIEW VIs that are available in the Functions palette are based on this concept: the Data Communications palette and contained subpallet, most of the Instrument I/O VIs (VISA, GPIB, GPIB 488.2), and some Data Acqui-sition and File I/O VIs use error I/O. The variable introduced in LabVIEW 8 also uses the error cluster. By using these VIs and wiring in the error clusters, much of the error detection work is already done for the programmer. These built-in VIs provide the detection needed in the lower-level operations. When wiring these VIs on the code diagram, you will notice that the Error In terminal is on the lower left side of the VI, whereas the Error Out terminal is on the lower right side. This is a convention followed by all VIs developed by National Instruments, and is also recommended when creating drivers.

Figure 6.3 is an example of how the error clusters can be used. The VI uses GPIB Write and GPIB Read from the Instrument I/O palette. It is a simple instrument driver that can be used to write data to and read data from an instrument. To perform error detection, the programmer only has to use the Error In and Error Out clusters and wire them accordingly in the code diagram. The error detection work is left to the Instrument I/O VIs. When this driver is needed as part of a larger application, the error I/O concept is used. Figure 6.4 uses two drivers with the Error In and Error Out wired. The second VI in the diagram will not execute if an error occurs during the execution of the first VI. Execution order is forced, causing the second driver to wait for the error cluster data from the first one. This approach can be applied successfully to larger applications.

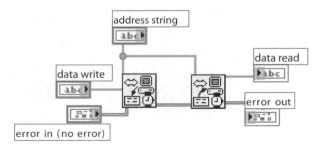

FIGURE 6.3

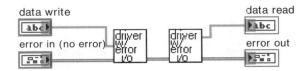

FIGURE 6.4

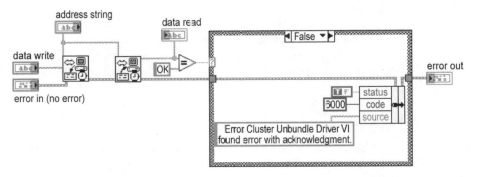

FIGURE 6.5

The error clusters can also be used to perform error checks other than those done by the available LabVIEW VIs. Suppose you are communicating to a device or application that returns acknowledgments when sending commands and data. An "OK" value is returned when the data is accepted and valid, and an "NOK" is returned if the data is invalid or the command is unknown. The LabVIEW VIs do not perform any check on instrument- or application-specific acknowledgments, only on general communication errors. Returning to the VI in the previous example, we can implement our own error check. Figure 6.5 shows how this is done.

The Bundle by Name was used from the Cluster palette to accomplish this. If the acknowledgment returned does not match "OK," then the error cluster information is altered. The Boolean is made true, the code assigned is 6000, and the source description is also wired in. LabVIEW reserves error codes 5000 to 9999 for user defined errors. If the acknowledgment returned matches the expected value, we wire the error cluster through the "true" case directly to Error Out without any alterations. The error detection for the correct acknowledgment will now be performed every time this driver is called.

Figure 6.6, Extra Source Info.vi, shows an example of how to get more information out of the error cluster for debugging and error-handling purposes. This VI adds extra information to the source string of the error cluster. First, the error cluster is unbundled using Unbundle by Name. The extra pieces of information that will be added include the time the error was generated and the call chain. Call Chain, available on the Application Control palette, returns the VI's call chain all the way to the top level in string format. The call chain information is useful for user-defined errors to indicate where the error was generated. These two pieces of data will then be bundled together with the original source information generated by the error cluster. You can put any other type of information you would like returned with the

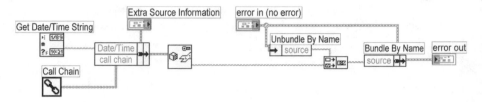

FIGURE 6.6

error cluster in a similar manner. It can be used to give the programmer more facts on the error that may be helpful for debugging. The errors can then be logged in a text file or database for reference. Error logging is demonstrated in Section 6.4.6 through a basic example.

6.3.2 ERROR CODES

A list of possible LabVIEW-generated errors is accessible through the Online Reference in the Help menu. The errors are listed by the error code ranges and the types of possible errors. Error codes can be either positive or negative values, depending on the type of error that is generated. When a zero error code is returned, it indicates that no error has occurred. Warnings are indicated with a code that is nonzero, whereas the status returned is "false." Table 6.1 contains a list of the error code ranges.

A handy tool for looking up error codes is also available through the Help menu in LabVIEW Version 5.0 and later. When Explain Error is selected, a new window appears with the error cluster on the left side and a text box on the right side. The error code can be input either in hexadecimal or decimal format. An explanation of the error will be provided for the error code in the text box. This tool provides a quick way to get additional information on an error for debugging purposes.

6.3.3 VISA ERROR HANDLING

VISA is a standard for developing instrument drivers and is not LabVIEW-specific. It is an Application Programming Interface (API) that is used to communicate with different types of instruments. VISA translates calls to the lower-level drivers, allowing you to program nonsimilar interfaces with one API. See Chapter 5 on instrument drivers for more information on VISA.

VISA is based on the error I/O concept, thus VISA VIs have both Error In and an Error Out clusters. When an error occurs, the VIs will not execute. There is a set of VISA-specific error codes that can be found in LabVIEW Help. The VISA Status Description VI can be used in error-handling situations. This VI is available in the VISA subpalette of the Instrument I/O palette.

When you are using instrument drivers that utilize VISA, there are some additional errors you may encounter. The first may be the result of VISA not being correctly installed on your computer. If you choose the typical install, NI-VISA is selected for installation by default. If you have performed the custom install, you must make sure the selection has been checked. You will not be able to use any

TABLE 6.1
Error Codes

Error Type	Code Range
Networking	-2147467263 through -1967390460
Instrument driver	1074003967 through -1074003950
VISA	-1073807360 to -1073741825
Report generation	-4105 to 41000
Formula parsing	-23096 to -23081
Mathematics	-23096 to -23000
Signal processing	-20999 and -20337 to -20301; -20115 to -20001
Point by point	-20207 to -20201
Regular expression	-4644 to -4600
Waveform	1820; -1811 to -1800
Apple Event	-1719 to -1700
Instrument driver	-1300 to -1210; 1073479937 to 107347994
Timed loop	-823 to -800
Windows registry access	-620 to -600
Signal processing, GPIB, instrument driver, formula parsing, VISA	0
GPIB	1 to 20; 30 to 32; 40–41
General	1 to 52; 67 to 91; 97 to 100; 116-118; 1000 to 1045; 1051 to 1086; 1088 to 1157; 1174 to 1188; 1190 to 1194; 1196 to 1198; 1307 to 1320; 1362
Networking	53 to 66; 108 to 121;1087; 1191
Serial	61 to 65
Windows connectivity	92 to 96; 1172; 1173; 1189; 1195; 1199; 14050 to 14053
Instrument driver	102, 103
MATLAB and Xmath	1046 to 1050; 1053
Run-time menu	1158 to 1169
Waveform	1800 to 1809
SMTP	16211 to 16554
Signal processing	20001 to 20353

VISA VIs unless your system has this option installed. Another error can be related to the lower-level serial, GPIB, or VXI drivers that VISA calls to perform the instrument communication. For example, if you have a GPIB card installed on your computer for controlling instruments, make sure the software for the card has also been installed correctly to allow the use of VISA VIs. You can use NI-Spy to monitor calls to the installed National Instrument drivers on your system. NI-Spy is briefly explained in Section 5.5.

When using VISA in your application, remember to close all VISA sessions or references you may have opened during I/O operations. Leaving open sessions can degrade the performance of your system. You can use Open VISA Session Monitor.vi to find out the sessions that you have open, and to close the ones that are not being used. This VI is available in the following directory: \LabVIEW\Vi.lib\Utility\visa.llb. This VI can be helpful while you are debugging an application.

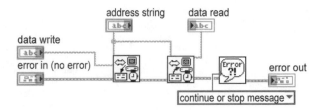

FIGURE 6.7

6.3.4 SIMPLE ERROR HANDLER

The Simple Error Handler can be found in the Time & Dialog palette in the Functions menu. This VI is used for error reporting. It is used with LabVIEW VIs that utilize error I/O and the error cluster. The purpose of the Simple Error Handler is to notify the operator that an error has occurred, but it can be customized for added functionality. It takes the error cluster as input and determines if an error was generated. If an error has been generated, the VI displays a dialog box with the error code, a brief description of the error, and the location of the error. The Simple Error Handler utilizes a look-up table to display the description of the error based on the error code.

As mentioned, one of the uses of the Simple Error Handler is for error notification purposes. The programmer can select the type of dialog box to display by wiring the corresponding integer or enumerated constant. A value of 1 displays the dialog box with only the OK button for acknowledgment. A value of 2 displays a button dialog box with Continue and Stop buttons. This allows the operator to stop execution of the program. A value of 0 gives no notification to the operator, even when an error has been generated. This might be used when exception handling is to be performed elsewhere by using the error?, code out, or source out, outputs from the Simple Error Handler.

You must keep in mind that this VI will halt execution until the operator responds to the dialog box. If your intention is to start the program and walk away, the program will not continue if an error is generated. Dialog boxes should be used only when the program is being monitored. Consider using e-mail for notification using the SMTP package. Chapter 8 also shows you how to incorporate the e-mail feature using .NET, or there are SNMP utilities as part of the Data Communications pallete.

Figure 6.7 shows how the Simple Error Handler can be used. This is the same VI shown in Figure 6.3. Notice that the Simple Error Handler has been merely added as the last VI in the flow. The value of 2, which corresponds to the two-button dialog box (Continue and Stop), is being passed to the VI. If an error is detected in either GPIB Read or GPIB Write, the dialog box will appear displaying the error code, description, and the source of the error.

6.3.5 GENERAL ERROR HANDLER

The General Error Handler essentially performs the same task as the Simple Error Handler. The Simple Error Handler offers fewer choices when used in an application. The Simple Error Handler is a wrapper for the General Error Handler. The General

Error Handler can be used in the same situations as the Simple Error Handler, but as the General Error Handler has a few more options, it can be used for other purposes where more control is desired.

The General Error Handler allows the addition of programmer-defined error codes and corresponding error descriptions. When these arrays are passed in, they are added to the look-up table used for displaying error codes and descriptions. When an error occurs, the possible LabVIEW-defined errors are searched first, followed by the programmer-defined errors. The dialog box will then show the error code description and specify where it occurred.

The General Error Handler also offers limited exception handling options. The programmer can set the error status or cancel an error using this VI. An error can be canceled by specifying the error code, source, and the exception action. Set the exception action to Cancel Error on Match. The look-up tables are searched when an error occurs. When a match is found, the error status is set to "false." In addition, the source descriptor is also cleared and the error code is set to zero in the output cluster. Similarly, when the status of the Error In is "false," it can be set to "true" by passing the exception action for the error code and source.

6.3.6 FIND FIRST ERROR

The Find First Error VI is a part of the Error Utility Package located in \Lab-VIEW\vi.lib\Utility directory as part of the error.llb package. The purpose of this VI is to create an Error Out cluster. It takes the following inputs: Error Code Array, Multiline Error Source, and Error In Cluster. When the Error In status is "false" or is not wired in, the VI tests to see if the elements of the error code array are nonzero. The VI bundles the first nonzero element, the source, and a status value of "true" to create the Error Out cluster for passing back out. As the source is a multiline string, the index from the array of error codes is used to pick the appropriate error source for bundling. If an Error In cluster is passed in, then a check is first performed on the cluster's status. When the status is "true," the Error In cluster will be passed back out and the array check will not be performed.

Find First Error is practical for use with LabVIEW VIs that do not utilize error I/O but pass only the error code value out. Some VIs that output only the error code are the original serial I/O VIs and some Analysis VIs. The Find First Error VI can be used to convert the error code from these VIs to a cluster. The error cluster can then be used in conjunction with other VIs that utilize error I/O.

Figure 6.8 is an example of how the Find First Error can be used. Both the Bytes at Serial Port.vi and the Serial Port Read.vi pass an error code out. An array is built with the two error codes that are passed out. A multiline string for the source is also created in the example. The source will give information on the origin of the error. The Find First Error.vi assembles the error cluster and passes it to Error Out. If an error has occurred, the first error that occurred will be sent to the Error Out cluster. If no error was generated, the Error Out cluster will contain a "false" status Boolean, no error code, and an empty source string. The error cluster can then be passed to the General Error Handler or the Simple Error Handler to display a dialog box if needed.

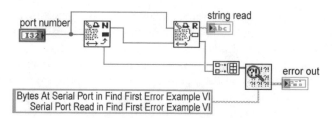

FIGURE 6.8

6.3.7 Clear Error

This VI is located in the dialog palette and as the name implies, it clears the error cluster of all values. Essentially, it takes an error cluster in and outputs a new error cluster with no information. This VI is useful when an exception has been success-fully handled.

If the application is expected to log errors seen or if the error cluster is being used to carry information in the error string, then this VI should not be used — it will clear the error string and code.

6.4 PERFORMING EXCEPTION HANDLING

Exception handling encompasses both the detection of errors and the treatment of the errors once they have been found. The previous sections presented several types of errors that can occur, as well as the built-in LabVIEW functions that are available for exception handling. This section will illustrate different approaches that are effective for managing errors. The effectiveness of an error handler can be improved by building it into your application during the early stages of development. It will support the readability and maintainability of your code, as well as code reuse. When error handling is not considered while you are architecting the application, the handling code will consist of patches for each exception.

You may have some questions about the implementation of exception handling code in order to make the handler both efficient and effective. When should error detection, reporting, and handling be performed? What should the application do when an exception is detected? Where and how should it be implemented? The following subsections will address the where, how, and what on exception handling approaches for your application.

6.4.1 When?

The question of when to implement is a little bit trickier and depends on the specific situation or application being developed. This may vary depending on the objective of the application, the amount of time available, the programmers' intent, and several other factors. Some areas of an application that need handling may be easier to identify than others. You may be able to identify areas where errors cannot be tolerated, or where errors are prone to occur, through past experience. These are definite targets for error detection, reporting, and handling.

To answer this question as completely as possible, you must also look at specific instances in an application to determine what alternative scenarios are foreseeable as well as their possible consequences. To illustrate this point, consider an example in which you must open and read or write to a file using an I/O operation. To answer if exception handling code is needed, and maybe even what is needed, think about the following scenarios and consequences. What will need to happen if the file that is being written to cannot be opened? What happens if the read or write operation fails because the drive is full? What happens if the file cannot be closed? Answering these questions will help put the need for handling into perspective for the application. It will also help you look at the application and determine where the exception handling activities are needed by asking similar questions. Error handling will definitely need to be implemented if the file I/O operation is crucial to the application, and if other parts of the program are dependent on this activity's being successful.

6.4.2 EXCEPTION-HANDLING AT MAIN LEVEL

To answer the "where" question, exception handling should be managed at the Main Level or Test Executive level. The Main Level controls and dictates program flow. By performing exception handling at the Main Level, the program execution and control can be maintained by the Top Level. This is important because the exception handler code may alter the normal flow of the program if an error is detected. You may want the code to perform several different actions when an error has occurred. When exception handling is performed at lower levels, program control must also be passed to the lower levels. This is a good reason why the implementation of an exception handler should be considered when architecting the application. Application structure and processes for application development are discussed in Chapter 4. Reading Chapter 4 will help you get a better perspective on how to approach the development of an application and other topics that must be considered before you begin.

Performing exception handling at the Main Level also eliminates the need for duplicating code in several subVIs. This permits the error handler code to be located in one place. The separation of error handler code from the rest of the code reduces confusion and increases readability and maintainability. Logical flow of the program will be lost in the clutter when error handling is performed with the rest of the code. This is explained further in Section 6.4.5 on exception handling with state machines.

The suggested style is similar to other programming languages where Error Information is sent to a separate piece of code for handling purposes. As mentioned earlier, both Java and C++ have a separate section that performs the error handling after the evaluation of an error is completed. There is no such mechanism inherent in LabVIEW, but this approach resembles it.

6.4.3 PROGRAMMER-DEFINED ERRORS

Defining errors was briefly discussed in Section 6.3.1 along with the error cluster. The ability to define errors is significant because LabVIEW leaves application-specific error handling to the programmer. As mentioned earlier, error codes 5000-

9999 are dedicated for use by the programmer. The programmer must perform error checking in circumstances where faults cannot be tolerated, as was shown in Figure 6.5. An error code must then be assigned to the error check as well as a source string to indicate the origination.

When implementing a programmer-defined error in a subVI or driver, you must make sure that an error was not passed in. Simply unbundle the error cluster and check the value of the status Boolean. If an error was passed in, but you fail to check the status, you may overwrite the error cluster with the new Error Information that you implemented. This will make it nearly impossible to find the root of the problem during the debugging phase. You must also make use of shift registers when using error clusters within loop structures to pass data from one iteration to the next. If shift registers are not used, error data will be lost on each iteration.

Records must be kept of the error codes that have been assigned by the user. A look-up table can be created that contains all of the error codes and sources assigned. This can then be used with the General Error Handler or with other exception handling procedures. It may be a good practice to maintain a database or spreadsheet of user-defined error codes. A database facilitates the management as the number of codes grows in size.

When you are assigning error codes, you can group similar errors into specified ranges. This is helpful when deciding the course of action when errors occur. For instance, you can set aside error codes 6000-6999 for incorrect acknowledgments from instrument I/O operations. When an error in this range occurs, you can identify it and decide how to deal with it easily. LabVIEW-generated errors are grouped in a similar manner to facilitate their identification and management.

User-defined warnings can also be assigned codes to indicate that an undesired event has occurred. You can use these to signal that the data taken may not be entirely valid due to the occurrence of some event during application execution. The user can investigate the source of the warning further to determine the validity of the data. Multiple errors can be reported and handled by unbundling the error cluster and appending the new information.

6.4.4 MANAGING ERRORS

Once you have a list of the errors that you want to deal with that can be detected, you have to decide what to do with them if they occur. When an error occurs it should be passed to the exception handling code. The exception handling code can deal with the errors in different ways. Expanding on the idea of grouping similar errors, the code can check to see what range the error has fallen in to determine the course of action. Figure 6.9, Error Range Example.vi, is an example of grouping ranges of error codes for handling purposes. When a set of exceptions is considered to be logically related, it is often best to organize them into a family of exceptions.

The easiest way to deal with an error is to simply display a dialog box to notify the user that an error has occurred. This dialog box can be as simple as the one displayed by the General Error Handler. You can create your own VI to display a dialog box to include more information, including what the user can do to trouble-shoot the error. This usually results in halting execution of the program.

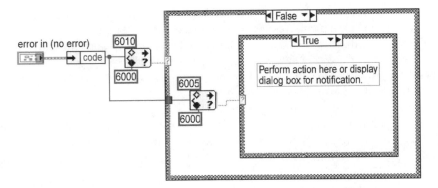

FIGURE 6.9

You can get more involved by attempting to correct or recover from an error in the exception handling code. In this case, the more general range checking technique will not suffice because the exact error code will be used to determine how to correct it. It also requires detailed knowledge of the error and exactly how it can be corrected. Suppose, for example, that you get a specific error telling you that the device under test did not respond to the commands sent to it. You also know that this happens when the device is not powered-on or has not been initialized properly. You can then attempt to correct this error by power cycling the device and initializing it. Then you can retry the communications and continue with the program if successful.

Figure 6.10 illustrates a technique for dealing with specific error codes as an alternative to the general range-checking method. This method needed to be used in LabVIEW 4.1 or older because the default state was not defined in these versions. If the error code did not exist in the array of error codes, the search 1-D array function would return -1. There is no '-1' case in the case statement and you would have had a problem pre LabVIEW 5. Current versions of LabVIEW have a default case permitting you to wire the code directly to the selector terminal of the case structure. For case statements using integers, such as this error array, set the default case to 0. This case will then execute for error codes for which no case has been defined.

The method displayed is similar to a look-up table described earlier. An array that contains all of the error codes is used with the Search 1D Array VI. The error

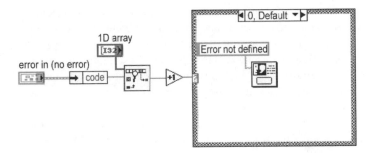

FIGURE 6.10

code is passed to it and the index of the error code is searched for. The index drives
the case statement, which takes the right course of action for the error code. If there
is no match for the error code, the Search 1D Array returns a value of −1. By adding
1 to the result, Case 0 is selected from the structure. This case will serve as the
default case when no match is found. In the example shown, a dialog box is displayed
indicating that the error code was not defined.

Another alternative is the use of strings to drive case structures. You can implement the previous example by unbundling the cluster to retrieve the source information. This string can then be used to determine the course of action by wiring it
to the case selector terminal.

6.4.5 STATE MACHINE EXCEPTION HANDLING

The use of a state machine offers several advantages for exception handling code.
One advantage is that the exception handling code can be located in one place. This
is done through the use of an Error state. The Error state is responsible for all
exception handling in the application. This eliminates the need for exception handling code in several places. Maintaining the code becomes easier when the code
resides in one location. Using a state machine also facilitates exception handling
management at the Main or Test Executive Level. The Error state is part of the Main
Level, so control is maintained at the upper level.

Another advantage is that duplication of error handling code is reduced when
the code is placed in one location. Similar errors may be generated in different parts
of your code. If you do not perform exception handling in one place, you may have
to write code in several places for the same type of error.

Conditional execution of code can be implemented without creating a complex
error handler through the use of a state machine. Exception handling code determines
program execution based on the severity of the error that was generated. You may
want your code to skip execution of the parts of the code that are affected by the error,
and continue execution of the rest of the program. For example, suppose you have a
series of ten different tests you want to perform on a device under analysis. If an error
occurs in Test 1 and that same error will affect Tests 5, 6, and 7, you may still want
to execute Tests 2, 3, and 4. In this case, using a queued state machine will simplify
the procedure for performing this task. The Error state can parse out the states that
correspond to Tests 5, 6, and 7 from the list of states to execute. In cases where the
error can be corrected, the program needs to remember where execution was halted
so it can return to the same location and continue. The use of state machines facilitates
implementation of this feature into exception handling code. Proper logic for diag-
nosing state information must be kept to make this possible. In addition, proper logging
and saving routines should be incorporated to ensure that data is not lost.

The conditional execution can also be applied to tests that fail. You can design
the application to execute a test depending on the outcome of another test. If Test
1 fails, you may want to skip Tests 2 and 3 but continue with the remaining tests.
Again, you can parse the tests that should not be executed. Chapter 3 discusses the
various state machines in depth. The example in Section 6.4.9 will help demonstrate
the implementation of exception handling in a state machine context.

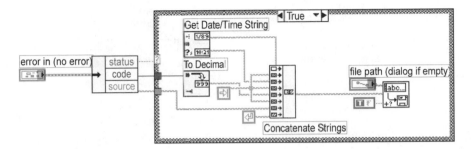

FIGURE 6.11

6.4.6 Logging Errors

Error logging is useful for keeping records of faults that have occurred during the execution of a program. The error log should report the code, the origin, a brief description, and when the error occurred. Upon the occurrence of an error, the log file is opened, written to, and closed. If further exception handling code exists, the error can be dealt with in the appropriate manner.

Error logging is beneficial in cases where the exception handling code has already been implemented and when there is no exception handler in the application. When exception handling has been implemented, error logging gives the programmer insight into the types of errors that are being generated and whether the code is handling them properly. The log can be used as a feedback mechanism to determine areas of the exception handling code that are unsatisfactory. These areas can then be enhanced to build a more robust application.

In instances when the exception handling code has not yet been developed, the error log can be used in a similar manner. The log can serve as a basis for developing the error handling code. The errors that occur more frequently can be addressed first. This method attempts to strike a balance in the amount of effort spent in developing an exception handler. The concept here is to gain the maximum benefit by attacking the most common errors.

Figure 6.11 is an example of a VI that logs errors. First, the status in the error cluster is checked to determine whether an error has occurred. If an error has been generated, the date, time, error code, and source are written out to a file that serves as the error log. The Write Characters to File VI is used to perform the logging. This VI can be used in multiple places where logging is desired, or in a central location along with other exception handling code. As the error information has been converted into a tab-delimited set of strings, it can be imported into Excel for use as a small database.

6.4.7 External Error Handler

An exception handler that is external to the application can be written to manage the errors that are generated during program execution. The application must then make a call to the external error handler. This can be beneficial when using the NI Test Executive. The error handler VI will be loaded when it is referenced in the

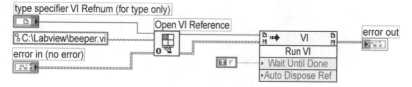

FIGURE 6.12

application. The error handler VI can be written to perform all the relevant tasks, similar to carrying out exception handling within an application.

If the error handler is written to accommodate general exceptions, it can be called in as many applications as needed. Figure 6.12, Load External Handler.vi, shows how a VI can be loaded and run from an application. First, a reference to the VI must be opened using Open VI Reference. This VI can be accessed through the Application Control palette. You must specify the path or directory in which the error handler resides. Set the VI Server Class to "Virtual Instrument" by popping up on the VI Refnum. The Invoke node is used to run the external VI. The Invoke node is also available in the Application Control palette. When the VI reference is passed to the Invoke node, the VI Server Class will automatically change to Virtual Instrument. Then, by popping up on "Methods," you can select the Run VI method from the menu. Data can be passed to the error handler VI using the Invoke node and selecting the Set Control Value method. The functions available on the Application Control palette are described in Chapter 2.

EXAMPLE:

An example of how an external exception handler is implemented is shown in Figure 6.13. This code diagram demonstrates the steps involved in using an external handler: opening a VI reference, passing the input values, running the external VI, and closing the reference. Opening a VI reference and running an external VI has already been described. In this example, the error cluster is passed to the external exception handler which determines the course of action.

First, a VI reference is opened to External Handler.vi as shown in the VI path. Then, the error cluster information is passed to External Handler.vi using the Set Control Value method on the Invoke Node. This method requires the programmer to specify the Control Name, the Type Descriptor, and the Flattened Data. The error cluster is passed to this method by flattening it using Flatten to String from the Data Manipulation subpalette in the Advanced palette. The flattened data string and the type descriptor are then wired directly from Flatten to String to the Set Control Value method. The Control Name is a string that must match identically the control name on the front panel of the VI to which the data is being passed. The name specified on the code diagram is Error In (No Error), as it appears on the front panel of the External Handler.vi. The VI is run using the Run VI method, and, finally the reference is closed.

Figure 6.14 illustrates the code diagram of External Handler.vi. This VI is similar to an exception handler shown previously. It takes the error cluster information and

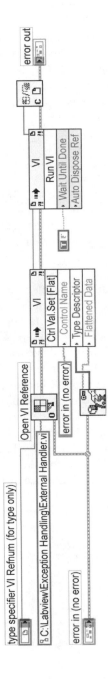

FIGURE 6.13

FIGURE 6.14

decides the course of action based on the error code. The Error Information is logged using Error Log.vi, and the case structure is driven by the error code. Case 0 is used as the default for error codes the handler is not prepared for.

In this example, the error cluster data was passed to the external VI. Similarly, data can be retrieved from the controls or indicators from the VI if it is desired. The Get All Control Values method can be used to perform this action. This method will retrieve all control or all indicator values from the external VI. The data is returned in an array of clusters, one element for each front panel control or indicator. The cluster contains the name of the control or indicator, the type descriptor, and the flattened data, similar to the way the values were passed to the External Handler VI in the example.

6.4.8 PROPER EXIT PROCEDURE

In situations where fatal or unrecoverable errors occur, the best course of action may be to terminate execution of the program. This is also true when it is not reasonable to continue execution of the program when specific errors are generated. However, abnormal termination of the program can cause problems. When you do decide that the program should stop due to an error, you must also ensure that the program exits in a suitable manner.

All instrument I/O handles, files, and communication channels must be closed before the application terminates. Performing this task before exiting the program minimizes related problems. Consider, for example, a file that is left open when a program terminates. This may cause problems when other users or applications are attempting to write to the file because write privileges will be denied.

Upon the occurrence of an error, control is passed to the error handler. Therefore, it is the responsibility of the error handler to guarantee that all handles, files, and communication channels are closed if the error cannot be recovered from. The easiest way to implement this is to have the error handler first identify the error. If the error that was generated requires termination of the program, code within the handler can perform this task. Figure 6.15, Close Handles.vi, is an example of a VI that is used solely to close open communication channels. A VISA session, file refnum, TCP connection ID, and an Automation Refnum are passed to this VI, which then proceeds to close the references.

A program should be written to have only one exit point, where all necessary tasks are executed. The best way to implement this is to utilize a state machine. By using a state machine, only one exit point is needed and will serve as the Close state. Correspondingly, there is only one place where all exception handling is

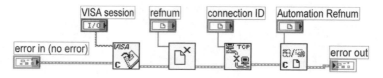

FIGURE 6.15

performed: the Error state. When an error is identified as fatal, the Error state will force the state machine to the Close state. The Close state will be responsible for terminating the program in the appropriate manner. All handles, files, and communication channels will be closed in this state. As only one Close state is needed, it will also be the last state executed during normal execution of the program when no error exists. This style makes the code easier to read and maintain.

6.4.9 EXCEPTION HANDLING EXAMPLE

Several methods of performing exception handling were provided in this section. A closing example that utilizes some of the topics that were discussed is presented in Figure 6.16. The example utilizes the state machine structure with an Error state for error handling.

The purpose of Next State.vi is simply to determine which state will be executed next. The Next State VI is also responsible for checking if an error has occurred after the completion of each state. When an error has occurred, the next state that will be executed is the Error state. The Error state first logs the error using the Error Log VI. The error code is checked to determine if it falls in a certain range that corresponds to instrument driver errors. If the error code is within that range, it is considered as unrecoverable or fatal in this example. When a fatal error is detected, the Close state is wired out to the Next State VI to execute the proper exit procedure.

If the error code does not fall in the range specified, the code is again compared to an array of user-defined error codes. This drives the case structure, which will take the action that is appropriate depending on the error that was generated. When no match results from this comparison, Case 0 is executed as illustrated in Figure 6.17.

When a match results for Case 1, the Remove States VI will remove the states that cannot be executed due to the error that was generated. Then, the program will continue with the states that can be executed according to the elements in the states array. This is shown in Figure 6.18.

Figure 6.19 shows the Close state of the state machine. This state is executed during normal termination of the program, and also when a determination is made that a fatal error has occurred. As shown in Figure 6.16, the Error state will force the Close state to execute when an unrecoverable error has been found. The only task of the Close Handles VI is to close any references and communication channels that have been opened. This will minimize problems when the application is run again.

This example demonstrates the ideas presented in this section. First, exception handling was performed at the Main Level so that program control did not have to be passed to lower levels. Second, the error handler code was separated from the rest of the code to increase readability. Not only does this reduce confusion, it also reduces the need for duplicating code in several places. Next, the use of a state machine allowed the placement of exception handling code in one location to increase maintainability and conditional parsing of tests. Error logging was performed to keep a record of exceptions that occurred. Finally, a proper exit procedure for the application was implemented. Following good practices in the creation of an exception handler will lead to sound and reliable code.

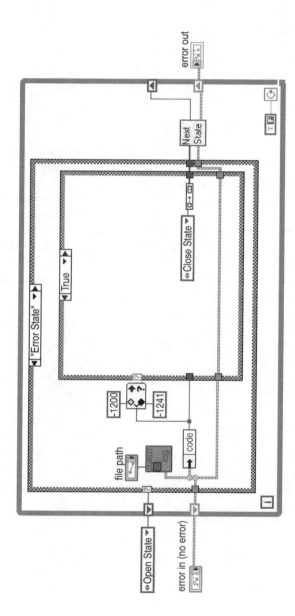

FIGURE 6.16

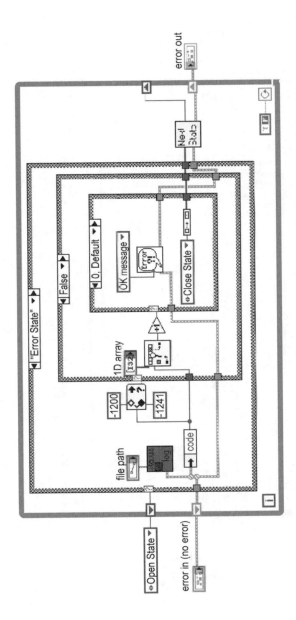

FIGURE 6.17

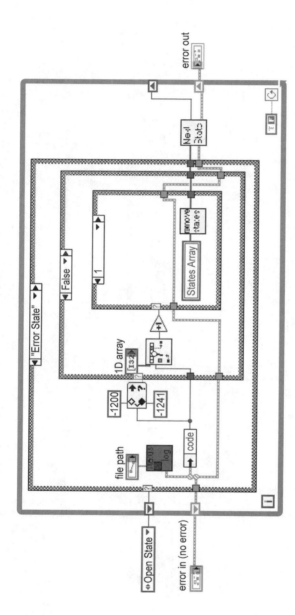

FIGURE 6.18

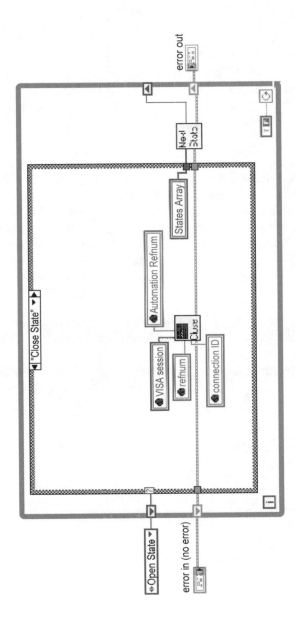

FIGURE 6.19

6.5   DEBUGGING CODE

The techniques described in the previous sections for exception handling can be utilized for debugging LabVIEW code. Error detection is very valuable during the testing phase of code. Detection assists in finding where and why errors occurred. Bugs are faults in the code that have to be eliminated. The earlier bugs are found, the easier they are to fix. This section covers some LabVIEW tools that facilitate the process of debugging VIs. First, broken VIs and the error list will be discussed. A description on how to utilize execution highlighting along with the step buttons will follow. Then, the probe tool, the use of breakpoints, and suspending execution will be described. Data logging and NI Spy will then be presented. Finally, tips on utilizing these tools to debug programs will be provided.

6.5.1   ERROR LIST

A broken Run button indicates that a VI cannot be executed. A VI cannot be run when one or more errors exist in the code. Errors can be the result of various events such as bad wires or unwired terminals in the code diagram. You may also see a broken Run button when you are editing the code diagram. However, when you are finished coding, the Run button should no longer be broken. If the Run button is broken, you can find out more information on the errors that are preventing the VI from executing by pressing the Run button. Figure 6.20 shows the Error List window that appears.

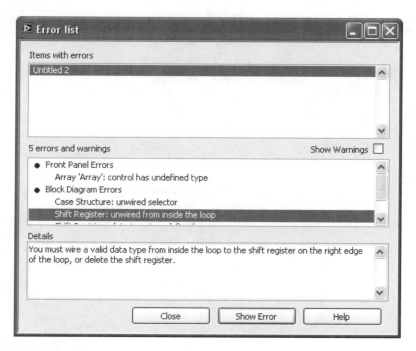

FIGURE 6.20

At the top of the Error List window is a list box that lists all of the VIs that contain errors and warnings. A box that lists all of the errors in each VI can be found just under this. Both front panel and block diagram errors will be listed. The list describes the nature of the errors. When an item in the list is selected, a text box below the list gives more information on the error and how it can be fixed. The Show Error button will find and highlight the cause of the error that is selected. There is also a checkbox, Display Warnings, which will list the warnings for the VI. The warnings do not prevent the VI from executing, but are recommendations for programming. You can set it to display warnings by default by selecting the corresponding checkbox in your Preference settings in the Edit menu.

Using the Error List, you can effectively resolve all of the errors that prevent the VI from running. Once all of the errors have been dealt with, the Run button will no longer be broken. The Error List provides an easy way to identify the errors in your code and determine the course of action to eliminate them.

6.5.2 Execution Highlighting

The Error List described above helps you to resolve the errors that are preventing a VI from running. But it does not assist in identifying bugs that are causing the program to produce unintended results. Execution Highlighting is a tool that can be used to track down bugs in a program. Execution Highlighting allows you to visually see the data flow from one object to the next as the VI runs. The data, represented by bubbles moving along the wires, can be seen moving through nodes in slow motion. The G Reference Manual calls this "animation." This is a very effective tool that National Instruments has incorporated into LabVIEW for debugging VIs. As LabVIEW is a visual programming language, it makes sense to incorporate visual debugging tools to aid programmers.

If you do not see data bubbles, perhaps your Preference settings have not enabled this option. By default, this option is activated. Select Preferences from the Edit pull-down menu, and choose Debugging from the drop-down menu. Make sure the box is checked to show data bubbles during Execution Highlighting.

Pressing the button with the light bulb symbol, located on the code diagram toolbar, will turn on Execution Highlighting. When the VI is run, the animation begins. Execution Highlighting can be turned on or off while the VI is running. Highlighting becomes more valuable when it is used in single-stepping mode. The speed of execution of the program is greatly reduced so you can see the animation and use other debugging tools while it is running.

6.5.3 Single-Stepping

Single-Stepping mode can be enabled by pressing the Pause button. This mode allows you to utilize the step buttons to execute one node at a time from the code diagram. Additionally, when Execution Highlighting is activated, you can see the dataflow and animation of the code while executing one node at a time. The Pause button can be pressed or released at any time while the VI is running, or even before it starts running. You can also press one of the step buttons located next to the

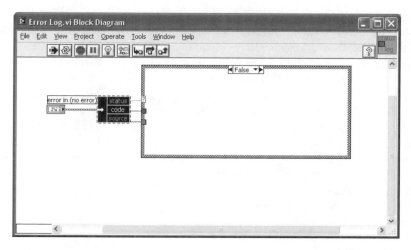

FIGURE 6.21

Execution Highlight button to enter Single-Stepping mode. The Pause button will become active automatically when these are used.

When the VI is in Single-Stepping mode, the three step buttons on the code diagram toolbar are used to control execution of the program. Depending on the code diagram, the step buttons will perform different actions. Use the Simple Help to determine what each button will do at a specific node on the code diagram. Simple Help can be accessed through the Help menu. When the cursor is placed over the step buttons, a description of their function will pop up. Figure 6.21 shows the Error Log VI in single-stepping mode with Execution Highlighting activated. The three step buttons can also be seen in this diagram.

The first step button on the toolbar is used for stepping into a particular structure or subVI. The structure or subVI will also be in Single-Stepping mode. You must then use the step buttons to complete the structure or subVI. The second button is used for stepping over objects, structures, and subVIs. If this button is pressed, the structure or subVI will execute and allow you to begin stepping again after its completion. The third button is used to complete execution of the complete code diagram. Once pressed, the remaining code will execute and not allow you to step through single objects unless Pause is pressed again.

6.5.4 PROBE TOOL

The Probe Tool can be accessed through the Tools palette or through the menu by popping up on a wire. The Probe Tool is used to examine data values from the wires on the code diagram. When a wire is probed, the data will be displayed in a new window that appears with the name of the value as the title. The probes and wires are numbered to help keep track of them when more than one is being used. You can probe any data type or format to view the value that is being passed along the wire. For example, if a cluster wire is being probed, a window with the cluster name appears displaying the cluster values. The values will be displayed once the

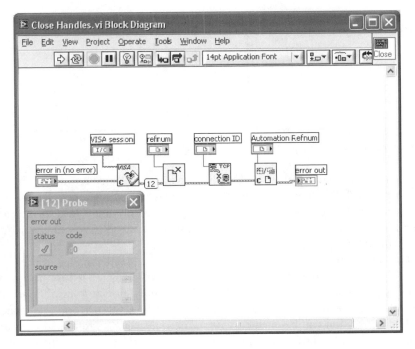

FIGURE 6.22

data has passed the point on the wire where the probe was placed when the VI was running.

The Probe Tool is very valuable when debugging VIs because it allows you to examine the actual data values that are being passed along the wires. If you are getting unexpected results or errors, you can audit values to ensure that they are correct. This tool helps you find the root of the problem. Figure 6.22 illustrates a probe on the error cluster between the VISA Close VI and the File Close VI. The wire is marked with a number, as is the window displaying the cluster values.

By default, auto probing is active in Execution Highlighting mode. This causes LabVIEW to display data values at nodes while Execution Highlighting is on. However, the complete data cannot always be viewed in this manner and is only useful for simple verification purposes. The Probe Tool will still be needed for data types such as clusters and arrays. Auto probing can be enabled or disabled from the same Preferences window as the data bubbles discussed earlier.

The conditional probe is one of the best debugging tools that have been added to LabVIEW. A conditional probe for a Numeric double precision value is shown in Figure 6.23. The data tab contains the typical information a probe has always contained — the value on the wire. The condition tab, however, is the new feature that adds a tremendous amount of value to the use of the probe in debugging.

Programmers that have used Visual Studio for C++ programming have had one very useful tool, called ASSERT. This function is a macro in C++ and part of the debugging object in Visual Basic, and this function is essentially a sanity check. The ASSERT function uses a Boolean expression as an argument. If the argument

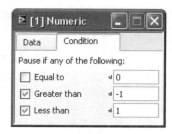

FIGURE 6.23   FIGURE 6.24

provided evaluates to true, then nothing happens. If the argument evaluates to false, the macro stops the program and provides a popup box telling where the assertion failure occurred so a developer can examine what issues could lead to the sanity check failure. This type of tool has become available to LabVIEW with the addition of the conditional probe.

The conditions available on conditional probes vary with the type of probe; not all wire types have conditional probes. For example, wires related to the communications pallet or the matrix control do not have conditional probes. Figure 6.24 shows the conditionals of a numeric conditional probe set to perform range checking. Each condition can be enabled with a different value. Figure 6.24 shows the probe configured to verify the range is between –1.0 and 1.0. If the greater or less than condition resolves to true, the program will pause and the developer has the ability to poke around the code diagram. This would allow a developer to set the conditional breakpoint to verify exception handling code catches issues and processes them as expected.

Conditional probes also function as a conditional breakpoint which makes it arguably the most flexible debugging tool in the LabVIEW toolbox.

6.5.5 BREAKPOINT TOOL

The Breakpoint Tool is another debugging device accessible through the Tools palette. As the name suggests, the Breakpoint Tool allows you to set a breakpoint on the code diagram. Breakpoints can be set on objects, VIs, structures, or wires. A red frame around an object or structure indicates a breakpoint has been set, whereas a red dot represents a breakpoint on a wire. Breakpoints cause execution of the code to pause at the location where it has been set. If it is a wire, the data will pass the breakpoint before execution is paused. A breakpoint can be cleared using the same tool that is used to set it.

Breakpoints are valuable because they let the user pause the program at specific locations in the code. The program will execute in its normal manner and speed until it reaches the breakpoint, at which point it will pause. The code that is suspect can then be debugged using Single-Stepping mode, Execution Highlighting, and the Probe Tool.

Once a breakpoint has been set, the program will pause at the break location every time it is executed. You must remember to clear the breakpoint if you do not want the program to pause during the next iteration or execution. If you save the VI while a breakpoint has been set, the breakpoint will be saved with the VI. The next time you open the VI and run it, execution will pause at the break location. You can use the Find function to locate any breakpoints that have been set.

Breakpoints are non-conditional, meaning whenever a breakpoint is encountered, the program is stopping. If conditional breakpoints are desired, the conditional probe is preferable to a breakpoint.

6.5.6 SUSPENDING EXECUTION

You can force a subVI to suspend execution, for debugging purposes, when it is called. This can be done using one of the following three methods. The first method is to select Suspend when Called from the Operate menu. The second method is to pop up on the subVI from the code diagram of the caller and select SubVI Node Setup. Then, check the box Suspend when Called. Alternatively, you can pop up on the icon while the subVI is open and select VI Setup. Then check the box Suspend When Called.

When you cause a subVI to suspend execution, its front panel will be displayed when it is called. The subVI also enters a special execution mode when it is suspended. The Run button begins execution of the subVI. When a subVI is suspended, it can be executed repeatedly by using the Run button. To the right of the Run button is the Return to Caller button. Once suspended, you can use Execution Highlighting, Single-Stepping, and the Probe Tool to debug the subVI. When you use Single-Stepping while a subVI is suspended, you can skip to the beginning and execute the VI as many times as needed.

6.5.7 DATA LOGGING

Data Logging is another LabVIEW built-in tool that can be used for debugging purposes. Front panel data can be logged automatically by enabling Log at Completion from the Operate menu. When the VI is run the first time, a dialog box will appear, prompting the user to enter a filename for storage. Alternatively, a log file can be selected before running the VI, by selecting Log from the Data Logging submenu in the Operate menu. When the filename is selected prior to running the VI, the dialog box will not appear. The front panel data is entered into that log file after the VI executes.

The Data Logging feature is a method for saving data from tests, similar to a database. LabVIEW enters a date and time stamp, along with the data for the indicators and controls from the front panel. The data can then be viewed by selecting Retrieve from the Data Logging submenu. Figure 6.25 illustrates how the data appears when data is logged and retrieved using this feature. This is a simple front panel with two controls and two indicators. The multiplication and addition results of the two integer controls are displayed in the indicators. This is how the data will be displayed when it is retrieved from the log file. The time and date stamp appears at the top, along with controls for scrolling through the records and deleting records.

Data Logging is useful for saving data values from tests and for debugging VIs. It serves as a mechanism for quickly saving data from specific VIs that are being

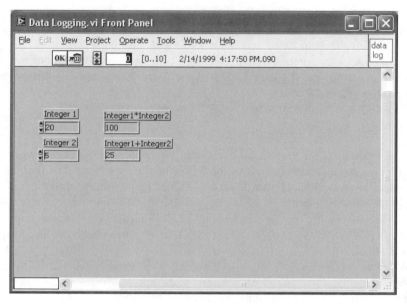

FIGURE 6.25

debugged. The saved data log can then be reviewed for suspect values. The data log is also useful for monitoring intermittent problems with VIs. The front panel data can be saved, retrieved, and purged as needed.

6.5.8 NI SPY/GPIB SPY

These two utilities are very similar and are both used as debugging tools on Windows operating systems. NI Spy monitors the calls that are made by applications to NI-488.2, NI-VISA, IVI, and NI-VXI drivers. Similarly, GPIB Spy tracks any calls that are made to GPIB drivers. They are useful for determining the source of communication errors, whether they are related to general communication problems or are application specific. They help you verify that communications with an instrument are correct. However, when either of these applications are running, they will degrade the speed of your application. Use them only when you are debugging your program to free up system resources, especially if execution time is a consideration for the application.

NI Spy displays the index number assigned to the call, a description of the operation and parameters, and the time that it occurred. The tool displays the calls as they are made during the execution of your application. Errors are immediately highlighted to indicate failures. NI Spy also allows you to log the activity for review at a later time.

GPIB Spy monitors calls to the Windows GPIB driver by Win32, and displays them while your application is executing. All errors or failures are highlighted for quick identification. You can view each call as it is made and see the results, including any timeouts. This utility can be used to verify that your application is sending the right calls to the Windows GPIB driver. GPIB Spy lists an index number of the call, the names of the GPIB calls, output of the status word ibsta after the call, output of the error word iberr, output of the count variable ibcntl, and the time of each call. All of these contain useful information on the performance of the application. You can view detailed information by using the Properties button on the toolbar.

Familiarization with GPIB, NI-488.2, and the ANSI/IEEE 488.2 communication protocol may be necessary to fully utilize and understand the debugging features on both GPIB Spy and NI Spy. A discussion of IEEE 488.2 is beyond the scope of this book.

6.5.9 UTILIZATION OF DEBUGGING TOOLS

The Error List, Execution Highlighting, Single-Stepping mode, Probe Tool, Breakpoint Tool, and suspending execution were described in the previous sections. These built-in LabVIEW features are very effective for debugging code when they are used in conjunction with on-line Help. Each one is a weapon the programmer can use for tracking down and resolving problems. These tools are summarized in Table 6.2 which lists the tool, its application or use, and how to access or enable it.

The software process model being followed determines when the debugging or testing phase for the code begins. In an iterative model, debugging is involved in each cycle of the process. In the Waterfall model, debugging is done only during one phase of the development cycle. In either case, the first action is to eliminate the errors that prevent the VI from running. As already described, the Error List will assist in removing these errors to allow the VI to run. This part of debugging should

---

TABLE 6.2
Debugging Tools

Tool	Application	Accessing
Error List	Used to list, locate, and resolve errors that prevent a VI from running.	Press broken Run button.
Execution Highlighting	Used to animate, visualize data flow along wires on code diagram.	Press highlight button with bulb.
Single-Stepping Mode	Allows execution of one node at a time.	Use Pause button.
Probe Tool	Displays data values passed along wires.	Available from Tools palette.
Breakpoint Tool	Halts execution of program at specific location.	Available from Tools palette.
Suspending Execution	Suspends subVI for repeated execution during debugging.	Use Operate menu, SubVI node setup by popping-up on icon or VI setup while VI is open.
Data Logging	Enables front panel data logging to file.	Use Operate menu and Data Logging submenu.
GPIB Spy/NI Spy	Monitor calls to Windows drivers.	Run application.

be performed as the application is being developed, regardless of the model being used. Getting rid of errors that prevent VI execution should be considered part of the coding phase in LabVIEW. This is analogous to syntax errors in traditional languages that are pointed out to the programmer during coding. The Error List makes this easy for even the novice programmer. It guides the programmer in resolving errors quickly.

If it is possible, try to test one VI at a time. Test the individual drivers and subVIs separately before attempting to run the main or executive. You may be overwhelmed when you try to debug a large program with many subVIs. Not only is it easier to concentrate on smaller parts of the program, but you reduce the errors that may be caused through the interaction of the subVIs with each other. A modular design approach with VIs that are specific and self-contained simplifies testing. This interaction through data flow may make it appear that more errors exist. You may also be able to create a simulator for I/O or other portions of the code that have not yet been prepared. Again, this will help in isolating problems related to the specific code at hand without having to deal with I/O errors.

Once the VI can be executed, the next step is to run it with Execution Highlighting enabled. The animation helps you see the data flow on the code diagram. Execution Highlighting will help you find bugs caused by incorrectly wired objects. While the VI is running, make sure that the code executes in the order that was intended, which can be identified with highlighting.

You may also want to probe certain wires with Execution Highlighting and make sure that the values are correct by using the Probe Tool. For instance, probing the error cluster between two objects or VIs will help narrow down where the error is generated. You will see the value of the Probe Tool for debugging once you begin to use it. The Probe Tool and Execution Highlighting can be used in Single-Stepping mode. Single-stepping mode lets you look at a section of code in even more detail to find the problems that exist.

If problems persist, a few suggestions are offered here for you to consider. These might seem basic, but they are the ones that are easy to overlook. First, make sure that the input values provided by the user controls are valid. The Probe Tool can be used to perform this check from the code diagram. When these input values are out of the acceptable range, the code will not execute as intended.

If you are performing communications with an external device, file, or application, check the commands or data sent. The device may not respond to unexpected commands. During this process, also check for correct file names, handles, and addresses. Examine the external device to see if it is functioning properly, and manually perform the actions you are trying to take through automation. Consider using delays in the program if the external device is not responding quickly. Investigate the execution order of your code to ensure that the correct sequence of events is occurring. Race conditions can result if the code is not executing as intended.

Inspect arrays for correct usage of indices. Arrays, lists, rings, and enumerated types all start off at zero and can cause potential problems if not accounted for. During this inspection, check case structures that are driven by these values to see if they correspond. Also make sure that you have a default case set up to ensure the correct code is executing. You should also examine loop structures to make proper use of shift registers so data is not lost. This includes proper initialization of the shift registers.

Set personal time limits for how long you will attempt to determine where an error exists in code. It becomes very frustrating to attempt to debug a section of code for hours. When your time limit expires, a second opinion should be brought in. This second perspective will see the programming problem differently and may well propose a solution or at least ask questions that may lead you to a solution.

6.5.10 EVALUATING RACE CONDITIONS

LabVIEW has always had inherent multitasking capabilities. In LabVIEW 5.0, multithreading became available. As the combination of hardware, operating systems, and software advances capabilities, new and exciting programming problems evolve in step with the newest technologies. This section will outline steps that can be taken to understand race conditions that can occur in LabVIEW code that leverages LabVIEW's parallel execution. Chapter 9 discusses multithreading in detail; this section will cover a more generic set of exceptions which are race conditions.

A race condition is simply when two branches of code are attempting to use the same set of data. Generally, a race condition involves one branch of code getting access to data in an order that was not intended. As an example, one branch of code might be coordinating instrument communications and is in the process of reading a trace from an instrument. A second branch of code is attempting to access the trace for analysis. The intention of the developer is to have the trace be read, stored, and then signaled to the analysis branch. A race condition exists if it is possible for the analysis branch to read the array before the communications branch has had the ability to write the array in memory.

Race conditions create special issues that may not be identified by typical debugging tools, and the reason is race conditions are timing related; most debugging tools do not give the developer a view of what is going on without altering the timing of the application. Single-stepping or execution highlighting most certainly changes execution timing in very profound ways. It will be almost impossible to identify a race condition in an application when the code is being stepped through.

Application logging can help identify some issues. In very high performance applications, it may disguise some issues though.

One way to view race conditions is they are not entirely deterministic. There are some probabilities that become involved. Essentially every VI, or VI subsystem, has a probability that it will be executing on a processor core at a given period of time. When application logging is used such as writing debugging information to a file, the amount of code running in a VI has been changed relative to the rest of the application. The code which is used to write the log file is the "new" code which is impacting application timing. This timing change can make some race conditions less likely to appear.

One way that we have successfully used to troubleshoot multithreaded C++ code that is applicable to LabVIEW is to draw out trees of the call stacks. Examine the LabVIEW application. Are the branches of code running in different subsystems?

If so, draw out the hierarchy of each subsystem. Record which VIs are called, and what data the VI is accessing. As the programmer, you will know which VIs have fairly small execution time, and which VIs have longer execution times. The smaller execution time of a VI, the lower the probability it is running concurrently with another specific VI.

Once you have the hierarchies drawn out for each branch or subsystems, examine the data accessed. Any data that is accessed by multiple branches is suspect of causing a race condition. Some common data may only be read by all branches. For example, at application start up, initial configuration information may be read in and stored. This type of data is unlikely to cause issues after application startup. Any data that is commonly accessed that can be written to by any of the branches may be a race condition.

In general, this level of analysis is not required for strictly LabVIEW code. If an application using external code such as .NET objects external code segments is exhibiting behavior such as crashing, examine which branches of code are accessing the external code. It is entirely possible, especially if LabVIEW's subsystems are being used, that there is a thread-related race condition. A semaphore can be used to restrict access to the code or object. Every branch or subsystem will need to use the semaphore to access the suspect code object.

In general, semaphores should be used to protect very specific areas — not entire sets of VIs. If entire sets of VIs are blocked off with a semaphore, then there is a very strong chance that application performance is going to be degraded. For example, code surrounding direct read or writes to a suspect object should be blocked off. Processing data that has been read or preparing data that is about to be written should be done outside the confines of the semaphore.

In many cases, access restrictions such as semaphores or notifications will resolve these types of problems. On occasion, a set of troublesome code will show up that access restriction does not fix. A last resort would be to set the suspect code to run in the User Interface subsystem of the application. This will force all calls to the code to be done by specific threads of the LabVIEW runtime engine. Forcing particular threads to be used is discussed in Chapter 9, multithreading.

6.6 SUMMARY

When you are developing an application, it may be easier to just omit code to perform error detection and handling because it requires extra work. However, exception handling is needed to manage the problems that may arise during execution. An exception is something that might occur during the execution of a program. These unintended events, or exceptions, must be dealt with in the appropriate manner. If exceptions are left unattended, you can lose control over the program, which may result in more problems.

An exception handler allows programmers to deal with situations that might arise when an application is running. It is a mechanism to detect and possibly correct errors. LabVIEW provides some built-in tools to aid the programmer in error detection and handling, but it is the responsibility of the programmer to implement the exception handling code. Several methods for dealing with errors were described in this chapter. The topics discussed will assist the programmer in writing more robust code through the implementation of exception handlers.

Exception handling should be considered at an early phase of application development. It is appropriate to take exception handling into account when the structure or architecture of the application is being decided upon. Better applications can be developed when exception handling is a forethought, not an afterthought. Exception handling, when built into an application, will lead to sound and reliable code.

BIBLIOGRAPHY

G Programming Reference, National Instruments, Austin, TX, 1999.
Professional G Developers Tools Reference Manual, National Instruments, Austin, TX.
LabVIEW Function and VI Reference Manual National Instruments, Austin, TX.
LabVIEW On-line Reference, National Instruments, Austin, TX.

7 Shared Variable

This chapter discusses the shared variable concept released in LabVIEW 8. The shared variable is a standard feature for the language that provides a service for distributed data throughout LabVIEW. The Data logging and Supervisory Control module extends the level of functionality. Shared variables are available for use on any operating system and real time (RT) targets. The shared variable engine itself requires a Windows installation or an RT target.

This chapter is devoted to a discussion of the shared variable and includes a discussion of the shared variable engine and communications employed to distribute data. On the surface, it may seem questionable to devote an entire chapter to a distributed data system. As this chapter progresses, it will become clear that the shared variable is not a simple service.

7.1 OVERVIEW OF SHARED VARIABLES

Shared variables are primarily a mechanism for distributed networking support. In general, distributed applications have processes that perform specific tasks with information that needs to be shared with other elements of the application. The shared variable provides a mechanism to the LabVIEW programmer that abstracts away the complexities of distributed programming. Distributed applications do not make life easier on the developer as they add-in a number of exception cases that need to be considered. For example, how does the application know if messages were lost, what if messages are late, the data is old, or what if it is necessary for IP networks to check that multiple copies of the same message have not been received?

The shared variable engine absolves LabVIEW developers from the vast majority of these issues. From a development perspective, what we primarily need to be concerned about is dragging and dropping the correct shared variable. Networking issues are handled by the shared variable engine and the client. It is possible from a shared variable to determine if the information is "fresh" or if the variable has not been updated in awhile. Network security issues are also addressed in configurable fashions through the shared variable engine and supporting NI security services. Distributed application development is roughly as complex as working with global variables.

Adding shared variables to projects and basic information are provided in Chapter 1 and Chapter 2. This section expands on shared variable mechanics and the benefits and penalties for using them.

Shared variables support essentially every data type that can be found in Lab-VIEW. Pull-down selection menus in the shared variable configuration screen provide what might be considered an incomplete list until you arrive at the last entry: custom control. Complex clusters can be configured as custom controls which, as stated, make the shared variable capable of supporting any data type that would be needed for distribution.

7.1.1 SINGLE-PROCESS VARIABLES

When creating a shared variable, one of the configuration items is to determine if the variable is network-published or single process. Single-process shared variables do not sound so intuitive. What is the benefit of using a shared variable that is not shared? It would seem not of much benefit at all. Essentially, a single-process shared variable functions much like a global variable. In fact, there is a considerable amount of common code used between the shared single-process variable and the standard LabVIEW global variable. Any place a global variable is used, a single-process shared variable will also work — such as sharing data between two parallel loops.

Using a single process shared variable does have two advantages: The first is the ability to network-publish the variable with a simple configuration change; going from a global variable to a network-published variable requires code rework. The second advantage is the ability to use the time-stamp functionality of the shared variable, which does not exist on the global variable.

There is one significant disadvantage to using the shared variable in a single process. Performance benchmarking performed by National Instruments shows that the reading of a shared variable and global variable are roughly equivalent. The time required to write data to a shared variable, however, is more than for a global variable.

The buffering feature of the shared variable can also be used to emulate the performance of a queue. Similar to replacement for a global variable, this is useful for RT targets and should be used in any queues that are expected to be distributed at a later time.

One key difference between the single process variable and network-published variable is that historical data logging is not available to single-process variables.

7.1.2 NETWORK-PUBLISHED VARIABLE

There are two supported formats for network-published variables. The shared variable configuration screen only allows the selection of a network-published variable. Data socket bindings are still supported and usable; however, new development should focus on the use of shared variables.

Unlike single-process variables, network-published variables can be buffered. The nomenclature of buffering is somewhat misleading; the variable's buffering functions as a FIFO queue. When using the queue, time-stamping data corresponds to when the particular copy was added into the queue. The use of the buffering feature needs to be carefully designed. All designs involving a queue have one exception case that must be considered: queue overflow. In the case of the shared variable, overflowing the queue is not allowed. Any writes to the queue when it is full get silently discarded. LabVIEW does not receive an error notification that overflow errors are occurring.

One disadvantage of a network-published variable, compared to its single-process counterpart, is the setting of initial values. Default values for published shared variables are not defined unless the DSC module is installed. When designing an application without the DSC module, a VI needs to step up and write data to that variable at system startup. Otherwise, the nodes subscribed to the variable will be receiving undefined data.

7.2 SHARED VARIABLE ENGINE

Shared variables are published through the shared variable engine. The engine itself is available only on Windows-based machines or RT targets. On Windows, the engine runs as a service on the machine. As a service, there are a few items to the engine that are worth noting. Services do not require a user to be logged in. The service is running as a part of the operating system and is not tagged to a user in the task manager. This allows for the engine to run regardless of who is or is not actively logged into the machine. Unfortunately, services are not allowed to have their own user interfaces; external applications need to run to communicate with the service.

7.2.1 Accessing the Shared Variable Engine

As the shared variable engine is a service, it does not present a direct user interface. There are two ways for an administrator to access the engine: The first interface is through the shared variable manager. The manager is a NI provided application that permits configuration of the engine. The second is the Windows Event Viewer. As a service, the engine logs events to the operating system. The event viewer does not permit configuration of the engine but does allow an administrator to evaluate events the service is generating.

7.2.1.1 Shared Variable Manager

The shared variable manager is shown in Figure 7.1. The manager will be the primary tool for accessing the engine as it gives the administrator the most control over the engine. The manager window consists of three frames which are (1) a list of items, (2) the watch list, and (3) the alarms window. The items list provides a listing of all shared variable libraries available on the system, including RT targets. Each library is listed as an independent process and can be stopped and started independently. This is a core feature on large distribution points which may have libraries for multiple applications running concurrently.

From the variable manager you can add or remove libraries, assuming you have appropriate permissions. By default, LabVIEW installation sets the shared variable services to operate without security restrictions. In Section 7.7, it will be pointed out that leaving this configuration without security is, in general, not advisable.

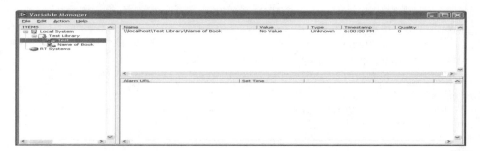

FIGURE 7.1

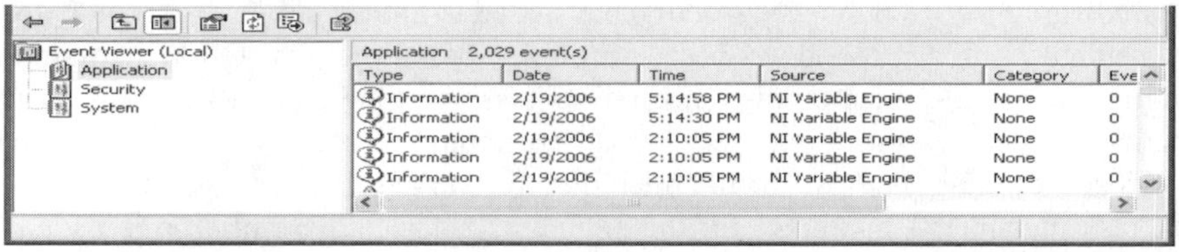

FIGURE 7.2

The watch list pane allows for viewing the status of individual variables. Specific elements that can be seen in the watch list are the name of the variable, the current value, its type, the time stamp of last value change, and the quality.

The variable manager allows for a user to add and configure variable libraries and variables without running LabVIEW. It also allows for libraries to be created from outside the context of a LabVIEW project.

7.2.1.2   Windows Event Viewer

The second tool for accessing the health of the engine is the event viewer. From the Windows Control panel, access administrative tools. The events viewer shows system recorded events for applications, security, and the system. Engine events are recorded in the application window. An example of the application counter is shown in Figure 7.2. The events recorded are fairly generic in nature, but will give an administrator knowledge of when the service was started or stopped.

7.2.1.3   Windows Performance Monitor

Another tool that can be used for monitoring the health or usage of the shared variable engine is the performance monitor. The performance monitor does not show directly items involving the shared variable engine but does provide some insights into how the engine and its host machine are performing. Section 7.4 discusses the networking aspects of shared variables, but for now one hint will be provided: the shared variable uses UDP as its transport protocol. The performance monitor is a Windows operating system tool that provides access to a large number of statistics about the health of the computer. Different elements of the operating system are broken up into "objects" such as the CPU, the hard drive, and networking statistics. One of value to us here

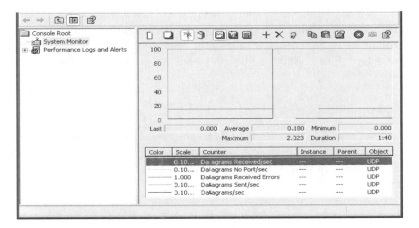

FIGURE 7.3

is the UDP performance. Figure 7.3 shows the performance monitor reporting on the volume of UDP traffic being processed by the computer. If a distributed application is experiencing problems, the performance tools can help identify what types of issues the networking may be experiencing. For example, if the error count is high, then there may be a problem with cabling or a switch in the network. As a reference, bit error rates on Ethernet are fairly low; on order of 1 in 1 million bits or less should be corrupted by a network. If you have hundreds of UDP messages per second arriving with errors, the problem probably exists in the network. The task manager's performance tab would supplement the performance counters. If the packet rates are large, compare this with the task manager's report of network utilization. A high utilization would suggest additional network capacity is needed.

One counter configuration that may be useful in determining the health of a networked application is shown in Figure 7.4. The performance monitor is tracking counters for the IP object, TCP object, and UDP object. Each object is displaying the Send and Receive counters. If there is suspicion that the host machine is very busy on the network — for example, it is hosting a web server for VI control and the shared variable engine — it can be expected to show the TCP and UDP Send/Receive counters running large numbers. It should not be surprising that the IP counters will run even larger numbers; in fact, it should be approximately equal to the sum of the UDP and TCP counters as IP is the underlying protocol of both TCP and UDP. The reason why the IP counter would only be approximately equal

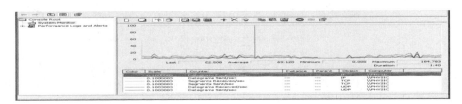

FIGURE 7.4

is because there is another commonly used protocol that also uses IP — ICMP. Ping messages and notification of message delivery failure are often sent through ICMP. It is possible for an administrator to monitor counts on multiple computers at once, so the counters shown above can be monitored for the shared variable engine and its clients.

7.2.1.4 Windows Task Manager

The Windows task manager is not really a tool to evaluate the health of the shared variable engine, but one tab on the task manager can yield some misleading information about the network utilization. When examining the Task Manager's Network Performance tab, it is reporting utilization of what the link is rated for, not what the network is capable of supporting. Most Ethernet links are 100 megabit or faster, but this does not mean the router the cable is plugged into is going to support this data rate. For example, the backbone of Rick's home network is a Nexland ISB SOHO router/firewall. The router supports 10/100 megabit operation on the links but is only capable of 6 megabit throughput internally. It would be very unlikely to see the Network Performance tab report over 6% utilization! Now this might sound clunky and old but consider the simple fact that it still has a higher throughput than many DSL and cable modem installations. When looking at the task manager's network utilization it is important to know what the other side of the Ethernet cable is capable of supporting. A remote client with a 128 kilobit DSL link may show 1.2% of its network being utilized, but that is all the DSL link has to offer. This may not be a factor in a typically modern engineering facility, but remote locations with high throughput may need to consider what their link to the system supports.

7.3 SHARED VARIABLE PROCESSES AND SERVICES

This section discusses the processes and services used in operating the shared variable. Several services are in use and provide different functional elements. Figure 7.5 shows some of the Windows services involved. Top of the services list is the Lookout Citadel server. The Citadel server is not dependent on the rest of the shared variable engine. It is installed with LabVIEW, but the DSC is needed to take full advantage of Citadel.

With the DSC module, additional capabilities available to distributed data include datalogging through the Citadel database, data alarms, data scaling, and initial values. In the event the shared variable engine is being used without the DSC

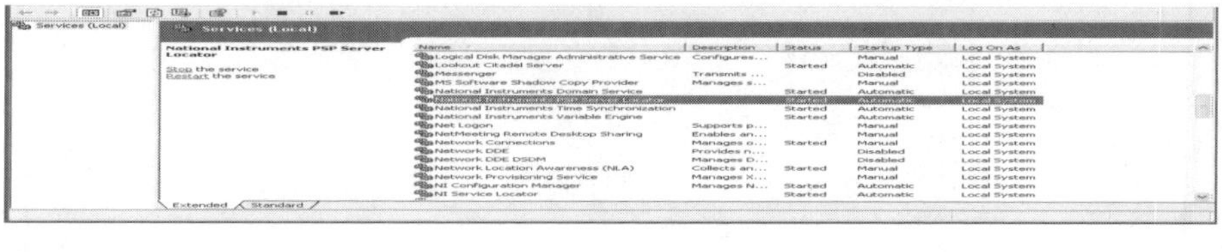

FIGURE 7.5

module, one of the first steps the application needs to do is set the shared variable's initial values on system startup.

Citadel is a streaming database and uses a feature called "deadbanding" for logging. If a value of a variable does not change, or does not change significantly, Citadel can be configured to not log the value update, so it is possible to have Citadel only log value changes of significance.

The DSC Module's alarm capabilities make this add-on desirable in many monitoring applications. The alarms have a significant level of configuration so the alarming feature can be tailored specifically to a scenario. Standard alarm configurations include values over high and low thresholds and a rate of change alarm. Rate of change alarms are useful in process control; for example, if a temperature sensor is reporting a mixture is heating too quickly or too slowly.

The National Instruments Domain Service is responsible for coordinating National Instruments security for the shared variable engine. This service is responsible for user logins and access to features of the shared variable such as who is allowed to stop library processes. Usage of this process is discussed in more detail in Section 7.7.

The National Instruments PSP Locater Service is responsible for coordinating the Publish-Subscribe-Publish Protocol transactions across system elements. It works with the NI security features that are part of the domain service. This service handles the detection of remote domains.

The National Instruments Time Synchronization Service is used primarily with Lookout or the DSC Module in LabVIEW. This module helps synchronization of time between remote machines; in general, there is no need to modify this service. In fact, it can be difficult to modify the time synchronization of a remote client without having either Lookout or the DSC Module installed. In cases where neither Lookout nor the DSC Module is installed, to configure this module to synchronize time to another machine, such as the one hosting the variable engine, use the following steps (note that this only works on Windows-based machines):

```
Start up a DOS window

enter: lktsrv -stop

enter: lktsvr -start <update interval> <IP Address>
```

Now this will work and cause the local time service to synchronize with the remote machine — but only until the computer is restarted; then this setting will be lost. To make a change like this "permanent," you can change the service Startup command in the Windows Services configuration. Go to the control panel and open up Administrative Tools. Open up the Services Applet and double click on the National Instruments Time Synchronization Service. A dialog window will open up and is shown in Figure 7.6. Stop the service, and add the second line shown previously into the start parameters box at the bottom of the window. Every time the computer is restarted, it will bring up the time service with the address specified. It is also possible to configure the service to synchronize time with multiple machines — separate the IP addresses using a space-delimited list.

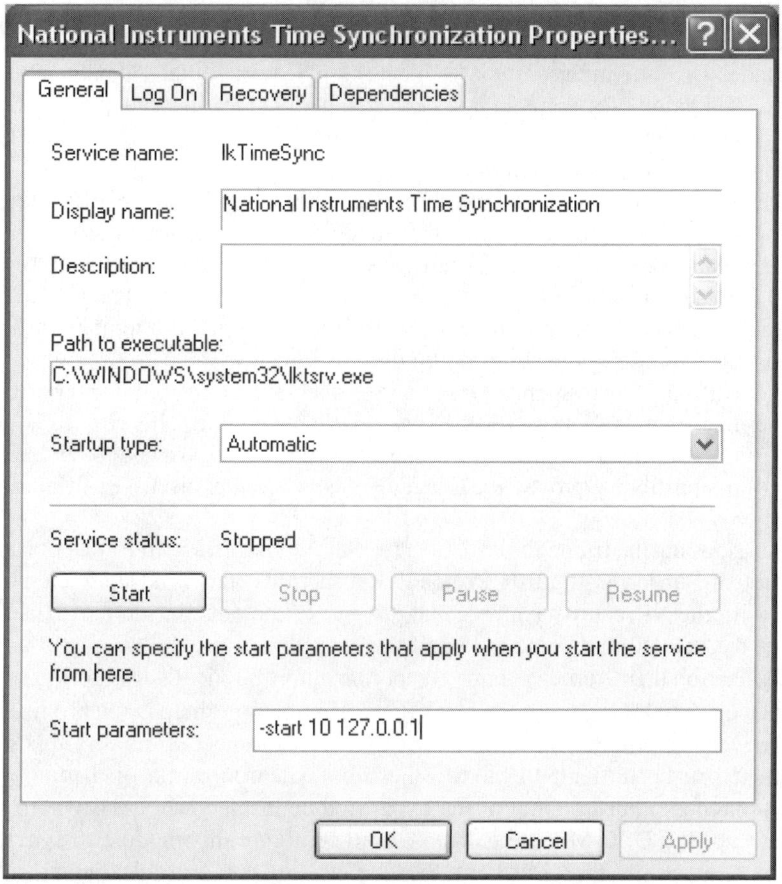

FIGURE 7.6

The National Instruments Shared Variable Engine is the last service to be mentioned, and it obviously implements the shared variable engine. This service is dependent on the NI Configuration Manager.

7.4 SHARED VARIABLE NETWORKING

Shared variables are transported across networks via a protocol called the NI Publish-Subscribe Protocol (PSP). The PSP protocol is proprietary in nature, so it is not possible to go into significant detail of message exchanges. There are several aspects of the protocol that can be discussed and we will begin with the underlying transport. The PSP protocol itself uses UDP/IP as the underlying transport protocol. UDP is not a reliable protocol; it is possible for PSP packets to be lost during transmission. The advantage to using UDP for the transport protocol is the lack of TCP-related overhead. TCP is reliable, but its windowing algorithm and associated acknowledgments can create what is called *network jitter*. When a TCP packet is lost, almost

every TCP stack in commercial use makes a fundamental assumption: the packet
was lost due to network congestion. The TCP stack will then perform a delay and
then attempt to retransmit the packet. The trouble with this mechanism is that the
application sending the message is unable to control the retransmission delay.
Retransmission delay is called *jitter*. Jitter is undesirable in shared variable applications because retransmissions end up delaying applications. If a front panel control
is bound to a shared variable, the goal is to attempt to acquire a current value for
the control. TCP-based communications could leave the application in a position of
waiting around for the TCP windowing to push data up to the application.

Another disadvantage to using TCP is the requirement that the connection be
open and maintained. The TCP socket needs to stay open, which requires the variable
engine to maintain a port for the specific variable. A large, distributed application
with dozens of shared variables would end up putting a load on the networking end
of the shared variable engine. The choice of UDP as the transport protocol allows
the engine to use a few ports across any number of remote applications. This is far
more efficient from the perspective of the engine and allows the engine to concentrate
on pushing data instead of maintaining TCP sockets.

As mentioned, UDP is not a reliable protocol because UDP itself does not
provide notification that the message was received. The NI-PSP protocol implements
acknowledgments and essentially resolves the unreliable aspect of UDP by implementing proprietary reliability functions on top of UDP. This may sound like considerable work, but, as stated above, TCP is not a desirable protocol when time can
be a critical factor in network communications.

The engine and clients use sockets in the low 6000 range — specifically, ports
6000 to 6010 in addition to 2343. Basic network security requires the use of firewalls
on every machine; these ports will need to be made available through the firewall.
Applications using both shared variables and network communications need to avoid
using these port ranges as they will be refused. The shared variable engine on the
server side is a service, which means it will be up and running before the application;
the application needs a user to launch it whereas the service comes on line before
a user has logged in. This means the server will always win the race to request the
ports from the operating system.

The format of data exchanged in the UDP packets is proprietary to National
Instruments, meaning the exact message formats and field definitions are not public.
What is known is that the PSP protocol uses uniform resource locators (URLs) to
identify information about the shared variables available on a given engine. The
engine will also forward out a lot of information about itself to a client during initial
"conversation." This startup conversation includes which processes are responsible
for different functions, including the process name, location, and version number
down to the build number. Distributed timing is also a common issue in distributed
applications; the engine notifies the client which process is creating time synchronization information.

The application in particular that is performing most of the server work is called
"Tagger" and is located in Program Files\National Instruments\Shared\Tagger. In
the event you are working on a machine using a software firewall such as Zone
Alarm, you may receive mysterious requests about an application called Tagger

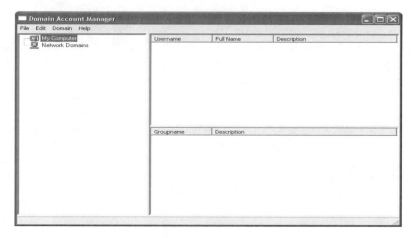

FIGURE 7.7

attempting to access the Internet. This would be the shared variable engine attempting to make itself available. Do not be alarmed that your machine has acquired a virus of some sort.

7.5 SHARED VARIABLE DOMAINS

The shared variable service uses the concept of domains to control access to shared variable services. The NI Security Service coordinates user logins and access to NI Services. It is a strong recommendation that the NI Security Services be used when deploying shared variables across any network.

The NI security apparatus is built on the concept of users, groups, and domains. A domain is a set of users and account policies. It is possible for users to have different policies applied in different domains. In order to create a domain, from LabVIEW select Tools-> Security-> Domain Account Manager to bring up the Manager. Figure 7.7 shows a machine with a fresh installation of LabVIEW — no domains exist on the local machine. The domain manager displays domains specified on the local machine and any known domains on the network. In order to create a domain, right click on local computer and create a domain.

The first dialog box to appear when a new domain is created is the Administrator password. Administrator is a unique account name, and its usage should be kept to a minimum. A machine can have a single domain. Once the password is created, the Administrator account is established, and the domain name is entered; you have to log in to the domain to do anything else. Creation of the machine's domain will be done as a "Guest" user, and the Guest user does not have authority to do anything with the machine's new domain. Log in as Administrator and add accounts for the user base. Once the domain is established, only accounts with administrator privileges will be permitted to modify anything in the Domain Account Manager. Figure 7.8 shows a local machine with a new LabVIEW Advanced Programming Techniques domain created. Only two predefined accounts exist for the domain, as mentioned,

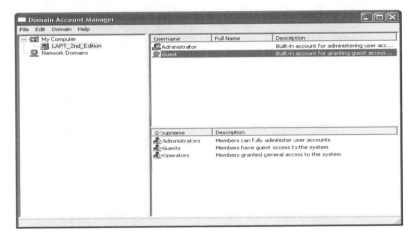

FIGURE 7.8

Administrator and Guest. In general, it is poor security practice to allow for only these two accounts. Every user should have a unique user account for the domain. In the case of the two predefined accounts, leaving everybody as a user allows everybody access to the entire NI security system which largely defeats the purpose of the system.

Once the domain is created, right click on it; selecting Properties shows the base configuration options for a domain. Figure 7.9 shows the general properties that can

FIGURE 7.9

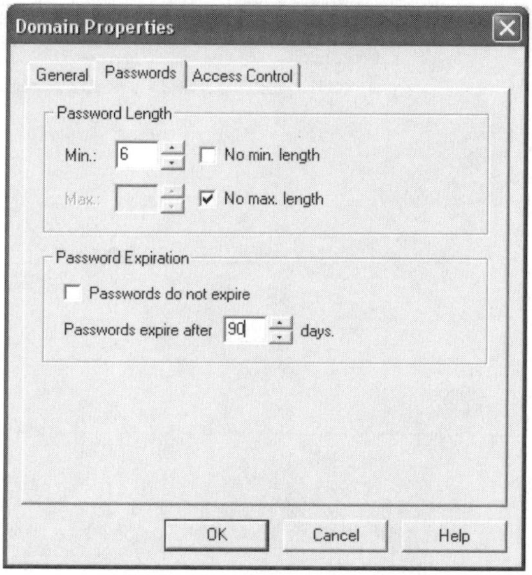

FIGURE 7.10

be configured consisting solely of the name of the domain. The hosting machine cannot be changed from this tab; to move the domain to a new machine, you will need to destroy it and recreate it on the target machine. The middle tab shows password policy configuration for the entire domain. Figure 7.10 shows the policy options available. It is recommended that password expiration policies be set for all domains. Depending on your employer or customer, guidelines on password expiration may already be defined. NI security settings should be configured to match any local IT requirements.

The last tab on the domain properties dialog is access control and is shown in Figure 7.11. This tab provides two lists, a list of machine-granted access and a list of machines denied access. Only one access list can be active at a time, so either this configuration is granting access to specific machines or denying access to specific machines. In general, we do not use this tab. If there is a security concern, it should be resolved with the machine's firewall, which will deny access to everything on the system instead of just NI-specific items.

In order to add a user to the domain, right click on the domain and select New User. Figure 7.12 shows the dialog box that is displayed. Once the user is given a name and description, a new password must be created before the user can be assigned membership to any groups. A base install of LabVIEW has a minimal set of security settings. The DSC Module permits for further security policy applications. Users are fairly self-explanatory. Any individual entity interfacing to the system is a user. Two built-in users are defined: the Administrator and Guest. Typically, each person using the system should be given a unique user account to the system.

Groups have three defined categories: Administrators, Operators, and Guests. It is possible to define additional groups. Additional groups with a basic LabVIEW

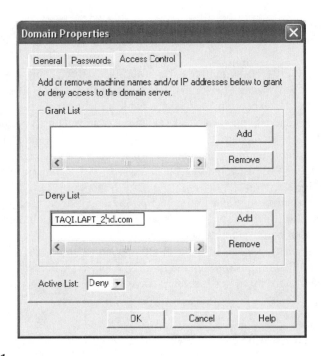

FIGURE 7.11

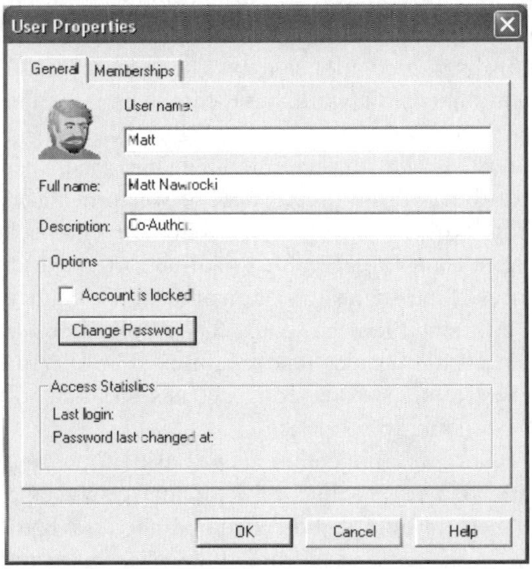

FIGURE 7.12

install do not add much value other than to provide logical groupings of users. The DSC Module adds some features that allow for additional security policies to be configured.

The security differences between the groups are largely minor. Administrator group members have access to security elements of the system. Operators can do everything with the exception of security, which includes starting and stopping variable libraries and adding variables to existing libraries. Guests surprisingly have a fair amount of permissions in the system by default, which include adding variables to the shared variable engine. There is no security configuration available in a base installation of LabVIEW. Tightening up the system would require either Lookout or the DSC Module.

Okay, so we have gone out and created a domain, added users, and assigned group membership to the users. The tie-in to the shared variable engine is now regulated by user permissions. Opening up the variable manager now requires a login and the security policies of the user are applied to limit what the user can do. Publishing variable libraries through the LabVIEW project will also use the security polices defined. Note that guest accounts have permission to publish variables to the engine by default.

7.6   PITFALLS OF DISTRIBUTED APPLICATIONS

This section will provide some information on issues that can be encountered when developing a distributed application; it is not entirely specific to the shared variable itself, but as shared variables can be used across machines on different networks, you may encounter some of the same issues.

This section is not intended to scare off development of a distributed application, but it is intended to raise a level of awareness as to what can be expected in the general realm of distributed applications. The shared variable abstracts a great deal of complexity around the network, but it cannot entirely absolve the developer of everything.

First and foremost, do not expect communications between computers to be 100% reliable every time, all the time. Links between machines do drop, packets get lost, multiple copies of the same packet may arrive, and packets may not arrive in the order they were sent. In general, a small network with a few switches will not see many instances of message duplication or out of order arrival; larger networks distributed across different ISPs, however, can. Applications built for wide area networks (WANs) need to consider link reliability. The shared variable engine is going to provide "best effort" in delivering updates on its distributed data, but the network is not always going to cooperate.

In the WAN scenario, the application must consider how sensitive the data is to time. If the application is not very time sensitive, then remote applications can work on stale data, and there is little to be concerned about. When time sensitivity is important, and data is not arriving on time, then errors need to be logged, alarms need to be sent out, and exception handling routines need to be engaged. One minor detail to consider on reporting errors: If the remote application is unable to communicate with the shared variable engine, then it probably can't send a network message to that server with an error message. A machine that is collecting error information should not be the one running the shared variable engine.

The shared variable engine on a remote machine cannot notify a remote application that it cannot communicate with. If the remote application must have current data, timestamps need to be evaluated. Real time environments that are not validating correct information in the required time frames should be logging errors.

The shared variable engine uses Netbios, and Netbios does not always work well when machines are not regulated by a domain controller. In the event applications are operating in an environment without a domain controller, register the shared variable engine by IP address or name.

Distributed networks that are split across an internet service provider (ISP) may run into particular problems. Many ISPs use Dynamic Host Configuration Protocol (DHCP) to assign IP addresses to nodes. A DHCP system assigns addresses for set periods of time — called an address license. Once this license time expires, the node needs to renew its address license, and it frequently will be assigned a new address. In such cases, one can use domain name services such as DynDNS to keep a constant name for a shared variable server. The firewall or router interfacing to the ISP needs to be configured to use a domain name service. Once the domain names are established, it will be possible to register the variable engine by domain name on the remote computer.

7.7 SHARED VARIABLES AND NETWORK SECURITY

Everybody needs a hobby, and some people choose these better than others. For reasons beyond the scope of this book, there is a large community of people with apparently nothing better to do than break into computer networks. Shared variables bring a lot of the power of networking to LabVIEW applications, yet at the same time, they also bring in some of the vulnerabilities. A full discussion of network security is beyond the scope of this chapter, but this section will outline some of the concerns a developer needs to be aware of when deploying a networked application.

Network security is essentially a "negative deliverable." Users will figure out when network security has failed, but proving that network security is working can be much harder. The simple fact that everything works may mean network security is preventing access, or it could mean that nobody has attempted to break into the network. The first assumption a developer needs to make is that attempts to violate the network will be made, and machines need to be configured to accept this reality.

First steps are fairly obvious and have nothing to do with LabVIEW. Every machine in the network should have antivirus applications with current virus definitions. Software firewalls should be operating on each machine, and a hardware firewall should restrict access at each location your network connects to the Internet. For example, a distributed network application gathering data at several manufacturing facilities may be exchanging shared variables over the Internet. Each facility needs to be firewall-protected at the Internet access point. In general, we never recommend using shared variables across the general public Internet. If your employer or customer has an IT department, they should be consulted on a distributed application. The IT organization will probably want to know the nature of the data being exchanged and the volume of data that will be exchanged, so they can verify that the bandwidth available in the network is sufficient for your application and other applications running in the system.

Next on the list is the use of wireless Ethernet. Almost nothing gives external access to a system better than an improperly configured wireless Ethernet router. The two main configuration items for security on wireless routers is encryption and MAC layer filtering. Both should be used.

Shared variables do not use encryption; any data sent over the Internet will be sent "in the clear," meaning any third party that can see the message exchange can see the data. If remote facilities need to be interconnected, they should be connected through a virtual private network (VPN). VPNs use encryption for data transfer between two remote networks and makes the two remote networks appear to be one private network. VPN selection, configuration, and operation is a bit large in scope to present in this book. The IT department of your employer or customer should have already installed a VPN; as mentioned above, it would be to your benefit to make sure they are prepared to support any bandwidth you require.

In some cases, the customer may not be network savvy. It is your responsibility as the developer to clearly communicate security-related requirements. In the event customers lose data to hackers, they will be upset, and it is unlikely they will blame themselves when it happens! In the event you find yourself in this situation, a statement of work and breakdown of costs should include basic security items such as virus software and definition update subscriptions, firewalls, and, where required, VPN gateways.

As an example, consider a scenario where a distributed application is simply monitoring and controlling the coolers for a dozen flower shops. The average florist does not have many impacts on national security, and it would seem that security restrictions should be fairly minimal. Customer records and financial information are not handled by this system, which would seem to put this system a bit lower on the security priority. However, security on this system is not negligible. If somebody managed to hack into the system, they could potentially wreak havoc. Potential attack scenarios would be denial of service and playback attacks.

One type of simple attack is to flood the variable engine or the hosting machine with excess packets. If the attacker identified the port range of 6001 to 6010 correctly, he or she could hit the engine itself with a flood of packets. There is also a range of packets in the 5001 to 5005 range used that can be identified with a packet sniffer. If a system startup is monitored, the port 2343 would also be noticed. When the variable engine receives a malformed message, it will silently discard it but still has to process the message to determine it needs to be discarded. This would still amount to a denial of service. Flooding ports 2343, 5001 to 5005, and 6001 to 6010 would tie down NI's security service which could prevent machines from registering to the domain.

As PSP is not encrypted, somebody with unauthorized network access could start off by monitoring data exchanges. Packets could be resent to the server in order to try to define their purpose and monitor the responses. In the event the server engine is not set to any level of security, the hacker could try writing data to the system! Results of this could range from late night pages claiming the coolers are running at 150° as an alarm condition to changing the set point of the cooler. The first scenario may sound somewhat amusing; the second scenario could result in the loss of inventory.

Before rendering a decision that the shared variable is too risky to deploy in a system, some common sense needs to be applied. It remains to be seen if the PSP protocol is hackable at this point. The PSP system has a lot more security built into it than many systems users would deploy. For example, if the example floral cooler control system were being built without the shared variable engine and instead was using custom written UDP packet exchanges, would the application validate the IP address where messages are coming in from? If not, then it would arguably be far less secure than a properly configured shared variable engine. The advice on this subject is pretty straightforward: Do not set a system up so that you will wind up finding out the expensive way. The shared variable engine greatly simplifies the act of distributing data, but security and access control need to be properly set up.

In order to minimize the potential of attacks, the following steps can be taken: First, designing distributed applications with central control may be undesirable. Remote clients should be designed such that they operate without the need of a central machine providing instructions. The cooler controls in this scenario should run the coolers regardless of the state of the network.

Firewall configurations could be set up to discard traffic from any machine not on a given list. The stock firewall in Windows XP Service Pack 2 allows for custom port permissions to be set which includes the ability to set a "scope." The scope can be set to allow only specific machines to pass traffic through the firewall on given ports. This is desirable because it eliminates any need to test the security of the NI system and reduces the processing requirements on the server. If the firewall rejects a packet at the entrance of the IP stack, the processing time required to route the doomed packet through the IP stack and into the NI security system will be eliminated.

Lastly, NI domains should be used. The standard LabVIEW installation does not provide much flexibility in the shared variable engine and domain security, but it can require somebody to log into the domain in order to interact with the engine.

If there are concerns that a system is under some type of attack, the tools mentioned to monitor the health of the engine can be used. If there are a dozen remote systems updating shared variables every minute, then the UDP traffic load is going to be low. The performance monitor can give an accurate count of how many UDP packets are arriving at any of the machines. If the counter values are through the roof, further investigation may be warranted.

Ethernet sniffers can be used to capture live data exchanges over a network for diagnosis. Ethereal is one off-the-shelf open source tool that does packet captures. This tool can be useful for monitoring traffic seen at an Ethernet interface. One issue to be aware of with Ethernet sniffers is they can generate very large trace files as networks tend to pass a lot of traffic that most users are not aware of. Systems with many traffic exchanges will generate tremendous amounts of traffic on their own; parsing an Ethernet trace file can become very time consuming.

In the event it becomes necessary to capture and analyze traffic, first start by identifying which IP addresses are the shared variable engine servers and which the clients are running on. Filters on the trace can be applied to minimize traffic displayed. For example, if the server were also running a Web server, filtering out HTTP traffic could eliminate a large volume of packets. Netbios exchanges will be common between the clients and servers; there is no need for concern when you see these. A lot of short messages will be exchanged between clients and servers, and most of it will be in nonprintable characters. Things that should be looked for are IP addresses not on the known list attempting to access NI related ports (2343, 5001 to 5005, and 6001 to 6010).

7.7.1 LabVIEW Specific Security Issues

LabVIEW provides a domain system for regulating users and access to network resources. LabVIEW domains should not be confused with Windows Domains; they are not the same. LabVIEW domains define three groups of users; Administrators, Operators, and Guests. Each group definition grants a fairly broad level of access to the shared variable engine. The security settings on the variable manager should be among the first things to be modified after installation. Figure 7.13 shows the security settings panel. This panel is accessed from the Edit menu on the Variable Manager. The only settings that can be configured on a basic LabVIEW installation are the ability to modify security settings and to manage variables. Our recommendation is to add each group to the list and lock down engine access. Administrator group members should be the only ones able to modify the security settings and manage the variables. Guests should be set to deny access to both security and variable management. The operators group may need some flexibility. In a fully deployed system, they do not need to manage variables, whereas if a system is in development, this group may need the ability to add or remove variables and libraries from the system.

With a configuration set to deny both security and variable management, guests can still add items to the watch list, start the shared variable engine, and can acknowledge alarms. Guests cannot change security settings, add or remove vari-

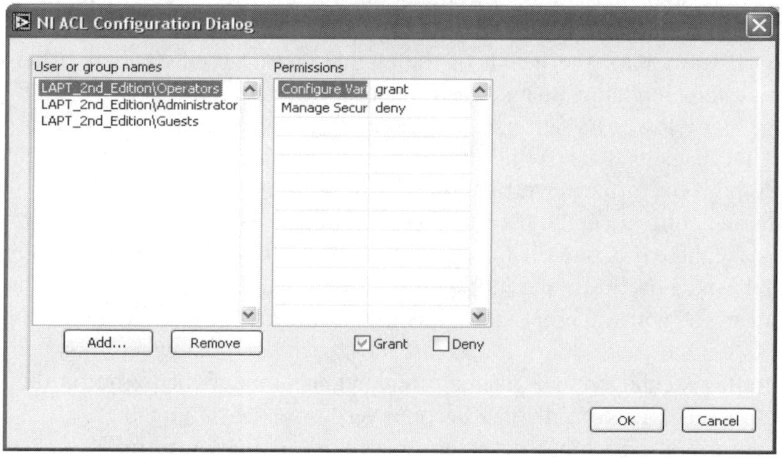

FIGURE 7.13

ables, start or stop processes, or add/remove libraries from the engine. The ability to start the shared variable engine is not much of a security risk; guests cannot stop the engine which would be a denial of service to other applications.

From a security perspective, most users need guest access once a system is deployed. Even better from a security perspective would be to use the DSC Module to tighten down permissions on the guest account. Individuals with administrator access should be kept to a minimum with a clear reason as to why they need the ability to configure the security settings of the system. Developers with a role that requires adding items to the system can be put into the Operator category, but if there is no particular need to start and stop processes or modify the libraries, they should be located into the guest group. It is possible to change a user from one group to another as required, and this is generally an advisable path to take. When a deployed system is running, there needs to be only a few administrators, and many guests. In the event an operator password is compromised, libraries can be removed from the shared variable engine; this would be a denial of service attack on the system.

BIBLIOGRAPHY

http://zone.ni.com/devzone/conceptd.nsf/webmain/
    5B4C3CC1B2AD10BA862570F2007569EF#4.

8 .NET, ActiveX, and COM

As many readers have known and suspected, there is much more to .NET and ActiveX than a LabVIEW interface. .NET and ActiveX are relatively new technologies that wrap around several other technologies, namely OLE and COM. This chapter gives a general outline of .NET, ActiveX, COM, and OLE technologies and how they may be applied by the LabVIEW programmer.

Component Object Model (COM) is a programming paradigm for software development. Microsoft is providing a path for developers to practice component-based development. The mechanical and electrical engineering communities have long practiced this. When designing a circuit board, resistors (components) are chosen from a catalog and purchased for use on the circuit board. Software developers have never truly had a component-based development environment.

Libraries, both static and dynamically linked, have been available for quite some time but were never well suited for component development and sale. Many companies are unwilling to distribute information about code modules such as header files or the source code itself. Dynamically linked libraries have proven difficult to work with because no mechanisms for version control are provided. Libraries may be upgraded and older applications may not be able to communicate with the upgraded versions. The ActiveX core, COM, is a standard that attempts to address all the major issues that have prevented software component development and sale. The result is a large pool of components that programmers may assemble to create new applications. Ultimately, this chapter will provide the background needed for Win32 LabVIEW developers to take advantage of the new component architecture.

LabVIEW itself may be called as an ActiveX component and may call other components. The VI Server has two implementations, one as an ActiveX control for Win32 usage, and a TCP-based server that may be used by LabVIEW applications on nonWindows platforms for remote control. The VI Server feature will also be covered in this chapter.

Microsoft.NET is a general-purpose development platform that is used to connect devices, systems, and information through Web services. .NET relies on communications between different platforms across the Internet using standard protocols. .NET provides for security, exception handling, threading, I/O, deployment and object lifetime management.

Following the overview of these technologies there will be a section of examples. These examples will show how to use .NET, ActiveX, and VI server. The examples will help illustrate the concepts described below.

different threads of execution take over the CPU. This procedure generates additional overhead. This does not happen with in-process COM objects. The fact that out-of-process communication occurs should not dissuade a LabVIEW programmer from using COM servers. This is something that developers will need to be aware of when developing time-critical applications.

One advantage out-of-process servers have is stability. When an application is started, it is given its own address space and threads of execution. In the event that one process crashes, the other processes will continue to execute. These other processes may experience problems because a peer crashed, but good exception handling will keep other processes alive to accomplish their tasks. It is possible to crash peer processes, but defensive programming can always mitigate that risk. Out-of-process servers can also be run on different machines using DCOM, which allows for the process to continue executing even if one machine crashes. Good examples that can take advantage of distributed computing are monitoring and production software.

8.1.1.5 The Variant

Programmers with experience in Visual Basic, Visual Basic for Applications, and Visual Basic Script should have heard of a data type called the *variant*. Variants do not have a defined type; these variables are meant to support every data type. In high-level languages such as VBScript, the variant allows for simple, fast, and sloppy programming. Programmers are never required to identify a variable type. The variant will track the type of data stored within it. If a variant is reassigned a value of a different type, it will internally polymorph to the new type. ActiveX, COM, and OLE controls allow the usage of variants, and therefore they must be supported in LabVIEW.

From a low-level language such as C++, variant programming can be difficult. The variant type has a number of binary flags that are used to indicate what type of data it contains. For example, variants can contain integers or arrays of integers. A C++ programmer needs to check binary flags to determine which are set: primitive type (integers) or arrays. LabVIEW programmers go through similar problems; we need to know what type of data the variant is supposed to contain and convert it to G data. The Variant to Data function is explained later in this chapter. LabVIEW programmers do not need to check for unions and bit-flags to work with Variants. We do need to know what type of data an ActiveX control is going to pass back, however. If you choose the wrong type, LabVIEW will convert it and give you something back you probably do not want.

8.2 COM

The core of the Component Object Model is a binary interface specification. This specification is independent of language and operating system. COM is the foundation of a new programming paradigm in the Win32 environment. ActiveX and OLE2 are currently based on the COM specification. Neither technology replaces or supercedes COM. In fact, both OLE and ActiveX are supersets of COM. Each technology addresses specific programming issues, which will be discussed shortly.

Because COM is intended to be platform and machine independent, data types are different in COM than standard types in the C/C++ language. Standard C/C++ data types are anything but standard. Different machines and operating systems define the standard types differently. A long integer has a different interpretation on a 32-bit microprocessor from that on a 64-bit microprocessor. Types such as "char" are not defined in COM interfaces; a short, unsigned data type is defined instead. COM defines its types strictly and independently from hardware.

COM does define a couple of data types in addition to primitive types used by nearly every language. Date and currency definitions are provided as part of the interface. Surprisingly, color information is also defined in COM. OLE and ActiveX controls have the ability to have their background colors set to that of their container. The information identifying background colors is defined in the COM specification. COM also uses a special complex data type, the variant.

8.3 OLE

We will demonstrate in Section 8.4 that OLE interfaces are, in fact, ActiveX objects. The reverse is not always true; some ActiveX objects are not OLE interfaces. OLE is an interfacing technology used for automation, the control of one application by another. Many programmers subscribe to the myth that ActiveX has replaced OLE, but this is not true. OLE is a subset or specialization of ActiveX components. In general, ActiveX will be more widely used. LabVIEW applications that control applications like Microsoft Word and Excel are using OLE, not ActiveX.

A document is loosely defined as a collection of data. One or more applications understand how to manipulate the set of data contained in a document. For example, Microsoft Word understands how to manipulate data contained in *.doc files and Microsoft Excel understands .xls files. There is no legitimate reason for the Word development team to design in Excel functionality so that Word users can manipulate graphs in their word documents. If they did, the executable file for Word would be significantly larger than it currently is. OLE automation is used to access documents that can be interpreted by other applications.

8.4 ACTIVEX

As described earlier in this chapter, the component object model is the key to developing applications that work together even across multiple platforms. ActiveX, like OLE, is based on COM. This section will discuss what ActiveX is, why it was developed, and how it can be used to improve your applications.

8.4.1 DESCRIPTION OF ACTIVEX

ActiveX controls were formerly known as OLE controls or OCX controls. An ActiveX control is a component that can be inserted into a Web page or application in order to reuse the object's functionality programmed by someone else. ActiveX controls were created to improve on Visual Basic extension controls. It provides a way to allow the tools and applications used on the Web to be integrated together.

The greatest benefit of ActiveX controls is the ability to reduce development time and to enhance Internet applications. With thousands of reusable controls available, a developer does not have to start from scratch. The controls available to the developer also aid in increased functionality. Some controls that have already been developed will allow the developer to add options to the Website without having to know how to implement functions. The ability to view movie files, PDF files, and similar interactive applications is made possible through the use of ActiveX.

ActiveX is currently supported in the Windows platforms, as well as Web browsers for UNIX and the Mac. ActiveX, which is built on COM, is not Win32-specific. This provides the ability to be cross-platform compatible, making ActiveX available to the widest possible group of users.

ActiveX controls can be developed in a number of programming languages, including Microsoft Visual Basic and Visual C++. The key to compatibility is the COM standard that ActiveX is built with. Development time is reduced because a developer can use the language that is most convenient. The programmer will not have to learn a new programming language to develop the control.

Some of the definitions of ActiveX technology have roots in object-oriented (OO) design. COM is not completely OO; however, much of the OOP design methodology is used in COM. The main benefits of OO are encapsulation, inheritance, and polymorphism. For more information on these subjects, see Chapter 10.

8.4.2 ACTIVEX DEFINITIONS

First, we will discuss some of the main ActiveX technologies. The main divisions of ActiveX include automation, ActiveX documents, ActiveX controls, and ActiveX scripting. After the discussion of these technologies, we will discuss some of the terms used with ActiveX as well as COM. These terms include *properties, methods, events, containers, persistence, servers, clients, linking,* and *embedding*.

8.4.3 ACTIVEX TECHNOLOGIES

ActiveX automation is one of the most important functions of ActiveX. Automation is the ability of one program to control another by using its methods and properties. Automation can also be defined as the standard function that allows an object to define its properties, methods, and types, as well as provide access to these items. The automation interface, Idispatch, provides a means to expose the properties and methods to the outside world. An application can access these items through its Invoke method. Programs being able to work together is critical to software reuse. Automation allows the user to integrate different software applications seamlessly.

ActiveX documents (previously called OLE documents) are the part of ActiveX that is involved in linking, embedding, and editing objects. ActiveX documents deals with specific issues relating to "document-centric" functionality. One example is in-place editing. When a user wants to edit an Excel spreadsheet that is embedded in a Word document, the user doubleclicks on the spreadsheet. Unlike previous versions of OLE documents, the spreadsheet is not opened in Excel for editing. Instead, Word and Excel negotiate which items in the toolbar and menus are needed to perform the editing. This function of ActiveX allows the user to edit the sheet in Word while still having all the necessary functionality. Another example of ActiveX documents is Web page viewing. When someone viewing a Web page opens a file, like a PDF file, the file can be displayed in the Web browser without having to save the file and open it separately in a PDF reader.

ActiveX controls (which replace OCX controls) are reusable objects that can be controlled by a variety of programming languages to provide added functionality, including support for events. ActiveX scripting is a means to drive an ActiveX control. This is mainly used in Web page development. An ActiveX control is lifeless without code to operate it. Because code cannot be embedded in the Web page, another method of control is necessary. That method is scripting languages. There are two common scripting languages supported in Web pages that are ActiveX compliant. The first is JScript (a type of JavaScript) and Microsoft Visual Basic Scripting Edition (VBScript).

8.4.3.1 ActiveX Terminology

Simply put, a method is a request to perform a function. Let's say we are programming a baseball team. The baseball team is made up of players. A pitcher is a specific type of player. The pitcher must have a ThrowBall method. This method would describe how to throw the ball. A full description of method was provided at the beginning of this chapter.

A property is the definition of a specific object's parameters or attributes. For instance, in our baseball example, the player would have a uniform property. The user would be able to define whether the player was wearing the home or road uniform.

An event is a function call from an object that something has occurred. To continue the baseball analogy, an event could be compared to the scoreboard. When a pitch is made, the result of the pitch is recorded on the scoreboard. The scoreboard will show ball or strike, depending on the event that occurred. Events, as well as methods and properties, occur through automation mechanisms. Events are covered in more detail in the following section.

A container is an application in which an object is embedded. In the example of an Excel spreadsheet that is embedded in a Word document, Microsoft Word is the container. LabVIEW is capable of being a container as well.

When an object is linked in a container, the object remains in another file. The link in the container is a reference to the filename where the object is stored. The container is updated when the object is updated. The benefit of linking is the ability to link the same object in multiple files. When the object is updated, all of the files that are linked to the object are updated.

When an object is embedded in a container, the object's data is stored in the same file as the container. If an Excel worksheet is embedded in a Word document, that data is not available to any other application. When the worksheet is edited, the changes are saved in the data stream within the Word document. With both linking and embedding, a visual representation of the data in the container is stored in the container's file. This representation, called *presentation data*, is displayed when the object is not active. This is an important feature because it allows the file to be viewed and printed without having to load the actual data from the file or data stream. A new image of the data is stored after the object has been edited.

Persistence is the ability of a control to store information about itself. When a control is loaded, it reads its persistent data using the IPersistStorage interface. Persistence allows a control to start in a known state, perhaps the previous state, and restores any ambient properties. An ambient property is a value that is loaded to tell the control where to start. Examples of ambient properties are default font and default color.

8.4.4 EVENTS

An event is an asynchronous notification from the control to the container. There are four types of events: Request events, Before events, After events, and Do events. The Request event is when the control asks the container for permission to perform an action. The Request event contains a pointer to a Boolean variable. This variable allows the container to deny permission to the control. The Before event is sent by the control prior to performing an action. This allows the container to perform any tasks prior to the action occurring. The Before event is not cancelable. The After event is sent by the control to the container indicating an action has occurred. An example of this is a mouse-click event. The After event allows the container to perform an action based on the event that has occurred. The final event is the Do event. The Do event is a message from the container to the control instructing it to perform an action before a default action is executed.

There are a number of standard ActiveX control events that have been defined. Some standard events include Click, DblClick, Error, and MouseMove. These events have Dispatch Identifications and descriptions associated with them. DispIDs have both a number and name associated with them to make each event, property, and method unique. Microsoft products use the dispatch ID numbers, where LabVIEW uses the dispatch ID names. The standard events have been given negative DispIDs.

8.4.5 CONTAINERS

An ActiveX container is a container that supports ActiveX controls and can use the control in its own windows or dialogs. An ActiveX control cannot exist alone. The control must be placed in a container. The container is the host application for an ActiveX control. The container can then communicate with the ActiveX control using the COM interfaces. Although a number of properties are provided, a container should not expect a control to support anything more than the IUnknown interface. The container must provide support for embedded objects from in-process servers, in-place activation, and event handling. In addition to the container providing a way for the application to communicate with the ActiveX control, the container can also provide a number of additional properties to the control. These properties include extender properties and ambient properties.

The container provides extender properties, methods, and events. They are written to be extensions of the control. The developer using the control should not be able to tell the difference between an extender property and the control's actual property.

There are a few suggested extender properties that all containers should implement. These controls are Name, Visible, Parent, Cancel, and Default. The Name property is the name the user assigns to the control. The Visible property indicates whether the control is visible. The Parent property indicates what the parent of the control is. The Cancel property indicates if the control is the cancel button for the container. The Default property indicates if the control is the default button for the container. There are a number of additional extender properties that are available to be used.

Ambient properties are "hints" the container gives the control with respect to display. An example of an ambient property is the background color. The container tells the control what its background color is so the control can try to match. Some of the most used ambient properties are UserMode, LocaleID, DisplayName, Fore-Color, BackColor, Font, and TextAlign. The UserMode property defines whether the control is executing at run time or design time. The LocaleID is used to determine where the control is being used. This is mainly for use with international controls. The DisplayName property defines the name set for the control. The ForeColor and BackColor define the color attributes for matching the control's appearance to the container.

8.4.6 How ActiveX Controls Are Used

The first requirement for using an ActiveX control or a COM object is that it be registered in the system. When applications like Microsoft Word are installed on a computer, the installer registers all of the COM objects in the system registry. Each time the application is started, the information in the registry is verified. There is a slightly different mechanism when the COM object or ActiveX control is contained in a Dynamic Link Library (DLL). The specifics will not be mentioned here; the important concept is that the item is known in the system.

The next issue is the interfaces to the control being used. The program using the control needs to be able to access the control's interfaces. Because the ActiveX control is built using COM objects, the interfaces are the common interfaces to the COM object. Some of the most common interfaces are mentioned in the section on COM. These include IDispatch and IUnknown. Most of the interfaces can be created automatically by programming tools without having to do the coding directly.

8.5 .NET

Microsoft .NET is a general-purpose development platform. .NET's main purpose is connecting devices, systems and information through Web Services. Web Services allow access to reusable components or applications by exchanging messages through standard Web protocols. The use of standard protocols like Extensible Markup Language (XML) and Simple Object Access Protocol (SOAP) enable computers from many different operating systems to work together. In addition to providing a means for cross-platform interoperability, .NET allows the user to run distributed applications across the Internet.

There are some additional benefits of .NET applications. The .NET Framework, which is one building block of the .NET platform, includes components for security, exception handling, threading, I/O, deployment, and object lifetime management. The integration of theses services in the .NET Framework makes application development faster and easier.

8.5.1 DESCRIPTION OF .NET

Microsoft .NET platform consists of numerous components and subcomponents. There are many books dedicated to individual pieces of the .NET platform, so we will just touch on some of the basics. In order to utilize .NET in your LabVIEW applications, very little knowledge of the nuts and bolts of .NET is necessary.

The .NET platform has six main components. The six components are the operating system, .NET Enterprise Servers, the .NET Framework, .NET Building Block Services, and Visual Studio .NET. The operating system is at the lowest level of .NET. The operating system includes desktops, servers, and devices. The inclusion of devices reflects the continuing trend to distribute applications and functionality to personal devices such as cell phones and PDAs. At the center of the .NET platform is the .NET Framework. The .NET framework consists of base classes to support Web Services and Forms. .NET provides a set of framework classes that every language uses. This will allow for easier development, deployment, and reliability. The .NET framework is built on the Common Language Runtime (CLR). The CLR is described further in the next section. At the highest level is Visual Studio .NET (VS.NET). VS.NET supports Visual Basic, Visual C++, Visual C#, and Visual J# for application development.

You may be asking yourself if .NET is supposed to be easier to develop and distribute, and has additional security and networking functionality, why would anyone still use COM? The question of what .NET means to COM's future is a controversial question. Well, first, let me assure you that COM is not going to become unsupported tomorrow. Microsoft has stated that the next Windows operating system will still support COM. There has been significant development of COM objects and applications. That is not going to be discarded. That being said, Microsoft does recommend that new application development be done using .NET. In order to be able to support existing ActiveX controls as well as to make the transition to .NET easier, you are able to call ActiveX objects in .NET.

8.5.2 COMMON LANGUAGE RUNTIME

The Common Language Runtime (CLR) is the foundation for the .NET Framework. The CLR is a language-neutral development and execution environment that uses common services to perform application execution. The CLR activates objects, performs security checks on the objects, performs the necessary memory allocations, executes them, and then deallocates the memory used. The objects the CLR operates on are called Portable Executable (PE) files. The PE files are either EXE or DLL files that consist of metadata and code. Metadata is information about data that can include type definitions, version information, and external assembly references. The CLR also supports a standard exception-handling mechanism that works for all languages.

8.5.3 Intermediate Language

Intermediate Language (IL) is also known as Microsoft Intermediate Language (MSIL) or Common Intermediate Language (CIL). All .NET source code is compiled to IL during development. The IL is converted to machine code when the application is installed or at run-time by the Just-In-Time (JIT) compiler by the CLR. The IL supports all object-oriented features including data abstraction, inheritance, and polymorphism. The IL also supports exceptions, events, properties, fields and enumeration.

8.5.4 Web Protocols

There are several Web protocols utilized in .NET including XML, SOAP, HTML, and TCP/IP. XML provides a format for describing structured data. This is an improvement over Hypertext Markup Language (HTML) in that HTML is a fixed format that is predefined. XML lets you design your own markup languages for limitless different types of documents.

Simple Object Access Protocol (SOAP) is a simple, XML-based protocol that is used to exchange structured data on the Web. SOAP is the main standard for passing messages between Web services and the Web service consumers. SOAP defines the format of the XML request and responses used for the messaging. Through SOAP, objects can talk to each other across the Internet, even through firewalls. This helps eliminate an issue developers have had using distributed COM (DCOM). Because the IP address information was included in the messaging in DCOM, communications could not go through firewalls without configuration.

8.5.5 Assembly

The assembly was introduced in .NET. The assembly is similar to a compiled COM module. The assembly will be a DLL or EXE file. The assembly contains the IL code as well as a manifest that stores information on assembly name, version, location, security, files that are in the assembly, dependent assemblies, resources, exported data types and permission requests. A single module assembly has everything within one DLL or EXE. You can also use the assembly linker to create a multimodule assembly. The multimodule assembly can consist of many module and resource files.

An assembly can be private or public. The private assembly is a static assembly that is used by a specific application. A private assembly must exist in the same directory as the executable. A public or shared assembly can be used by any application. The shared assembly must have a unique shared name. All assemblies must be built with a public/private key pair. The public/private key pair will make the assembly unique. This can then be used by the CLR to ensure the correct assembly is used.

8.5.6 Global Assembly Cache

In the COM world, the shared library (DLL) used by the application is registered in the Windows Registry. When another application is installed on the computer, there is a risk of overwriting the DLL that is being used with another version that does not support all the features that are needed. This is a common issue for COM programmers. .NET eliminates this issue by ensuring that the executable will use the DLL that it was created with. In .NET the DLL must be registered in the Global Assembly Cache (GAC). The GAC is a listing of public assemblies on your system. This is similar to the registry for COM objects. In the GAC the DLL must have unique hash value, public key, locale, and version. Once your DLL is registered in the GAC, its name is no longer an issue. You can have two different versions of the same DLL installed and running on the same system without causing problems. This is possible because the executable is bound to the specific version of the DLL. A .NET assembly stores all references and dependencies in a manifest. You can install, uninstall, and list assemblies in the GAC using the assembly registration utility (gacutil.exe).

8.6 LABVIEW AND ACTIVEX

The previous sections discussed OLE, COM, and ActiveX to provide a better understanding of how they are used. The terminology, evolution, significance, and operation of these technologies were explained in order to establish the needed background for using ActiveX effectively. This section will go one step further to define how LabVIEW works with ActiveX. This includes discussions of how ActiveX can be used in LabVIEW, the ActiveX container, and the functions available for use in the code diagram. The goal of this section is to show LabVIEW's ActiveX interfaces by describing the tools that are accessible to programmers. This chapter will conclude with a brief review of the VI Server and the related functions. The ActiveX and VI Server functions are very similar, making it an appropriate topic to include in this chapter.

The next chapter provides numerous ActiveX examples to demonstrate how to utilize this feature in LabVIEW programs. It opens the door to an entirely new set of possibilities that may not have been thought about before. The examples will provide the foundation for successfully implementing ActiveX through the illustration of practical applications.

8.6.1 THE LabVIEW ActiveX Container

The ability to act as an ActiveX control container was first added to LabVIEW 5. By adhering to specifications for interacting with ActiveX controls, LabVIEW has become a client for these controls. It allows LabVIEW to utilize the functionality that the controls have to offer. To the user, an ActiveX control appears as any other control on the user interface. The operation of the control is seamless and unnoticeable to the end user. When used in conjunction with the available LabVIEW controls and indicators on the front panel, it becomes an integrated part of the application.

8.6.1.1 Embedding Objects

LabVIEW allows ActiveX controls and documents to be embedded on the front panel of a VI by using the container. Once the control or document has been placed

FIGURE 8.1

inside the container, it is ready for use and can be activated in place. Additionally, the objects' properties, methods, and events are made available to the programmer for use in the application.

A container can be placed on the front panel by accessing the ActiveX subpalette from the Controls palette. Then, to drop a control or document inside the container, pop up on the container and select Insert ActiveX Object from the menu. Figure 8.1 displays the dialog box that appears with the options that are available to the programmer once this is selected. Three options are presented in the drop-down box: Create Control, Create Document, and Create Object from File. The first option is used to drop a control, such as a checkbox or slide control, into the container. The other two are used to drop ActiveX documents into the container.

When Create Control is selected in the drop-down box, a list of ActiveX controls along with a checkbox, Validate Servers, will appear. The list consists of all ActiveX controls found in the system registry when the servers are not validated. If the box is checked, only registered controls that LabVIEW can interface to will be listed. For instance, some controls that come with purchased software packages may have only one license for a specific client. Any other client will not have access to this control's interfaces. Sometimes third-party controls do not adhere to ActiveX and COM standards. LabVIEW's container may not support these interfaces, which prevents it from utilizing their services. After a registered control is dropped into the container, its properties, methods, and events can be utilized programmatically. A refnum for the container will appear on the code diagram.

Selecting Create Document also produces a list of registered document types that can be embedded on the front panel. These are applications that expose their services through an interface, allowing LabVIEW to become a client. When Create Object from File is selected, the user must select an existing file to embed on the front panel with an option to link to it directly.

In both cases, the embedded object behaves as a compound document. Right-clicking on the object and selecting Edit Object launches the associated application.

332   LabVIEW: Advanced Programming Techniques

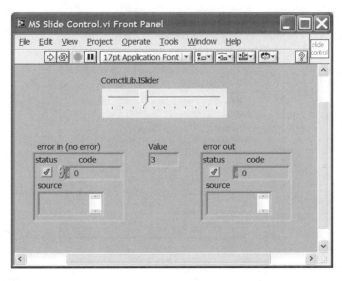

FIGURE 8.2

The user can then edit the document using the application. The automation interfaces that are supported by the application can be utilized from the code diagram after the object is placed inside the container. Once again, a refnum representing the container will appear on the code diagram.

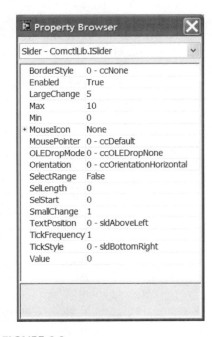

FIGURE 8.3

8.6.1.2 Inserting ActiveX Controls and Documents

This section will demonstrate how to insert and use ActiveX documents and controls within LabVIEW's container. Figure 8.2 shows the front panel of a VI with the Microsoft Slider Control, Version 6.0 (SP4), inserted into the container following the procedure described in the previous section. This control is similar to LabVIEW's built-in Horizontal Slide Control.

The properties of the slide control can be modified by right-clicking on it and selecting Property Browser. Figure 8.3 illustrates the window that appears for setting the slide's properties. There are options for altering the orientation, tick style, tick frequency, and mouse style of the control. Essentially, every

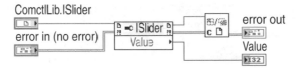

FIGURE 8.4

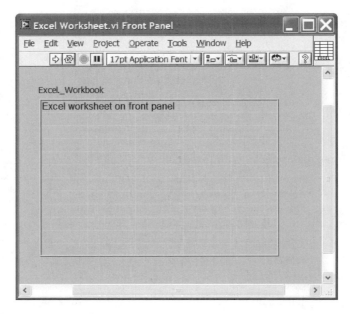

FIGURE 8.5

ActiveX control defines properties for which you can either set or retrieve the current value.

Figure 8.4 displays the code diagram of the same VI. Note that the ActiveX control appears as a refnum on the diagram. This is a simple example in which the Property node, detailed further in Section 8.6.2.2, is used to retrieve the value of the control set from the front panel. With LabVIEW's ActiveX container, numerous other controls can be integrated into an application. The programmer is no longer limited to the built-in controls.

ActiveX documents are embedded into LabVIEW's container by either inserting an existing file or creating a new document with an application that exposes its services. Figure 8.5 illustrates the front panel of a VI with a Microsoft Excel worksheet dropped into the ActiveX container. This was done by choosing Create Document from the drop-down menu, then selecting Microsoft Excel Worksheet from the list of available servers. The text shown was entered in the cell by right-clicking on the worksheet and selecting Edit Object to launch Microsoft Excel. Alternatively, you can pop up on the container worksheet and select Edit from the Document submenu. Figure 8.6 shows the same VI, but the Excel sheet is white with grid lines, which is the way you are probably used to looking at it. This was

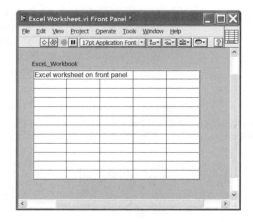

FIGURE 8.6

done by editing the object and filling the cells with white and selecting the gridlines. Once the Excel sheet is updated you can just close it, and the changes will be reflected on the VI front panel. It is not possible to just use the paintbrush to change the color as was done in earlier versions. The container refnum can now be utilized programmatically to interface with the worksheet's properties and methods. It is possible to use the properties to change the color programmatically as well.

8.6.2 THE ACTIVEX PALETTE

The Connectivity palette on the Functions palette holds the ActiveX subpalette. This subpalette contains the automation functions that can be used to interface with ActiveX servers. This section briefly describes these functions: Automation Open, Automation Close, Invoke Node, Property Node, and Variant to Data. With these functions, you have everything you need to work with and utilize ActiveX within LabVIEW. You really do not need to know the details of how COM works to use the services provided by servers. It is similar to using GPIB Read or Write in an application. These functions provide a layer of abstraction so programmers do not have to learn the intricacies that are involved.

8.6.2.1 Automation Open and Close

The function of Automation Open is to open a refnum to an ActiveX server. This refnum can then be passed to the other functions in the palette to perform specific actions programmatically. To create the Automation Refnum, first pop up on the Automation Open VI to choose the ActiveX class. Select Browse in the Select ActiveX Class submenu to see a list of the available controls, objects, and interfaces that LabVIEW has found on the system. Figure 8.7 shows the dialog box that appears with the drop-down menu for the type library, and the selection box for the object. LabVIEW gets information on the server's objects, methods, properties, and events through the type libraries. Additionally, a Browse button lets the programmer find

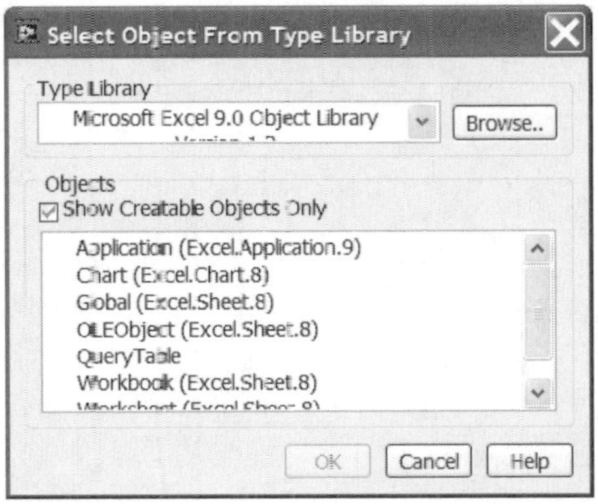

FIGURE 8.7

other type libraries, object libraries, and ActiveX controls that are not listed in the drop-down box.

Once you have found the object that you want to perform automation functions on, select OK on the dialog box, and the refnum will be created and wired to Automation Open. If you wire in a machine name, a reference to an object on a remote computer will be opened using DCOM. If it is left unwired, the reference will point to the object locally. DCOM objects can be instantiated only from the Automation Open VI.

As the name suggests, Automation Close is used to close an automation refnum. You should always remember to close refnums when they will not be used further or before termination of the application in order to deallocate system resources.

8.6.2.2 The Property Node

The Property node is used to get or set properties of an ActiveX object. A refnum must be passed to the Property node to perform the Get or Set action on the object. The automation refnum, either from Automation Open or the refnum created from inserting a control into the front panel container, can be passed to the node. Note that this is the same Property node that is available in the Application Control subpalette to read or write properties for an application or VI. When performing application control, however, a reference must first be opened to an application or VI using the Open Application Reference function.

Once a refnum has been passed to the Property node, you can pop-up on the node and select a property that you wish to perform a read or write action on. The list of properties the object supports will be listed in the Properties submenu. If you wish to get or set more than one property, simply add elements from the pop-up menu, or drag the node to include as many as are needed. Then, select a property for each element. To change, or toggle, the action from read to write (or vice versa), select the associated menu item by popping up on the node. Some properties are read only, so you may not be able to set these properties. In this case, the selection in the pop-up menu will be disabled.

FIGURE 8.8

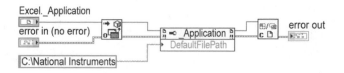

FIGURE 8.9

Figure 8.8 is an example in which the Property node is used to get the default file path that Microsoft Excel uses. Automation Open is used to open a reference to Microsoft Excel; Microsoft Excel 9.0 Object Library Version 1.3 was selected from the type library drop-down box and Application (Excel.Application.9) from the object list. DefaultFilePath is one of the items that is available for the application in the Properties submenu of the node. C:\Documents and Settings\Matt Nawrocki\My Documents was returned as the default file path that Excel uses when Save or Open is selected from the File menu. Alternatively, this property can easily be used to set the default file path that you want Excel to look in first when Open or Save is selected. First, you must pop-up on the Property node and select Change to Write from the menu. Then a file path can be wired into the node. This is depicted in Figure 8.9, where C:\National Instruments is set as the default path.

8.6.2.3 The Invoke Node

The Invoke node is the function used to execute methods that an ActiveX control makes available to a client. The number of methods that a particular control offers can vary depending on its complexity. Simple controls such as the Microsoft Slider Control, shown earlier in Figure 8.2, have only five methods. Complex objects like Microsoft Excel, may have upwards of several hundred methods available. Virtually all actions that can be performed using Microsoft Excel can also be performed using automation.

An automation refnum, either created from a control placed in the front panel container or from Automation Open, must be passed to Invoke node. Once an automation refnum has been wired to Invoke node, you can pop-up on it and select an action from the Methods submenu. The method selected may have input or output parameters that must be considered. Some inputs are required but others are optional. The data from output parameters can be wired to indicators or used for other purposes as part of the code.

Figure 8.10 shows the code diagram of an example using the Invoke node to execute a method in Microsoft Excel. As before, Automation Open is used to open a reference to Microsoft Excel. Then, the refnum is passed to the Invoke node, and

FIGURE 8.10

CheckSpelling is selected from the Methods submenu. When the method is selected, the input and output parameters appear below it. The CheckSpelling method checks the spelling of the word passed to it using Excel's dictionary. If the spelling is correct, a "true" value is returned. If the spelling is incorrect, a "false" value is returned. The only required input for this method is Word; the rest are optional. In a similar manner, you can utilize the services offered by any object that supports automation. This is a very effective way to realize code reuse in applications.

Complex ActiveX servers arrange their accessible properties and methods in a hierarchy. This requires the programmer to use properties and methods of a server to get to other available properties and methods. Applications such as Microsoft Excel and Word operate in a similar manner. After you have opened a reference to one of these applications, the Invoke node and Property node can be used to get to other properties and methods that are not directly available. The first reference opened with Automation Open is also known as a "creatable object." Note that when you are selecting the ActiveX class for Automation Open, the dialog box (shown in Figure 8.7) gives you the option to show only the createable objects in the selection list. When the option is enabled, LabVIEW lists those objects that are at the top of the hierarchy. What does this mean to a programmer who wants to use ActiveX and automation? For simple objects, a programmer is not required to know the details about the hierarchy to use them. They have relatively few properties and methods, and are simple to use programmatically. When complex objects or applications are being used, however, a programmer needs to know how the hierarchy of services is arranged to achieve the desired result in a program. This means that you have to refer to the documentation on the server, or help files, to use them effectively. The documentation will help guide you in utilizing ActiveX in your applications.

Figure 8.11 illustrates the hierarchy of Microsoft Word through an example. This VI opens a Microsoft Word document, named test.doc, and returns the number of total characters in the document. First, a reference to Microsoft Word Application is opened using Automation Open. Then, Documents is the first property selected with the Property node. Documents is a refnum with a set of properties and methods under its hierarchy. The Open method is then used to open the file by specifying its path. Next, Characters is a property whose Count property is executed to retrieve the number of characters contained in the document. Observe that all references are closed after they are no longer needed. The Microsoft Word hierarchy used in this example proceeds as shown in Figure 8.12. Microsoft Word application is at the top of the hierarchy with its properties and methods following in the order shown.

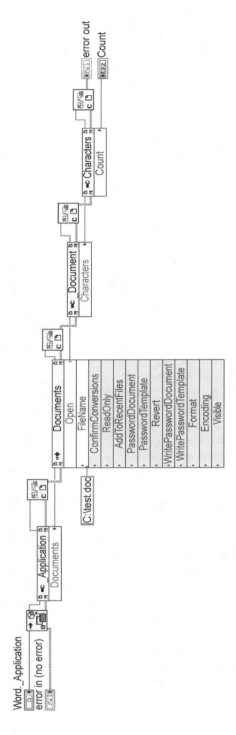

FIGURE 8.11

MS Word

↓

Documents

↓

Open

↓

Characters

↓

Count

FIGURE 8.12

8.6.2.4 Variant to Data Function

A variant is a data type that varies to fit whatever form is required of it. A variant
can represent strings, integers, floating points, dates, currency, and other types,
adjusting its size as needed. This data type does not exist within LabVIEW; however,
many ActiveX controls do make use of it. Variants are valid data types in Visual
Basic, which is often used to create ActiveX controls. Therefore, with the addition
of ActiveX support, LabVIEW must be able to deal with variants in order to pass
and retrieve data with objects.

LabVIEW supplies a control and a function VI to handle variants when working
with ActiveX because it does not interpret variants. The OLE Variant is a control
available in the ActiveX subpalette provided to facilitate passing variant data types
to servers. The Variant to Data function converts variant data to a valid data type
that LabVIEW can handle and display programmatically. To use this function, simply
wire in the Variant data to be converted and the type that you want it converted to.
A table of valid data types can be accessed from the online help by popping-up on
the function. You can wire in any constant value among the valid data types to which
you need the Variant converted.

Figure 8.13 shows an example using To G Data. The code diagram shown is a
subVI in which the value of the active cell in a Microsoft Excel workbook is being
read. Its caller, which opens a reference to Excel and the workbook, also passes in
the Application refnum for use. As shown, the data type for the cell value is a variant.
The actual data in the spreadsheet is a string, therefore, the variant passed from

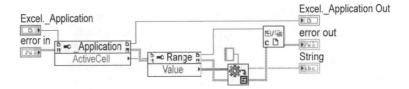

FIGURE 8.13

Excel must be converted to a valid LabVIEW data type using the Variant to Data function. An empty string constant is wired to the type input, and a string indicator is wired to its output. If the data in the workbook is a number, a numeric constant (integer or floating point) can be wired to the type input for conversion.

When an ActiveX object requires a variant data type as input, the programmer is usually not required to perform any conversion from the valid LabVIEW types. There is a function on the ActiveX subpalette that does convert to the Variant data type if needed. Look back at Figure 8.11, the example in which a Microsoft Word document was opened and the number of characters retrieved. The Open method required a file name of variant type. A string constant was wired to this input without any conversion. If you look closely, you will see a coercion dot, indicating that LabVIEW coerced the data to fit the type required.

8.6.3 USING THE CONTAINER VERSUS AUTOMATION

When you first begin to work with ActiveX in LabVIEW, it may be unclear whether to use the front panel container or Automation Open to utilize the services of an object. For instance, some servers can be used by either dropping them into the container or creating a refnum with Automation Open. The general rule of thumb, when using ActiveX is to utilize the front panel container when working with controls or embedded documents that the user needs to view. The CWKnob Control (evaluation copy that is installed with LabVIEW) is an example of a control that needs to be placed in the front panel container so that the user can set its value. If Automation Open is used to create the refnum for the knob, the user will not be able to set its value. This applies to controls, indicators, or documents that the user must be able to view or manipulate in an application. Once placed in the container, the Invoke node and Property node can be used to perform necessary actions programmatically, as demonstrated in previous examples.

When you need to use ActiveX automation to work with applications like Microsoft Word or Excel, it should be done from the code diagram, without using the container. Use Automation Open to create a refnum by selecting the ActiveX Class to make use of the available services. Then, use Invoke node and Property node to perform needed actions, as shown in Figure 8.11. A front panel container was not used in that example. You may need to refer to documentation on the objects' hierarchy of services. On the other hand, if a specific document's contents need to be displayed and edited, then you can embed the document on the front panel and perform the actions needed.

8.6.4 EVENT SUPPORT IN LabVIEW

Many ActiveX controls define a set of events, in addition to properties and methods that are associated with it. These events are then accessible to the client when they occur. In order to handle an event, the event must first be registered. The event can then be handled by a callback. This is significant because LabVIEW can interface to properties, methods, and events to fully utilize all of the services that ActiveX objects have to offer. This section will explain and demonstrate how to use the Event functions available in the ActiveX (or .NET) subpalette.

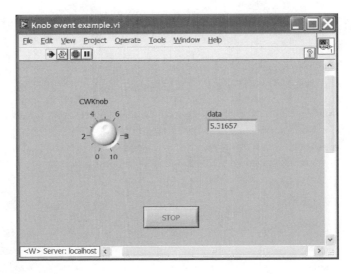

FIGURE 8.14

8.6.4.1 Register Event

For our example on using ActiveX events, we will put a knob control on the front panel of a VI and display the current value in a digital indicator. There will also be a stop button to stop execution. The front panel is shown in Figure 8.14. The first step in working with ActiveX events is to insert the control in a container or use the automation refnum to call an Active X object. For this example we will place a Component Works Knob in an ActiveX container on the front panel. This will create a reference on the code diagram. Now we need to insert the Register Event Callback Function on the code diagram. In order to be able to select an event to operate on you will need to wire the Knob control reference to the Event input on the Register Event Callback function. You might initially think that the reference should go to the Register Event Callback reference input, but the callback reference is a separate reference that does not need to be wired. Now that the reference is connected to the Event input, you will be able to click on the down arrow to select the desired event. For our example we want to display the value of the knob in a digital indicator as the knob is changed. To do this we need to select the PointerValueChanged event.

The Register Event Callback function should now have three inputs listed: the Event input, the VI reference, and an input for user parameters. The VI reference input, which is a link to the callback that will handle the event, will be discussed in the next section.

8.6.4.2 Event Callback

The Event Callback is a VI generated to perform any operations when the event occurs. For our example we want to display the digital value of the knob when it is changed. To do this we need to change the value of the data indicator in the main VI. Because this piece of data will need to be updated in the callback, the information

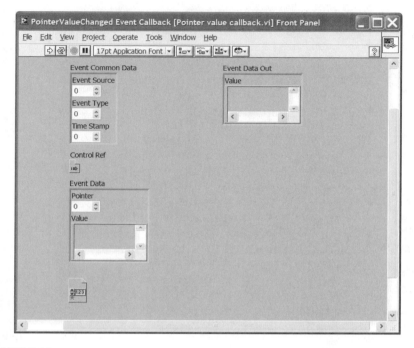

FIGURE 8.15

will need to be wired to the user parameters input on the Register Event Callback function. To do this you need to create a VI Server Reference for your indicator. This is done by right clicking on the indicator and selecting Reference from the Create option. This reference can now be used to access the properties and methods of the indicator.

Once you have wired the knob control reference and the user data input to the function, you can create the callback. To create the callback you right click on the Callback VI reference input and select Create Callback VI. A new VI will open with several controls and an indicator on it. What shows up on the callback will change based on the event selected and the information wired to the User Data input. The callback VI front panel is shown in Figure 8.15.

The Event Common Data control will provide a numeric value representing what the source of the event was (LabVIEW, ActiveX or .NET). The Event Type specifies what type of event occurred. For a user interface event the type is an enumerated type control. For all other event types a 32-bit unsigned integer is returned. The time stamp indicates when the event occurred in milliseconds. The Event Data control (and sometimes indicator) is a cluster of event-specific parameters that the callback VI handles.

To be able to update the value of the digital indicator in the main VI we need to first get the current value of the control. This is done by unbundling the value parameter from the Event Data control. What is contained in the value control is the current value stored as a variant. We must convert the variant to a number using the Variant to Data function. To be able to update the data indicator we will access

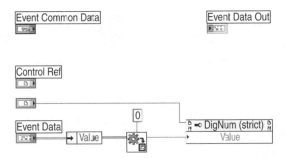

FIGURE 8.16

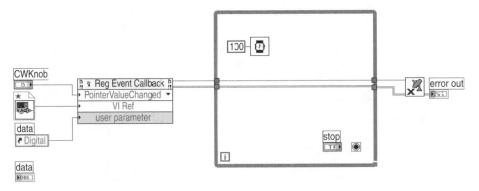

FIGURE 8.17

the value property from the data indicator reference and pass the current value. The final callback code diagram is shown in Figure 8.16.

Now that the callback has been created we can finish our main VI. In order to continually monitor the knob we insert a While loop (with a delay to free up system resources) on the code diagram. We also insert a stop button to exit the loop when we are done. Finally, we need to unregister our event. You notice that nowhere on the main VI do we do anything with the data indicator. This is all handled through the callback, the indicator will update whenever the event occurs. The code diagram for the main VI is shown in Figure 8.17.

8.6.5 LabVIEW as ActiveX Server

LabVIEW has both ActiveX client and server capabilities. This means that you can call and use external services inside LabVIEW's environment, as well as call and use LabVIEW services from other client environments such as Visual Basic. This functionality enhances the ability to reuse LabVIEW code by being able to call and run VIs from other programming languages. It also gives you flexibility to use the language that best suits your needs without having to worry about code compatibility.

When using LabVIEW as an ActiveX server, the Application and Virtual Instrument classes are available to the client. Consult the online reference for details on the methods and properties that these classes expose to clients. To use LabVIEW

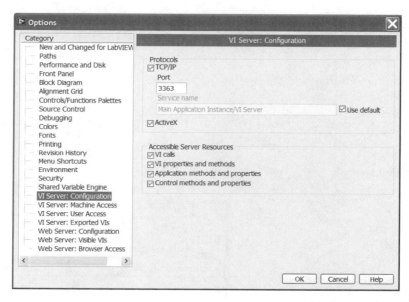

FIGURE 8.18

as an ActiveX server, you must first ensure that this functionality has been initiated. Select Options from the Tools menu, and VI Server: Configuration as shown in Figure 8.18. Check the ActiveX protocol selection and the appropriate Accessible Server Resources as needed. You can allow VI calls, VI properties and methods, Application methods and properties, and Control methods and properties. The online reference also has an ActiveX example to demonstrate how this feature can be used. It is an example of a Visual Basic macro in which a VI is loaded, run, and its results retrieved for further use.

8.7 LABVIEW AND .NET

As with using ActiveX in LabVIEW, you do not need to have in-depth knowledge of .NET to be able to utilize the controls and functionality that .NET provides. Now that you have some exposure to properties and methods, you will find that implementation of .NET is fairly easy. LabVIEW can be used as a .NET client to access the properties, methods and events of a .NET server, but LabVIEW is not a .NET server. Other applications cannot directly connect to LabVIEW using .NET.

8.7.1 .NET CONTAINERS

One part of .NET that can be used through LabVIEW is the .NET control. The .NET control is similar to an ActiveX control. LabVIEW allows you to embed the control on the front panel. The .NET control will appear like a standard LabVIEW control and will operate seamlessly when its attributes are controlled in your application. In order to use a .NET control on the front panel of your application you will need

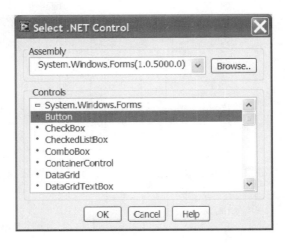

FIGURE 8.19

to use a container. The .NET container is available to place on the Front Panel from the .NET/ActiveX palette.

Once you have placed a container on the front panel you can right click on the container and choose Insert .NET control. This will cause a dialog box to come up that will allow you to search for the control you want to use in your application. The dialog box is shown in Figure 8.19. For this example we will select the radio button control. This control is contained in the System.Windows.Forms assembly.

Now that your control is embedded in your front panel you can start to program the application to utilize the control. One way to start is to use the class browser. Using the class browser you can view the properties and methods for an assortment of items including VI Server, ActiveX and .NET objects. To start you will need to launch the class browser from the tools menu. Once the browser launches you can select the object library. From here you select what type of object you are looking for and then either select an object that is already loaded or to browse for your object. Once the object is chosen, in this case the System.Windows.Forms, you can select the Class. For our example we choose the RadioButton class. Now you can view all the available properties and methods. The class browser is shown in Figure 8.20. Notice in the class browser that there are several methods with the same names. The Scale method and the SetBounds method are listed twice. In the case of the first Scale method the inputs are Single dx and Single dy. The second Scale method has a single input named Single ratio. In .NET you can have a single property or method name with different inputs. If you select a property or method you can place the item on the code diagram by selecting create or create write at the bottom of the dialog and then placing the node on the diagram.

For our example we are just going to have a loop that will monitor the radio button status. When the radio button is clicked the loop will stop. The value of the radio button will be displayed in an indicator. As we already have the radio button on the front panel, we will use the class browser to select the checked property. This property will output a Boolean value. Once the property button is placed on the code

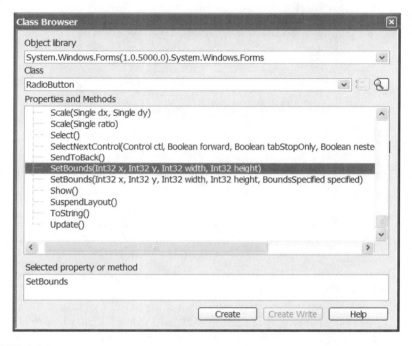

FIGURE 8.20

diagram you can wire the output to the conditional terminal of the While loop. The code diagram is shown in Figure 8.21. The front panel of the running VI is shown in Figure 8.22.

Obviously this is a simple example, but the same method would be used for larger applications.

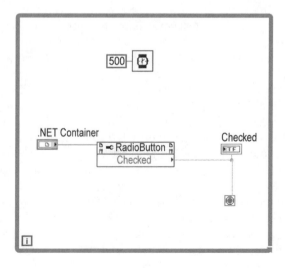

FIGURE 8.21

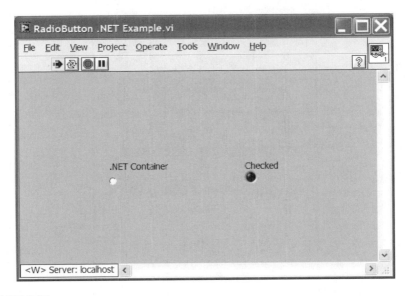

FIGURE 8.22

You must have the .NET framework installed. As discussed earlier, the .NET framework is the core of .NET, and is required to be able to be able to create and use .NET objects. The framework can be downloaded from the Microsoft Website.

8.7.2 .NET PALETTE

The .NET palette contains most of the same functions as the ActiveX palette. There is a Property node, Invoke node, register and unregister events, close reference, and static VI reference. There are a couple of new functions to the .NET palette. The first is the constructor node. This is similar to the automation open in ActiveX programming. The constructor is used to create an instance of a .NET object. There is also a pair of functions for converting a reference to a more specific or general class. The common properties have been discussed in depth in the ActiveX section; here, we will cover only the constructor node and the class conversion functions.

As mentioned, the constructor node is used to create an instance of a .NET object. The first step is to place the constructor node on the code diagram. Once the node is placed on the diagram, a dialog automatically opens. If you close the dialog box, or if you decide to select a different constructor, you can right click on the constructor node and choose Select Constructor. The top entry of the dialog is the assembly. For this example we will select the System assembly. Now you can go through the second window, which shows the creatable objects. We have scanned through the list to get to the System.Timers object. There are several subobjects under the main object. For this example the Timer object is selected. In the bottom window there are two available constructors. They are both timer constructors, but they have different inputs. Here, the Timer() constructor will be selected. The dialog is shown in Figure 8.23. Once the constructor is loaded on the diagram you can access the properties and methods.

FIGURE 8.23

The class conversion functions are similar to the typecast function. They are used to convert one type of reference to another. In some cases, a reference generated through a property or invoke node does not yield the properties or methods needed. The reference might be a general reference that needs to be more specific. I know this sounds confusing, but bear with me. Let us say that you go to a full service gas station to get gas for your car. You pull up to the pump, tell the attendant you want a fill-up. At this point you have opened a reference for the generic gas class. The problem is you don't want just any kind of gas; you want to get unleaded gas. You need to convert your existing reference to a more specific reference, the unleaded gas reference. Now you can access the properties of the unleaded gas reference such as grade. This is what you are doing with the To More Specific Class. You are getting access to the properties or methods of a more specific class than the reference you currently have. There is an example of using this function to control an instrument through .NET in the examples section.

To use the To More Specific Class function you need to connect the current reference. There is also an input for the error cluster. Finally, there is an input for the Target Class. You can use a constructor node to generate a reference for this input. The output is the type of reference you need for the selected object.

8.8 THE VI SERVER

VI Server functionality allows programmers to control LabVIEW applications and VIs remotely. The functions available for use with the VI Server are on the Application Control subpalette. You must first ensure that LabVIEW is configured correctly to be able to use this capability. Figure 8.18 displays the VI Server configuration dialog box. Activate the TCP/IP protocol, then enter the port and enable the

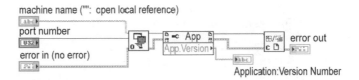

FIGURE 8.24

appropriate server resources you wish to use. In the VI Server: Machine Access menu item, you can enter a list of computers and IP addresses that you wish to allow or deny access to the VI Server. In the VI Server: User Access menu item, you can enter a list users and groups that you wish to allow or deny access to the VI Server. Finally, the VI Server: Exported VIs item allows you to create a list of VIs that you want to allow or deny access to remotely. The wildcard character (*) allows or restricts access to all VIs on the machine.

The Open Application Reference function is used to open a reference to Lab-VIEW on a remote machine. The machine name, or IP address, must be wired to the function along with a TCP port number. This should coincide with the TCP port number that was specified in the configuration for VI Server on the remote machine. If a machine name is not wired to the function, a local reference to LabVIEW will be opened. Once the reference is opened, the programmer can manipulate the properties and methods of the local or remote LabVIEW application. Use Close Application or VI Reference when you are finished using the reference in an application.

The Property node is used to get or set properties and the Invoke node is used to execute any methods. These functions are identical to the ones described previously while discussing ActiveX functions. The Property node and Invoke node from both ActiveX and Application Control subpalettes can be used interchangeably.

The code diagram in Figure 8.24 demonstrates how to make use of the VI Server. This example simply retrieves the version number of the LabVIEW application on a remote computer. First, a reference to the application is opened and the machine name and port number are wired in. Then the Property node is used to get the version number, and the reference is subsequently closed. The online help describes all of the properties and methods that can be utilized programmatically.

The Open VI Reference function is used to open a reference to a specific VI on either a local or remote machine. Simply wire in the file path of the VI to which you wish to open the reference. If the VI is on a remote computer, use Open Application Reference first, and wire the refnum to the Open VI Reference input. Figure 8.25 shows an example first introduced in the Exception Handling chapter. It opens a reference to External Handler.vi and sends it the error cluster information

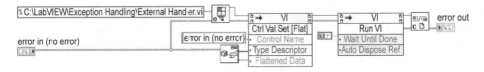

FIGURE 8.25

using the Set Control Value method. Then the External Handler is executed using the Run VI method. Finally, the reference to the VI is closed with Close Application or VI Reference.

8.9 ACTIVEX AND .NET EXAMPLES

In this section we will provide several examples that utilize ActiveX and .NET controls and automation. The material here serves two purposes. The first is to give you enough exposure to these technologies so that you will be comfortable using them in your own applications effectively. By employing the services offered by objects, code reuse is significantly enhanced. This is a key advantage that was gained by the introduction of COM. The second intent is to give you some practical examples that you can modify and utilize in your applications. Even if the examples are not directly applicable to your situation, they will give you a new way of thinking that you may not have considered before.

8.9.1 COMMON DIALOG CONTROL

The Common Dialog control is familiar to almost every Visual Basic programmer. Microsoft elected to provide a uniform interface to the dialog boxes for printing, opening, saving, and color selection. These were wrapped into a single dialog box and became known as the Common Dialog control. Every application you use, you can see the familiar Print, Open, and Save Boxes which are using this control. It was desirable for Microsoft to have standard mechanisms for users to perform common tasks through user interface elements. This allows the Windows operating system to provide a consistent look and feel to end users regardless of which company was developing software to use on the operating system.

We can use the Common Dialog control to keep the Windows look and feel consistent for our end users. This control is useful if you are using file saving and not using the high level interfaces provided by LabVIEW. If you are using the high level interfaces, you do not need to use this control; LabVIEW is using it for you.

This example uses the Microsoft Common Dialog Control Version 6.0. The Common Dialog Control is relatively simple to use and will serve as the introductory example. It is useful for prompting the operator to select a specific file for opening and saving purposes, while also allowing them to navigate through the Windows directories.

Figure 8.26 displays the code diagram of Common Dialog.vi. The Common Dialog control was placed in the front panel ActiveX container, and is represented

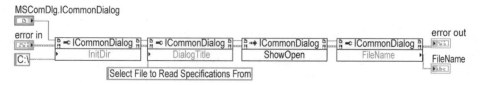

FIGURE 8.26

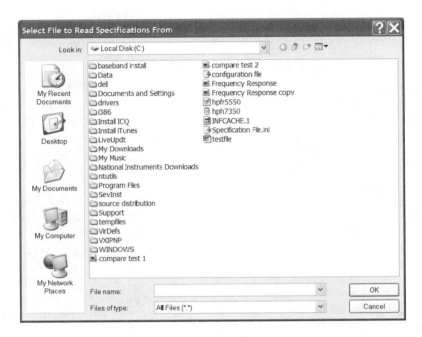

FIGURE 8.27

by the refnum on the diagram. The objective of this VI is to display the dialog box and instruct the user of the application to select a file. The name of the file that the user selects is then retrieved from the control and used elsewhere. In this example, the first action taken is to set the InitDir property to C:\. This causes the dialog box to display the contents of C:\ when it appears. Next, the DialogTitle property is set to prompt the user to select a specification file. Then the Show Open method is executed, which simply displays the dialog box. Finally, the FileName property is read back to find out the name of the file that was selected. The Common Dialog.vi can be used as a subVI whenever a similar function needs to be performed. Figure 8.27 shows the dialog box as it appears when the VI is executed.

The front panel container was used to insert the Common Dialog Control in this example, but we could just as easily have used Automation Open. The reason for using the container was to have the control displayed instead of the automation refnum. This allows the programmer to modify properties of the control quickly, by popping up on it. On the block diagram, the InitDir and DialogTitle properties could have been set in the same step, by popping-up on the Property node and selecting Add Element. This being the first example presented, two separate property nodes have been used to simplify the block diagram.

8.9.2 PROGRESS BAR CONTROL

The Microsoft Progress Bar can be used to keep a consistent look and feel in your applications. Programmers have been using slide controls for years to emulate the Microsoft Progress Bar. Now that the control is available through ActiveX and

.NET, we can use it without any workarounds. As we mentioned, a strong advantage to using this control is that it will always have the same look and feel of other applications running on Windows. In the event that Microsoft releases a new version of Windows containing a new look to the Progress Bar, the change will be automatically updated into your application if you replace the older control with the new one on your system. The way the Progress Bar looks could potentially change with a new release of Windows; however, the interface for this control is not likely to change.

As this is a user interface element, there are two different methods that we can use to create it: we can insert it into a container, or we can use the ActiveX Open. User interface elements are typically created with a container so there are no issues regarding the location of the control of the display. Therefore, we will not have an open VI for this control; users will need to place either an ActiveX or .NET container on the front panel. Once the container is placed on the front panel, right-clicking on the control allows you to insert an object. Insert a Microsoft Progress Bar control. You may notice that there are several different Progress Bar controls listed with different version numbers. Each release of languages like Visual Basic will have a new version of the Progress Bar control. Typically, we use the "latest and greatest" version of each control, but you may feel free to use any of the versions currently listed. According to the rules of COM, each version of the control should have the same interface. Additional functionality can only be added to new interfaces. This ensures that you will not have compatibility issues if another version of the control is updated on your system.

Once the control is placed on the front panel, you can resize it to dimensions that are appropriate for your application. The control itself has numerous properties and methods. Properties for this control include display concerns such as appearance, orientation, mouse information, and border style. If you want the typical 3-D look and feel, the appearance property should be set to 3-D. This control uses enumerated types to make it easy for programmers to determine which values the control will accept. All appearance properties for the control use enumerated types. Orientation allows programmers to have the control move horizontally or vertically. The default value is horizontal, but it is possible to have the Progress Bar move up and down, as is done in Install Shield scripts for hard drive space remaining.

The mouse information allows you to determine which mouse style will be shown when the user locates the mouse over the control. An enumerated type will identify which mouse styles are defined in the system; all we need to do is select one.

Minimum, Maximum and Value properties are used to drive the control itself. Minimum and Maximum define the 0% and 100% complete points. These values are double-precision inputs to allow for the most flexibility in the control. Setting the value allows the control to determine what percentage complete the bar should display. This simplifies the task of displaying progress bars over the old technique of using a LabVIEW numerical display type; the calculations are performed for us.

The driver for this control is fairly simple. We will have a few property-setting inputs for the configuration-related properties, and a VI to support setting the Value property. Default values will be assigned to the properties on the VI wrapper so

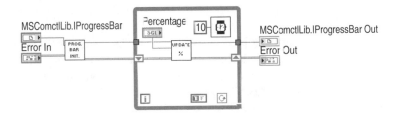

FIGURE 8.28

inputs do not need to be applied every time we configure a control. The code diagram
for this VI appears in Figure 8.28.

8.9.3 MICROSOFT CALENDAR CONTROL

This example uses the Microsoft Calendar Control, Version 9.0. In order to use this
control, you will need to place an ActiveX container on the front panel of your VI.
You can select the Calendar control by right-clicking on the container and selecting
Insert ActiveX Object. When the dialog box comes up you will need to select
Calendar Control 9.0. The properties and methods for this control are very manage-
able. There are 22 properties and 11 methods. The example used here will exercise
a majority of the properties. (Note that there is a .NET calendar control on the .NET
palette that has many of the same properties and methods.)

The first method displays the About box for the control. There is no input for
this method. Actually, none of the methods for the calendar control have inputs. The
actions are initiated by simply invoking the methods. There are methods for advanc-
ing the calendar control to the next day, week, month, and year. By invoking these
methods you will increment the Calendar control by the specified value. These
methods will also refresh the calendar display. There are methods for moving the
calendar control to the previous day, week, month, and year. There is a separate
method for refreshing the calendar control. Finally, there is a method for selecting
the current date on the control.

If you were implementing these methods in an application, you could make a
front panel selector that would drive a state machine on the code diagram. The state
machine would contain a state for each method. The user of the application would
be able to interactively move around the calendar through the use of these front
panel controls.

The background color is the first property available in the Property node. This
input requires a long number representing the RGB color for the background color
of the control. The valid range of typical RGB colors is 0 to 16,777,215. The next
set of properties relate to configuring the day settings. You can read or write the
value for the currently selected Day, set the Font, set the FontColor, and the Day-
Length. The DayLength property designates whether the name of the day in the
calendar heading appears in short, medium, or long format. For example, to set the
DayLength property to long, you would have to wire a 2 to the property input. This
would result in the day being displayed as Monday instead of Mon. or M. The

property after DayLength is First Day, which will specify which day is displayed first in the calendar.

The next set of properties relates to the grid style. The GridCellEffect property sets the style of grid lines. They can be flat, raised, or sunken. The GridFont property sets the font for the days of the month. You can also set the font color and line color for the grid. There are five properties relating to visibility. They allow you to show or hide items such as the title. The Value property allows you to select a day on the calendar or read the currently selected date from the calendar. The data type of the Value property is a variant. A Null value corresponds to no day being selected. The ValueIsNull property forces the control to not have data selected. This is useful for clearing user inputs in order to obtain a new value.

In the following example we will configure the Calendar control to use user-selected properties for a number of attributes. The user will be able to set the length of the day and month, as well as the grid attributes. After configuring the display settings, we will prompt the user to select a date. Using ActiveX events, the program waits for the user to click on a date. After the date has been selected, we read in the value of the control and convert the value to the date in days. The program will then calculate the current date in days. The current date will be subtracted from the selected date to calculate the number of days until the specified date. This value will be displayed on the front panel. The code diagram and front panel for this example are shown in Figure 8.29. The day, month, and year are all returned as integers. The month needs to be converted to a string by performing a Pick Line and Append function. This function selects a string from a multiline string based on the index.

8.9.4 WEB BROWSER CONTROL

This example utilizes the Microsoft Web Browser control. The Web Browser control can be used to embed a browser into the front panel of a VI, and has various applications. It allows you to navigate the Web programmatically through the COM interface and see the pages displayed on the front panel window. This control can be dropped into LabVIEW's front panel container. In LabVIEW 8, an icon for an ActiveX Web browser container is in the .NET & ActiveX palette. The resulting container on the front panel is equivalent to inserting the ActiveX object into an empty container.

The Web Browser control can be very useful as part of an application's user interface. This example will illustrate how to utilize the control in a user interface to display online documentation or technical support for an operator. A simplified user interface is displayed in Figure 8.30 that shows an enumerated control and the Microsoft Forms 2.0 CommandButton. Obviously, this would be only one control among several that you may want to make available in your user interface. The enumerated control lets the operator select the type of online support that is desired. When the Go!! button is pressed, another window appears with the Web page selected from the control. This is ideal for use in a company's intranet, where the Web-based documentation is created and saved on a Web server.

Figure 8.31 shows the code diagram of this user interface. After setting the Caption property of the CommandButton, an event queue is created for Click. The

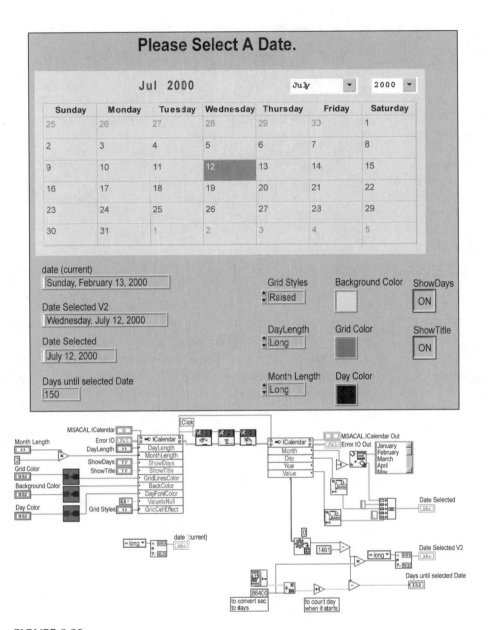

FIGURE 8.29

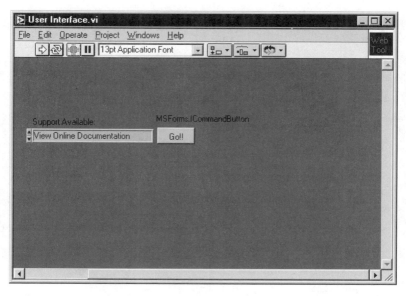

FIGURE 8.30

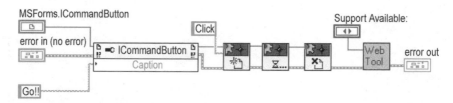

FIGURE 8.31

VI sleeps until Go!! is clicked, after which Web Browser.vi executes. The Web Browser VI was configured to display the front panel when a call is made to it. This option is made available if you pop-up on the icon in the upper right corner of a VI panel or diagram. Select VI Setup from the pop-up menu and the window shown in Figure 8.32 appears. Note that checkboxes have been selected to show the front panel when the VI is called and to close it after execution if it was originally closed.

The front panel window with the embedded browser appears loading the URL specified, as shown in Figure 8.33. National Instruments' home page is displayed in the browser window in this example. The Browser control allows the user to click on and follow the links to jump to different pages. When the operator is finished navigating around the documentation online, the Done CommandButton is pressed to return to the main user interface.

Figure 8.34 illustrates the code diagram of the Web Browser VI. Based on the support type selected from the user interface, the appropriate URL is passed to the Navigate method on the Browser control. As the front panel is displayed when the VI is called, the URL will begin loading in the browser window. Once again, the Caption property is modified, this time to display "Done" on the CommandButton.

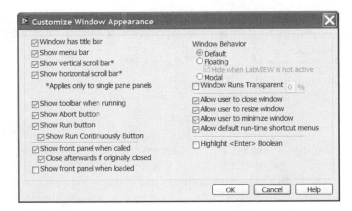

FIGURE 8.32

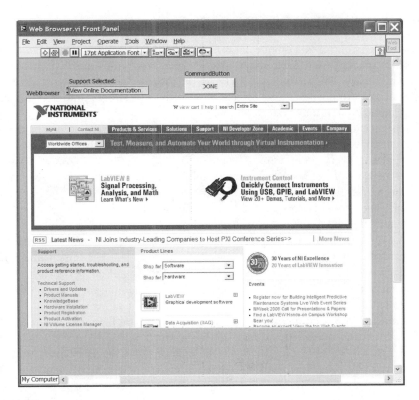

FIGURE 8.33

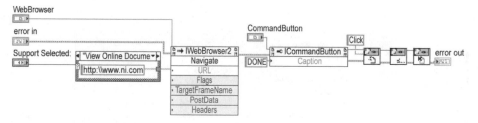

FIGURE 8.34

Next, an event queue is created and the VI waits for the Click event. When Done is clicked, the queue is destroyed and the window closes, returning to the user interface.

The Browser control does not have many methods or properties in its hierarchy of services, making it relatively simple to use. The CommandButton is also a simple control to use in an application. The button's support of several ActiveX events makes it an attractive control for use. The Click event is convenient because it eliminates the need for a polling loop.

8.9.5 MICROSOFT SCRIPTING CONTROL

The scripting control is a unique control that was used to add functionality to both Visual Basic and Visual Basic script. One of the internal components of this control is the dictionary. The dictionary is very useful for storing and retrieving data via a key. When working with objects such as clusters that store configuration information for an application, it may be desired to access the clusters by a key. The key can be a number such as an array index. It might be desirable to use a string for the key. Humans tend to identify strings better than numbers, which is why enumerated types are popular in every programming language.

As an example of why a dictionary might be useful, consider a very simple application that stores basic information about people, such as their names, phone numbers, and e-mail addresses. We tend to remember our friends' names, but their phone numbers and e-mail addresses may elude us. We want to be able to enter the name of one of our friends and have their name and e-mail address returned to us.

One solution is to store all the names and addresses in a two dimensional array. This array would store names, phone numbers, and e-mail addresses in columns. Each row of the array represents an individual. Implementing this solution requires that we search the first column of the array for a name and take the row if we find a match. Elegance is not present, and this approach has problems in languages such as C, where array boundaries do not just grow when we exceed the bounds. Also, consider that if we have a lot of friends and add one past the array boundaries, the entire array will be redimensioned, which will cause performance degradation. The big array solution does have a benefit; it can be easily exported to a tab-delimited text file for presentation and sorting in an application like Microsoft Excel.

It would be easier if we could use a cluster to store all the information, and search the clusters by the name to find the information. We can use clusters and a one-dimensional array, but we still have the problem of arrays and resizing. The

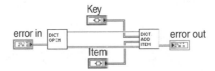

FIGURE 8.35

next problem we encounter is when we want to remove people from the array. This involves splitting the array and rebuilding it without the undesired row. Memory hits will again show up because we need to allocate two subarrays and one new array.

Linked lists are a possible solution, but searching the list is not particularly efficient. We would have to search the elements one at a time, from beginning to end, or use sorting algorithms to implement other search patterns. Regardless of the method chosen, we will spend time implementing code or will waste CPU time executing the code. Enter the scripting control and its Dictionary component. Microsoft describes the dictionary as an ActiveX equivalent to a PERL associative array. Associative arrays take two values one value you want to store and a value that you want to reference it with. We are not being very specific about the values that are stored and used for reference because the item types are very flexible. If we want to store clusters of information and refer to them by strings, we can. If we are interested in referring to strings by numbers, we can do that, too.

The interface to Dictionary is fairly simple, and a small collection of VIs to use the control are provided on the companion CD. First, we need an open Dictionary VI to create a reference to the ActiveX control. This VI will not be documented because it is a simple automation open command wrapped in a VI. A Close VI is provided; it is an automation close in a wrapper.

To insert a pair of items, called the "key and item," we invoke the Add Item method. This method is wrapped in a VI for ease of use. The control accepts variants, and we will use the variants with this control. The other option is to have wrappers that accept each data type and polymorph the LabVIEW type into a variant. This would require a lot of work, and the benefits are not very high. Either way, the types need to be polymorphed. Each cluster of personal information we want to insert into the dictionary is added to the map with the Add Item VI. The use of this VI is shown in Figure 8.35.

To verify that an item exists, the Key Exists VI is used. This VI asks the dictionary if a certain key is defined in the dictionary. A Boolean will be returned indicating whether or not the key exists.

Removing items is also very easy to accomplish; the Remove Item VI is invoked. The key value is supplied and the item is purged from the array. Removing all items should be done before the dictionary is closed, or memory leaks will occur. This VI calls the Remove All method and every element in the dictionary is deleted.

If we want to know how many items are in the dictionary, the Count property can be retrieved. Count will tell us how many entries are in the dictionary, but it does not give us any information as to what the key names are. A list of all key names can be obtained with the Get Keys VI. This VI will return a variant that needs to be converted to an array of the key type. For example, if you had a dictionary

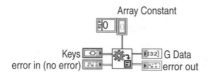

FIGURE 8.36

mapping integers to clusters, the output of Get Keys would have to be converted to an array of integers. An example of the conversion is shown in Figure 8.36. Our driver will not convert values for us because this would require a large number of VIs to accomplish, or would not be as reusable if we do not cover all the possible data types. In this case, variant usage is working in our favor.

The array returned by Get Keys can be used with a For loop to get each and every item back out of the dictionary if this is required. We will have an array of elements, which is generally undesirable, but there are cases where we will need to do this. Using the For loop to develop the array is efficient, because autoindexing will allow the For loop to predetermine the size of the output array. This array can then be dumped to a file for storage.

This example demonstrated a simple application of the Dictionary control. This control can be very useful in programming applications where an unknown number of data items needs to be stored for use at a later time. Applications that should consider using this control would have storage requirements for which arrays cannot be practical, and for which using a database would be overkill. Database programming can be challenging, and interfacing to the database may not be very fast. The Dictionary is an intermediate solution to this type of programming power, and variants allow this tool to be very flexible.

8.9.6 MICROSOFT SYSTEM INFORMATION CONTROL

It has always been possible through API calls to identify which version of the Windows operating system your code was executing. An easy interface to gather this information is presented by the Sysinfo control. This control has a variety of information regarding the operating system, power status, and work area. The Sysinfo control does not have any methods, but does have a list of interesting properties and events. The events are not necessary to use the control; all properties can be accessed without needing to respond to events.

Designing this driver collection will require four VIs for the properties. We will have one VI to open the control, one to handle power status, one to give operating system information, and the last to give work area information. No Close VI will be built. There is no obvious need to wrap the automation close in a VI; that will just add extra overhead.

NASA-certified rocket scientists would not be needed to verify that the code diagram in Figure 8.37 simply creates a connection to the control. We will not be asking for any properties from the control at this time. Opening ActiveX controls should be done as soon as possible when an application starts up. This allows us to

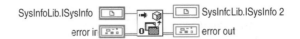

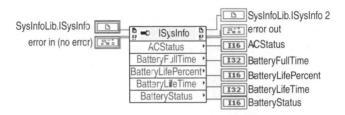

FIGURE 8.37

FIGURE 8.38

open the control at the beginning of a run and query information from the control
when it becomes necessary.

The Power Information VI code diagram in Figure 8.38 is nearly as simple as
the Open statement. The values that can be returned are AC Status, Battery Full
Time, Battery Status, Battery Percentage, and Battery Life. All return values are in
integers, and we will now explain why and what they mean. AC Status has three
possible values: 0, 1, and 255. Zero means that AC power is not applied to the
machine, 1 means that AC power is being supplied, and 255 means the operating
system cannot determine if power circuitry is active. Battery Full Time is actually
the number of seconds of life the battery supports when fully charged; a value of
−1 will be returned if the operating system cannot determine the lifetime the battery
is capable of. Battery Percentage is the percentage of power remaining in the battery.
A return value of 100 or less is the percentage of power remaining. A value of 255
or −1 means that the system cannot determine the amount of battery life left. Battery
Life returns the number of seconds remaining of useful power from the battery. A
value of −1 indicates that the system cannot determine this information.

This information is only useful for laptop or battery-powered computers. The
average desktop computer has two power states: AC on or AC off. If the desktop
is hooked up to an uninterruptible power supply (UPS), we will not be able to get
to the information from the Sysinfo control. Some UPS devices have interfaces to
the computers they support, and a different driver would be needed for that type
of monitoring. Consult the documentation with your UPS to determine if a driver
is possible.

The next set of properties, the operating system information, may be useful when
applications you are running are on different variants of Windows. Currently, there
are more variations of Windows releases than most people realize. If you are doing
hardcore programming involving ActiveX, OLE, or custom DLLs, the operating
system information may be useful for troubleshooting an application. A second option
is to dynamically change your code based on the operating system information.

Operating system properties available are the version, build number, and platform. "Platform" indicates which Win32 system is running on the machine. The

"Build Number" indicates which particular compile number of the operating system you are running. Actually, this property will not be particularly useful because the build number is not likely to change. Service packs are applied to original operating systems to fix bugs. The version number also indicates which version number of Windows you are currently running.

The work area properties are useful if you programmatically resize the VI. This information gives the total pixel area available on the screen considering the task bar. This way, panels can be scaled and not occupy the area reserved for the task bar.

LabVIEW users can take advantage of the events this control supports. Many of the commands are not going to be of use to most programmers. The operating system will trigger events when a number of things are detected. One good example is that when a PCMCIA card is removed from a laptop, the operating system will detect this (although not as quickly as many programmers would want). Once the system has detected the removal of a PCMCIA card, and you have an event queue for this control, you will be informed by the operating system that a PCMCIA card has been removed. This type of functionality will be useful for developers who need to support portable computer designs. PCMCIA card removal notification would allow an application to stop GPIB or DAQ card handling and would be useful for adding robustness to an application. For example, code that could be executed once this event happens would be to halt reads, writes, or commands that route to the PCMCIA card until a PCMCIA Card Inserted event is generated. The event handling for this control is left up to the reader to look up. The system control should be supplied with most systems.

8.9.7 MICROSOFT STATUS BAR CONTROL

The Microsoft Status Bar control is used to provide familiar status bar information to users. This control can be used in test applications to show the current status of processes currently running. The Status Bar control allows programmers to define different panels to supply basic information to users during execution. Many of the features of this control are useful for user interface designs, including concepts such as tool tips. When a user holds a mouse over an individual panel, a yellow text box will give basic information as to what the panel is describing. This type of information makes it easier for users to interface with applications. Most hardcore programmers consider this type of programming to be "fluff," but user interfaces are the only aspect of an application the end users work with. Giving users the ability to see and understand more about what the application is doing is a relevant topic in any design. This control is being presented to lead us into the next control, the Tree View control. The Tree View control uses more complicated nesting and relationships between data members, so the Status Bar is a beginning point that leads into more complex controls.

The Status Bar is yet another user interface element, and its strong point is the standard Windows look with little programming needed. Selecting the properties of the control can configure many of the details on the bar's appearance. Figure 8.39 shows the menu selection leading us to the status bar's properties. We can select

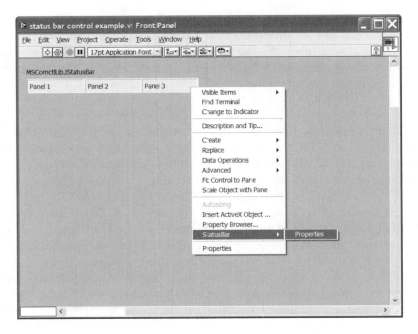

FIGURE 8.39

general properties to configure basic appearance information such as 2- or 3-D separators for the control.

The Panels tab allows us to configure the panels that we wish to display to the users. Panels may contain text, tool tips, and images. Figure 8.40 shows the Tabs Configuration window for this control. Each panel is flexible in the sense that

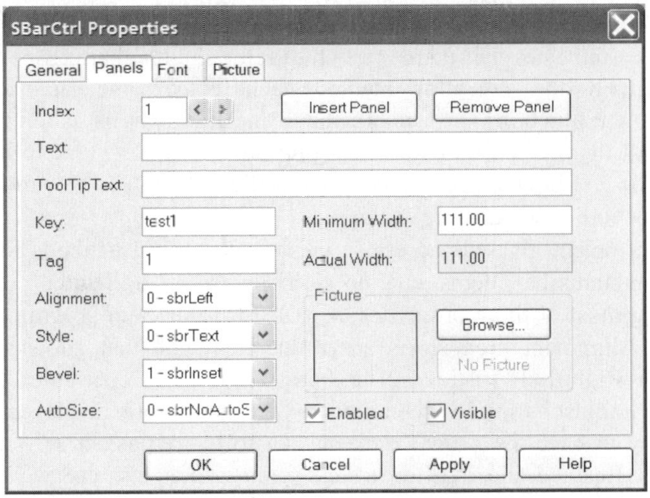

FIGURE 8.40

individual panels can be enabled and disabled without impacting the rest of the control. Panels can be beveled, which causes them to appear to "stick out" from the rest of the control, or inserted, which makes them appear to be "pushed in" relative to the other panels. Text can be centered, or right- or left-justified for appearances. All of these properties can be configured for startup and altered during runtime.

In order to work with the control at runtime, we will require some kind of driver to interface to the control. The Status Bar control has a few methods and properties. The only property of the status bar we are interested in is the Panels property. This property returns an ActiveX refnum that give us access to the complete list of panels of the control. The Panels property is accessed like an ActiveX control, and Panels itself has properties and methods we can work with.

The Panels object has a single property: Count. Count indicates that it is a read and write property. You can set the count to any number you desire; the control will simply ignore you. This is an error in the interface in terms of the interface description: Count should be an "out" property, meaning that you can only read the value. Count returns the number of panels the bar is currently holding.

Methods for the Panels object include Add, Clear, Item (Get), Item (Put Ref), and Remove. The Add method adds a new panel for you to place information. Clear removes all of the panels that are currently defined. The Remove method takes a variant argument similar to the Item method. This method removes the specified method from the status bar.

Item (Get) and Item(Put Ref) allow us to modify a single panel item. Both methods return a refnum to the specific panel. The Item (Get) method takes a variant argument that indicates which particular panel we want to work with. Our driver will assume you are referencing individual panels by numbers. The Item(Put Ref) method takes an additional parameter, a refnum to the replacement panel. For the most part, we will not be building new items to insert into the control, and we will not support the Item(Put Ref) method in our driver.

You may wonder why there are two different Item methods. The additional text in parentheses indicates that there are two different inputs that can be given to the method. ActiveX does not allow methods to have the same name and different arguments to the function. The function name and argument list is referred to as the "signature" of the method. ActiveX only allows the function name to have a single signature. The text in parentheses makes the name different, so we are not violating the "one function, one signature" rule.

The Item object gives us access to the properties and methods for individual panels in the status bar. Items have no methods and 13 individual properties. We will be using most of these properties, or at least make them programmable in the driver. The Alignment property is an enumerated type that allows us to select Centered, or Right/Left Justified. The Autosize property configures the panel to automatically adjust its size to accommodate what is currently displayed. Bevel gives us the ability to have the panel "stick out" or to look "pushed in" relative to the other panels. Enabled allows us to make specific panels disabled while the rest of the control remains enabled. Index and Key allow us to set the index number for the panel or a text "key" which identifies the panel. We can use the key in place of the index number to reference the panel. This is desirable because, programmatically,

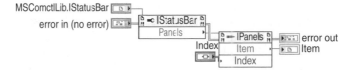

FIGURE 8.41

FIGURE 8.42

it is easier to work with descriptive strings than integers. The Picture properties allow you to hand a picture reference to the control to change the currently displayed image. We will not be using the Tag items in this driver.

The driver itself is going to consist of a number of VIs to set the individual tasks. It is not likely that programmers will want to set each property of a panel every time they access it. Therefore, the driver collection should be structured with each property to a VI in order to maximize flexibility.

One design decision we need to reach before writing the driver is how we access the individual items in the control. We will generally be dealing with the item property of the Panels item of the control. There are a number of steps in the control, and we may decide that we are too lazy to continuously reference the control to reference the panels to reference the item. It makes sense to write a wrapper VI to encapsulate the traversal of the ActiveX objects. This type of wrapper will save us a bit of coding time, but an extra VI will create some performance hits on execution time. What we can do is write a VI that takes the status bar refnum and the item and traverses the objects, returning a refnum to the specific key. This VI can be run with subroutine priority to eliminate a lot of the VI overhead that is incurred when we use subVIs. We will use the variant argument as the front panel control so programmers will be free to work with either the index or a key name. The code diagram for this VI appears in Figure 8.41.

The VI to set the Tool Tip text for an individual panel item is shown in Figure 8.42. We call our Get Item VI to return the refnum to the specific panel for us. This VI will be used in all of our driver VIs to obtain a reference to the desired panel item. We will not show the rest of the drivers because there is little additional information to learn from them, but they are available on the companion CD to this book.

8.9.8 MICROSOFT TREE VIEW CONTROL

Assuming that you have worked with Windows Explorer, you are familiar with the basics of what the Tree View control does. Tree View is a mechanism to present data to users in an orderly, nested fashion. Information is stored into the control in the form of nodes. A node is simply a text field and optional bitmap. The Tree View control is capable of displaying images in addition to simple text fields. Potential applications of this control to LabVIEW developers are displaying configuration or test result information in a nested format. This function is now built into LabVIEW as the Tree Control; we will also discuss the ActiveX implementation for this control. This control is part of the Windows common controls, and is arguably one of the most complex controls in this set.

Before we begin going into the details of how the Tree View control works, we need to get some background information on how the tree data type works. The Tree control stores data in a similar fashion to the tree data structure. As a data structure, trees store data in the form of leaves and branches. Each data point will be a leaf, a branch, or the root. Data items that point to yet other data items will be branches. The root is the first data element in the tree and is both the root and a branch. We will not display any illustrations of what a tree looks like; the easiest way to explain the data type is to instruct you to use Windows Explorer. This is a Tree View-based application and demonstrates well what the Tree View control looks like. At the root, you would have Desktop. All other elements in Explorer descend from the Desktop item. Desktop, in turn, has children. In the case of the computer on which I am currently typing, the desktop has six children. The children are My Computer, Network Neighborhood, Recycle Bin, My Briefcase and a folder titled, "LabVIEW Advanced Programming." Each of these children, in turn, are branches and contain other data items which may or may not point to children of their own. This is how a tree data structure is set up. A special case of the tree structure has two children for each branch. This is known as a "binary tree."

Each element in a tree contains some type of data and references to its parent and children objects. Data that can be stored in trees is not defined; this is an abstract data structure. Trees can be used to store primitive types such as integers, or complex data types such as dispatch interfaces to ActiveX controls. In fact, the Tree View control stores dispatch interfaces to ActiveX controls. The Tree View control uses the dispatch interfaces to store references to its children and parent. Data stored in the Tree View control is a descriptive string and a reference to a bitmap.

Now that we have an idea how a tree structure works, we may begin exploring the operation of this control. A Tree View control needs to be inserted in an ActiveX container on a VI's front panel. This control has a short list of methods that we will not be using ourselves. Most of the methods of this control are related to OLE dragging and dropping. We do not intend to perform any of this type of programming, and do not need to implement driver VIs to support it.

The property listing for this control has a number of configuration items for the control itself, such as display options and path separator symbols. The Windows Explorer uses the plus and minus signs as path separator symbols. We will use them by default, but we have the ability to change them to whatever we desire. The main property that we will need access to is the Nodes property. In the Tree View control, each of the leaves and branches has been titled a "node." The nodes contain all information that we need, such as the data we are storing and the location of parent, children, and siblings (nodes at the same level as the current node). Nodes serve as gateways to access individual data elements similar to the Items property in the Status Bar control. This is a fairly common theme to the architecture of many of Microsoft's controls. You have not seen this type of design for the last time in this chapter.

The Nodes property is itself an ActiveX refnum. This refnum has the methods Add, Clear, Control Default, Item, and Remove. This configuration, again, is very similar to the Status Bar control, which will make learning this control much easier for us. Item and Control Default have properties with the (1) following them because they have two different sets of arguments that go with them. We will "stick to our guns," and not use the methods that have the (1) following them because we are not going to track around different refnums for each of the nodes in this control. It is possible for us to have several hundred or even thousand items, and we will use the control itself to store the information.

Most methods use a variant argument called "Key." Key is used to perform a fast lookup for the individual node that we want. The command for Clear does not use the Key argument, it simply dumps all of the contained nodes, not just a particular one. The Add method requires a few additional arguments, most of which are optional. Add has the arguments Relative, Relationship, Key, Text, Image, and Selected Image. We need to specify only Text if we desire. The control will go ahead and, by default, stuff the element at the end of the current branch. Item allows us to gain access to individual data elements stored in the tree control. This method takes a variant argument called Index and returns an ActiveX reference to the item we are requesting.

The Item object has only two methods that are related to OLE: dragging and dropping. We will not be using this method in this driver. There are a number of methods, however, that give us complete control of navigating the tree. We can identify the siblings of this item. Properties for first and last sibling give us the top and bottom element of the branch we are currently on. The next property gives us a reference to the sibling that is below the current item in the list. This allows us to not track the keys for the individual elements in the tree and still work our way around it. The driver VI for this will be a single VI that has an enumerated type to allow a programmer to select which relationship is desired and return a refnum to that object. The code diagram is shown in Figure 8.43.

Individual elements can have text strings to identify their contents in addition to bitmap references. Users tend to appreciate graphical displays, and the bitmaps will make it easier for them to navigate the tree in search of the information they are interested in. For example, Windows Explorer uses bitmaps of folders and files to differentiate between listed items. Inserting images into the Tree View control is

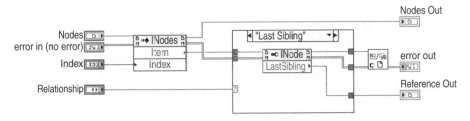

FIGURE 8.43

a bit of work. We need an associated Image View control that has the image, and we will pass an ActiveX refnum into the control. The Image View control is structured with the same item properties and methods as the Tree View and Status Bar Controls, and will not be covered in this book.

Now that we have mentioned how this control works, what properties and methods it has, and skipped out on the Image List control, we should begin to write and use the driver. We started this example by stating that this is only a user interface element. That is not really the case. You can use the Automation Open function to create the element, not display it and systematically store all configuration information into it. Functionally, it will operate very similarly to the Dictionary control we presented earlier. Flattening any configuration information into a string will allow you to use this hive structure to store information. Unlike the Dictionary control, the Tree View control will allow you to have a directory-like structure to the information. One of the biggest problems we have seen with large-scale applications is having a good design for basic information storage. This control offers an interesting solution to this problem.

8.9.9 MICROSOFT AGENT

Microsoft Agent is a free ActiveX control that we can use to generate animated characters that can speak and decode audio input. These characters provide another element of a user interface design and behave somewhat similar to the Help characters in Microsoft Office. Unlike the Office characters, Agent's characters do not reside in their own window; they are free-floating and may be set up anywhere on the screen. In addition, Microsoft Agent characters can be used by any application. This type of user interface element is not something we expect to appear in requirement documents, but the control itself is a great example of COM programming.

This section will provide an overview of how this control works, and develop a driver collection to utilize it. All of the details on this control cannot be presented; enough information exists that a separate book can be written on this control. In fact, a book has been written for programming Microsoft Agent. Microsoft Press published *Programming for Microsoft Agent,* and additional information is available on Microsoft's Website.

Microsoft Agent is a service, meaning that it uses an out-of-process server and a control to reference the server itself. The server is responsible for controlling and managing the characters for different applications. This makes it possible to control Agent as a DCOM object. The characters can be made to appear on different machines to provide up-to-the-second status information.

Before we begin the design of the Microsoft Agent driver, we need to examine the design of Microsoft Agent itself. Microsoft Agent uses aggregated controls, meaning there are multiple controls for us to work with. The Agent control is a base starting point to interface to the server object, but most of the operations we will perform will be handled through ActiveX interfaces created by the Agent control.

Each of the aggregated controls represents a subset of the functionality of Microsoft Agent. Programming is simplified because each of the methods and properties are contained in appropriately named controls. This design is similar to

SCPI instrument command sets, where commands are logically grouped into a hierarchy of commands. Agent follows the COM design methodology, and each of the branches in the hierarchy is a COM object.

We will first present the embedded controls with their methods and properties. Events will be covered after the main driver has been developed.

8.9.9.1 Request Objects — First Tier

The first object that is of use to most programmers, but not as useful to LabVIEW programmers, is the Request object. The Request object is analogous to the error cluster in LabVIEW. All Agent commands return a Request object to indicate the status of the operation. This object is not as useful for LabVIEW programmers because each method and property invocation returns an error cluster, which contains similar information as the Request object. Languages such as Visual Basic do not have built-in error cluster-type support, and objects such as Request are trying to make up for that limitation. None of the elements of our driver will use the Request object. Each time we are passed a Request object we will issue an ActiveX Close on it.

Programmers who are interested in using the Request object in their work will want to know that its properties include Status, Number, and Description. Status is an enumerated type that is similar to the status code in the stock LabVIEW error cluster. The Status property has four values instead of two, and includes information such as Successfully Completed, Failed, In Progress, and Request is Pending. The Number property is supposed to contain a long integer that contains an error code number. Description contains a text explanation of the current problem or status. As we mentioned, the Request object is an imitation of LabVIEW's stock error cluster.

8.9.9.2 Other First-Tier Controls

The next four objects that are mentioned for completeness, but not used in our driver, are the Speech Input, Audio Output, Commands Window, and Property Sheet objects. Each of these serves purposes directly related to their names, but are not necessary to successfully program Microsoft Agent. Additional information on them can be located on Microsoft's Website, or in the book *Programming for Microsoft Agent*.

Properties that are available at the Agent control base are Connected, Name, and Suspended. The Connected property is a Boolean that we will need to inform the local control that we wish to establish a connection to the Agent server. We will set this property in our Agent Open VI. Name returns the current name assigned to the ActiveX control. This may be useful to Visual Basic programmers who can name their controls, but LabVIEW programmers work with wires. We will not make use of the name property. The Suspended property is also a Boolean and will indicate the current status of the Agent server. We will not make use of this property, but programmers should be aware of it as it can be useful for error-handling routines.

8.9.9.3 The Characters Object

The first tier object in the hierarchy that is of use to us is the Characters property. Again, this is an embedded ActiveX control that we need to access in order to get to the individual characters. The Characters object has three methods, and we will need to be familiar with all of them. The Character method returns a refnum to a Character control. The Character control is used to access individual characters and is the final tier in the control. Most of our programming work will be done through the Character object. The Character method returns a refnum, and we need to keep this refnum.

The Load method is used to inform the Agent server to load a character file into memory. Agent server is an executable that controls the characters. For flexibility, the characters are kept in separate files and can be loaded and unloaded when necessary. We will need the Load method to inform the server to load the character we are interested in.

Unload is the last method of the Characters control. Agent server will provide support for multiple applications simultaneously. When we are finished with a character, we should unload it to free up resources. Not unloading characters will keep the character in memory. This is not a mission-critical error, but it does tie up some of the system's resources.

The last method is the Character method. This method simply returns a refnum to a character object in memory. An assumption is made that this character has already been loaded into memory. If this is not the case, an error will be returned. The refnum returned by this function is covered next.

8.9.9.4 The Character Control

We finally made it. This is the control that performs most of the methods that we need for individual character control. The Character control requires the name of the character as an argument to all properties and methods. This will become a consideration when we develop the driver set for this control. There are a number of embedded controls, and we need to track and keep references to the controls we will need.

The Character control supports methods for character actions. The Activate method allows the programmer to set a character into the active state. Active has an optional parameter, State, that allows the programmer to select which character is to be activated. We will assume the topmost character is to be activated in our driver.

Our driver will use the methods Speak, MoveTo, Play, Show, and Hide. These methods all cause the characters to take actions and add a new dimension to our application's user interface. The method names are fairly self-explanatory, but we will briefly discuss the methods and their arguments. Speak takes a string argument and causes the character to display a dialog bubble containing the text. The control can also be configured to synthesize an audio output of the text. MoveTo requires a coordinate pair, in pixels, for the character to move towards. An animation for moving will be displayed as the character moves from its current position to the new coordinate pair. Play takes a string argument and causes the character to play one of its animations. Animations vary among characters, and you need to know which animations are supported by characters with which you choose to develop. Each of the Play arguments causes the character to perform an animation such as smile, sad, greet, and others. Hide and Show require no arguments and cause the character to be displayed or not displayed by the server.

There are a number of properties that can be set for Agent, but we are going to need only two of them. Visible returns a Boolean indicating whether or not the character is visible. This is a read-only property. If we want the character to be displayed, we should use the Show method. The other property we will make use of is the Sound Effects property. This property determines whether or not the Agent server will generate audio for the characters.

Now that we have identified all the relevant methods and properties we need to use Microsoft Agent, it is time to make design decisions as to how the driver set will be structured. Unlike the other controls presented in this chapter, Agent uses aggregated controls, and it is possible to have several different refnums to internal components of the control. This is not desirable; requiring programmers (including yourself) to drag around half a dozen different refnums to use the control is far more work than necessary. We can pass around a refnum to the base control and the character name and rebuild the path of refnums back to the character as we use the control. Potentially, there are performance hits every time we go through the COM interfaces, but for user interfaces performance is not an issue. Users still need to drag around two pieces of information. We really only need one: the character name.

It stands to reason that the control should keep an ActiveX refnum as an internal global variable. This will allow programmers to keep names for multiple characters and run them through the same Agent control. This would be efficient on memory, because we need to instantiate the control only once. This also allows for different VIs that are running independently to use the same server connection, which is more efficient for the Agent server.

We will rebuild the paths back to aggregated controls for all calls, but as we decided before, performance is not a significant issue for this control. Most other controls do not support multiple connections; programmers need to make other instances of the control. Agent does not need multiple copies to run multiple characters. This example is going to show a different way of handling ActiveX controls. We will make this driver DCOM-enabled by allowing the programmer to supply a machine name for the control.

The first problem we encounter with using a global variable for a reference is multiple calls to our Open VI. We will use a technique called "reference counting." Each time a user calls the Open VI, an integer count stored in the global variable will be incremented. If the number equals zero before we perform the increment, we will call ActiveX Open. If the number is nonzero, we will increment the reference count and not call ActiveX Open. This VI is shown for both cases in Figure 8.44 and Figure 8.45. The Open VI returns only an error cluster; all other VIs will use the control through the global variable, and perhaps need to know the character name. When we need to open the control, we will set one property of the control. Connected will be set to "true."

The Close VI works in a similar fashion, except we use the decrement operator. The decrement is performed before the comparison. If the comparison shows the number is not greater than zero, then we know we need to close the control. When the comparison after the decrement is greater than zero, we know that we need to store the new reference count value and not to close the control. C++ programmers

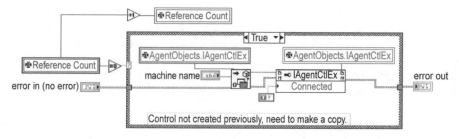

FIGURE 8.44

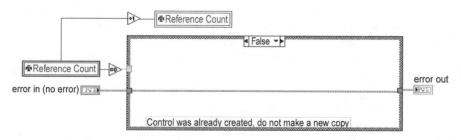

FIGURE 8.45

typically use reference counting, and this technique has applications for LabVIEW programmers with ActiveX controls.

The first method we will write is the Load Character method. This method requires the name of the character to load and the load key. The load key we will use is a path to a file on a hard drive. Agent defines this as the load key, so the control can be used in ActiveX-enabled Web pages. The Load key can also be a URL pointing to a character file. Agent character files use an ACS extension. This VI simply builds the Load key from the character name and supplied path. We use the global variable to access the agent refnum, and then we access the character's property. This property is used to gain access to the Load method. Once the Load method is performed, we close off the returned load value. Load's return value is a Request object. We do not need this object, because any errors generated would be reported back in the error cluster. Visual Basic programmers who do not have an error cluster to work with need Request objects. This VI is shown in Figure 8.46.

It is important to note that we closed off only the Request object's refnum. We are not finished with either the control or the character refnums; in fact, we have

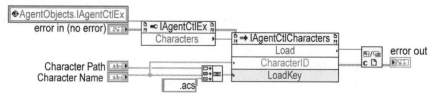

FIGURE 8.46

FIGURE 8.47

only begun to use both of these objects. Our next function to implement will be the Unload method.

The Unload VI will operate in a similar fashion to the Load VI. Again, we do not close off either the characters refnum or the agent control refnum; both of these objects are still necessary. The characters refnum is needed by other VIs that are making use of the Agent control, and the Agent control itself requires access to the Characters object. Dropping the reference count on this object could cause the buried control to be unloaded from memory, which would be a surprise to the main Agent control. The Unload VI is shown in Figure 8.47. Unlike the Load VI, Unload does not need the path to the character file, and does not return a Request object.

The last method of the Characters object is the Character method. We will not need a VI for this method; the driver will always handle it under the hood. Character is needed only when passing instructions to specific characters. Programmers will need to pass only the character name to our driver; the driver will handle the rest of the details.

Now that we have mentioned passing the character names in, it is time to start implementing the individual character controls for the driver. Characters need Show and Hide methods to enable them to be seen or hidden. The VI to perform this work is shown in Figure 8.48. Show and Hide will be implemented in a single VI to cut down on the total number of VIs that are necessary. Both commands return a Result object, and we promptly close this object.

Moving the Agent character around the screen will be easy to accomplish with the Move Character VI. We need three inputs from the user: the name of the character and the coordinates for the character to move to. You have probably noticed the cascading access to each of the elements. There is not much we can do about this; there is no diabolical plot by Microsoft to wreck the straight paths of our VIs. Visual Basic handles aggregated controls nicely. The LabVIEW method of tracking the error cluster clutters up our diagrams a bit. The Move To command in Visual Basic would look like this: result = Agent.Characters.Character("Taqi").MoveTo(100,100). This is easy to read in Visual Basic, and the SCPI-like structure of the commands makes it very easy to understand what the command is doing with each portion of the Agent control.

The next command that can be implemented is the Speak command. This is going to be one of the more widely used commands for this driver. Speak causes the character to display a bubble containing the text, and, if configured, to generate synthesized audio for the text. Speak.vi is structured very similar to Move To but requires two strings for input. This VI is shown in Figure 8.49.

The last animation command we are going to support is Play. Play is a generic command that all characters support. Agent allows for flexibility by not defining

FIGURE 8.48

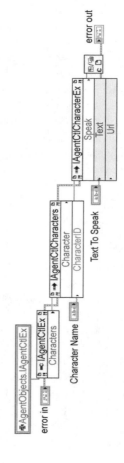

FIGURE 8.49

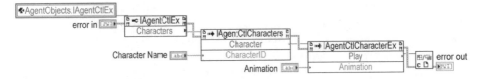

FIGURE 8.50

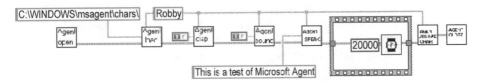

FIGURE 8.51

what animations the characters must support. Different characters can have animations that are related to the specific character. For example, a dog character could have an animation named Bark, while Bark would not be of much use to a President Clinton character. The VI implementing this function is shown in Figure 8.50. Programmers will need to know which animations their characters support.

Now that we have an intact driver for Agent, we shall put it to work. The VI shown in Figure 8.51 shows a simple application that opens Agent, loads the Robby character, and has it speak. The 20-second pause makes sure that the character has time to speak before we unload the character and release the control. It is important to let the character have time to complete its actions, or the server will cut out the character midsentence!

What makes the agent driver very easy to work with is that we encapsulated all details regarding ActiveX into wrapper VIs. The user's application looks like any standard LabVIEW VI. It is not obvious that the COM subsystem is in use, or even that Agent is running on another machine. Good driver development for ActiveX controls can make some of ActiveX's complexities a nonissue for programmers. When working with ActiveX, it is best to write a solid driver that makes the control look and feel like a standard LabVIEW VI.

This example may not be practical for many applications; dancing parrots may not be suitable for many production applications. We did manage to demonstrate programming techniques for advanced ActiveX controls. We saw a control that uses aggregation, and eliminated complexity from the driver for the benefit of the driver's users. Reference counting was presented and used to minimize the number of instances we need of the control.

8.9.10 Registry Editing Control

The Registry Editing control is perhaps the most dangerous control that will appear in this chapter. The Registry Editing control allows a programmer to get, edit, delete, and modify Registry keys. Only programmers who understand what the Registry is, and what not to do with it, should use this control. If you do not understand the

Registry or are not comfortable with editing and modifying its contents, feel free to skip over this control. Folks who like to tinker with ActiveX controls may want to backup the System Registry and some of their valuable data.

The Win32 Registry contains information necessary for booting the system and identifying the location of applications, components, and system services. Historically, most information needed to start an application was stored in an INI file. The System registry provides a new database storage mechanism to allow for a single file to handle major initialization of applications. Individual data items are stored in the form of keys. Each key takes on a particular value, such as a string or integer. ActiveX controls have several keys that are stored in the Registry. One of the parameters that each control must store is the location of the control itself. When LabVIEW uses an ActiveX Open, it has a name for a control, and that is about it. LabVIEW talks to the COM subsystem, which in turn contacts the System Registry to locate a component. Keys are grouped into directories called "hives." There are five base hives: HKEY_CLASSES_ROOT, HKEY_USERS, HKEY_CURRENT_USER, HKEY_LOCAL_MACHINE and HKEY_CURRENT_CONFIG. Application information is stored in the HKEY_CLASSES_ROOT hive, and generic user information is HKEY_USERS. Information specific to the currently logged-in user is stored in HKEY_CURRENT_USER. The current configuration and dynamic data hives store information that we really, really do not want to look at or modify. HKEY_CLASSES_ROOT stores information about all applications and components in the system. If you examine the registry using the Registry Editor (Regedit) you will see the five base classes. Expanding the root key will show a hierarchy of folders that start with file extensions. Each file extension is listed in the Registry, and the system uses these keys each time you double-click on a file. When you double-click on a file with a VI extension, the system searches the root classes for the .vi key and determines that labview.exe is the file that should be used to open it. Needless to say, when storing information about your particular applications, HKEY_CLASSES_ROOT is the place to store it. It is strongly recommended that you do not access other hives; it is possible to completely confuse a system and seriously confused systems need their operating systems reinstalled!!

Many large-scale LabVIEW applications can take advantage of the Registry to store startup information. Information that is typically stored in the Registry would be information that is needed at application startup and configuration information that does not change frequently. Examples of startup data that could be stored in a Registry is the name of the last log file used, which GPIB board the application is configured to use, and calibration factors if they do not change frequently during execution of an application. Data that should not be stored in the Registry are items that change frequently, such as a counter value for a loop. Accessing the Registry takes more time than standard file access, which is why it should be used at application startup and shutdown. Key values should be read once, stored locally in LabVIEW, and written back to the Registry when the application is exiting. During normal execution, Registry access would cause a performance hit.

The drivers for this control are amazingly simple. The Open Registry VI simply calls ActiveX Open and returns the handle to the control. It is perfectly safe to open this control; the refnum does not explicitly mean you are accessing the Registry. The Delete Key VI is one that should be rarely used, but it is included for completeness.

8.9.11 Controlling Microsoft Word

This example will give a brief overview of programming Microsoft Word through LabVIEW. Microsoft Office is designed to be programmable and extensible through Visual Basic for Applications (VBA). Complete coverage of all of Word's automation abilities is well beyond the scope of this book, but we will try to give you a starting point for using Microsoft Word to add to your current applications. Controlling any of the components of Microsoft Office is not true ActiveX programming; it is actually OLE Automation.

Microsoft Word 2000 (Version 9.0) has two main creatable objects that Lab-VIEW can use. The first is the Application object and the second is a Document object. The Application object alone has 91 properties and 69 methods. The Document object has a whopping 125 properties and 60 methods. Obviously, we will not be covering all the properties and methods of these controls in this book. Also consider that a number of these properties and methods are aggregated controls of their own, which leaves the possibility of another couple hundred methods and properties for each control!

This dizzying array of methods, properties, and objects seems to make controlling Microsoft Word impossible. It really and truly is not. One of the Document properties is "Password." Unless you are using password protection for your documents, you do not need to use this property. The Versions property is another almost-never used property; in fact, many Word users do not even know that Word has a version control system built in. What this proves to us is that we need to figure out which methods and properties we need to get up and running.

Both the Application and Document controls are available to perform different tasks. The Application object is intended to provide control for applications to the entire Word application, where the Document control is meant to allow a program the ability to work with a specific document. This control in particular tends to be fussy about when components are closed off. ActiveX Close does not necessarily remove the DLL or EXE from memory, it calls a function named Release. The Release function uses reference counting, similar to what we did in the Microsoft Agent driver, to determine if it is time to unload itself from memory. Microsoft Word itself will maintain reference for itself on all documents, but having multiple references open on some components seems to cause problems. Releasing references as soon as possible seems to be the best way to deal with this particular control.

To start with, we need to gain access to Microsoft Word. We do this by using ActiveX Open to create an instance of the Microsoft Word Application object. This will start Microsoft Word if it is not presently running (this is a strong clue that you are performing OLE automation and not ActiveX programming). We can make the application visible so we can see things happening as we go along. To do this, set the Visible property of Microsoft Word to "true." Word just became visible for us (and everyone staring at our monitor) to see. Word sitting idle without any documents being processed is not a very interesting example, so we need to get a document created with data to display.

There are a number of ways we can go about getting a document available. Possible methods are to create one from any of the templates that Word has to work with, or we can open an existing text file or Rich Text Format (RTF) document. This brief example will assume that we want to create a new document based on the Normal template. The Normal template is the standard template that all Word documents derive from.

After opening a communications link with Word, we need to gain access to its document collection. The documents commands for Word are located in the Documents property. Like Microsoft Agent, Word uses aggregated controls to encapsulate various elements of the application's functionality. Documents is analogous to the Characters property of the Agent control. Major document commands are available through this control, including the ability to select which of the active documents we are interested in working with. Methods available from the Documents object are Open, Close, Save, Add, and Item. Open, Close, and Save work with existing documents and Close and Save work with existing documents that are currently loaded into Word. The Item method takes a variant argument and is used to make one of the documents currently open the active document.

Our example will focus on creating new documents, so we will invoke the Add method to create a new document. Optional arguments to Add are Template and New Template. The Template argument would contain the name of a document template we would want to use as the basis for the file we just created. New Template indicates that the file we are creating is to be a document template and have a file extension of "DOT." We will not be using either of these arguments in this example. The Add method returns an ActiveX refnum for a Document object. This refnum is used to control the individual document.

To save the document, we invoke the Save As method of the document refnum. The only required argument to this method is the file name to save the document as. There are a total of 11 arguments this function can take, and we have only made use of the File Name property. Other methods are there to allow programmers complete control of the application. The operations we have just described are shown in the code sample presented in Figure 8.52.

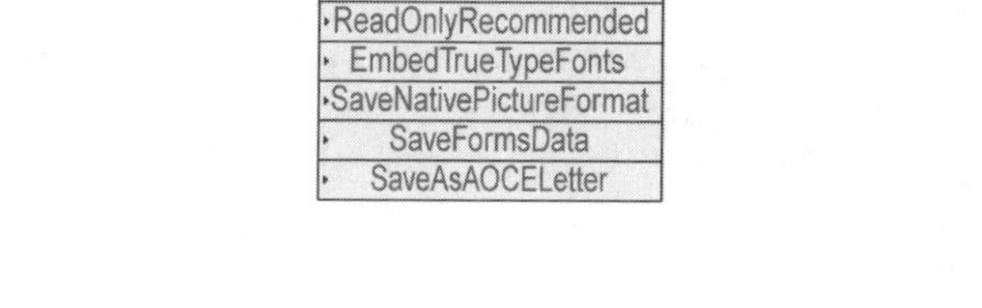

FIGURE 8.52

Thus far, we have managed to get control of Word, create a document, and save the document. We have yet to address the topic of putting text into the document to make it worth saving. Considering that the primary purpose of Word is to be a word processor, you would be amazed at how challenging it is to put text into a document. One issue that makes inserting text into Word documents difficult is the complexity of Word documents themselves. You are allowed to have images, formatting options, word art, and embedded documents from other applications — how do you specify where to put the text? Navigating through paragraphs is a trivial task when you are working with the application yourself, but programming the application is another story.

To insert text into the document we need the Range object. The range object is analogous to the mouse for applications. When users determine that they want to insert text at a particular location, they put the mouse at the location and click. This is much more difficult to do programmatically. Selecting the Range method with the Invoke node produces a new reference. Once you have a reference to the range object, you can access its properties. To insert text in a particular location, you would select the Start property. This indicates where to start the insertion. Now that you have programmatically clicked your mouse on the desired location, you need to insert your text. By using the Text property for the Range object, you can enter a string containing the text you want to insert. If you will be making multiple insertions of text, you can use the End property. This property provides the index to the end of the text insertion. The index read here could be used to start the next text insertion.

Let's assume that for some reason you need to perform a series of tests, take the results, and e-mail them to a group of users. We could use the MAPI control that we presented earlier, the SMTP driver we developed in chapter 3, the Internet toolkit, or Microsoft Word. We will insert the test results into a Word document and then attach a routing slip to it. This will allow us to send the other users a Word document containing the test results, and we would need to use only one ActiveX control to perform the task. Using the Routing Slip property gives us access to the document's routing recipient list.

If there is anything the preceding snippet shows us it is that anything we can do with Word on our own, we can do through the OLE interface. Typically, this is how application automation is intended to work. The VIs used to control Word are included on the companion CD.

8.9.12 MICROSOFT ACCESS CONTROL

This section will cover an example using ActiveX automation to write to a Microsoft Access database. Microsoft Access 8.0 Object Library Version 9.0 is used in this example. Access has many properties and methods in its hierarchy of services and can be complicated to use. This example will simply open an existing database, insert records into a table, and close the database. If you wish to perform additional operations with Access, you should find other reference material on controlling Access through automation. Because of its complexity, you can spend hours trying to accomplish a simple task.

Figure 8.53 displays the Upper Level VI, Insert Records into Access.vi, that is used to insert records into the Access database. The block diagram makes calls to

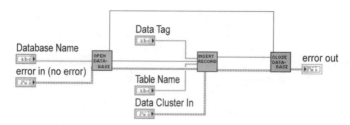

FIGURE 8.53

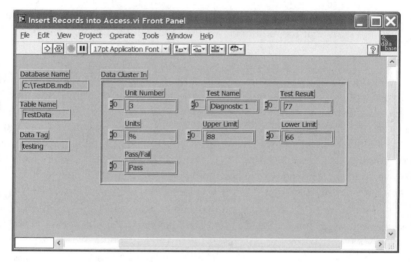

FIGURE 8.54

the following VIs: Open Database.vi, Insert Records.vi, and Close Database.vi. The code for this task was broken up logically into these subVIs. The required inputs include Database Name, Table Name, and Data Cluster In. The Data Cluster is simply a cluster containing one control for each field in the database. The front panel with this data cluster is displayed in Figure 8.54. Each control in the cluster is an array, and hence this VI can be used after collection of each data point or after accumulation of all data to be shipped to the database. It is more appropriate for use after all data is accumulated to maintain efficiency because the references to Access are opened and closed each time this VI is called. This VI can be used if you have an existing database, but you should be familiar with Microsoft Access. Table 8.1 defines the steps you can use to create a new database and table with the fields used in this example.

After the database and corresponding table have been created, you can use the Open Database.vi to open a reference. The block diagram for this VI is shown in Figure 8.55. Microsoft Access 9.0 Object Library Version 9.0 was selected from the Object Type Library list from the dialog window for selecting the ActiveX class. The Application object was selected from the list in the object library.

TABLE 8.1
Steps to Create a Database and Table in MS Access

Step 1. Launch Access application.

Step 2. Select New Database from the File pull-down menu.

Step 3. Select Blank Database, Name, and Save the Database using the file dialog box. TestDB.mdb is the name used in this example.

Step 4. Select the Tables tab, select New to create a table, and select Design View.

Step 5. Enter the field names as shown in Figure 8.56.

Step 6. Name and save the table. TestData is the name of the table used in this example. A Primary key is not needed.

Step 7. Close the database. LabVIEW will not be able to write to the database while it is open.

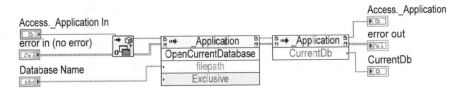

FIGURE 8.55

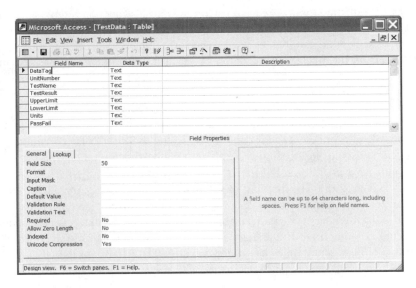

FIGURE 8.56

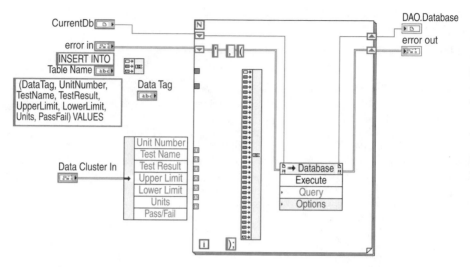

FIGURE 8.57

The Invoke node is then used to call the procedure to open the database. OpenCurrentDatabase is the name of the method selected from the long list of methods available. This method opens an existing database as the current database. The file path is the only required input. A database that resides on a remote machine can be specified if desired. The Exclusive input can be left unwired and defaults to "false" to open the database in shared mode.

The next method that is executed is CurrentDb. This function returns a database object which is essentially a reference to the database that is currently open. The reference, or pointer, can then be used to perform numerous operations on an open database. Both the Access._Application and CurrentDb refnums are passed out of this VI. The CurrentDb refnum is passed to the Insert Records.vi as a reference for the function. The block diagram of this VI is shown in Figure 8.57. This VI is responsible for generating and executing the SQL (Structured Query Language) string for inserting the records into the database.

The data cluster is sent to the database using the INSERT INTO command. This command is used to insert single records into the destination table. The syntax for this command is as follows: INSERT INTO table name (field 1, field 2, ..) VALUES ('data 1', 'data 2', ..). The first half of the statement is generated outside of the For loop because it is constant. The data values are added to the statement inside the For loop. Each of the elements of the cluster is an array. The loop is auto-indexed to execute as many times as there are elements. The second half of the command is generated only as many times as needed to get the data across to the table.

Inside the For loop is an Invoke node with the Execute method selected. The Execute method performs an operation on the database pointer passed to it. Query is the only required input for this method. Any valid SQL Query action expression wired to the method will be executed. SQL expressions are used by various databases to perform actions. Examples of some SQL statements include INSERT INTO,

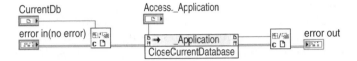

FIGURE 8.58

SELECT, DELETE, and ORDER BY. The CurrentDb reference is then passed out
of the VI.

Finally, the database and the refnums are closed in the Close Database.vi. The
block diagram for this VI is displayed in Figure 8.58. The CloseCurrentDatabase
method is executed to close the pointer to the current database that is open in
Microsoft Access.

Other procedures are available if you want to write to databases through Lab-
VIEW. One alternative is to utilize Microsoft DAO (Data Access Objects) in your
LabVIEW VIs. DAO objects are used to represent and access a remote database and
perform the necessary actions on the database. These objects use COM and can be
called through LabVIEW using Automation and the ActiveX container. ADOs
(ActiveX Data Objects) are also available to perform similar actions with databases.

Another alternative available is the SQL Toolkit, an add-on package that can be
purchased to perform database operations on various commercial databases. The
toolkit simplifies the process of creating databases, inserting data, and performing
queries. The SQL Toolkit comes with drivers for many popular databases. It requires
you to configure your ODBC (Open Database Connectivity) settings by specifying
a data source. This DSN (Data Source Name) is then used as a reference or pointer
to the database. The DSN is a user-defined name that represents the database that
will be used. This user-defined name is used in the SQL Toolkit VIs to perform the
necessary actions programmatically. The toolkit comes with several template VIs
that can be customized to get you up and running quickly with a database.

8.9.13 Instrument Control Using ActiveX

The examples to this point have shown how to insert objects in your code and to
control other applications. We will now discuss a method for controlling a piece of
test equipment using ActiveX. In reality, you are not really controlling the instrument
directly; you are controlling an application that controls the instrument. For this
example we will discuss the Agilent 89601A Vector Signal Analysis software.

The Agilent 89601A software is designed to demodulate and analyze signals.
The inputs to the software can come from simulated hardware, the Advanced Design
System (ADS), or from actual test equipment. The 89601A software can act as a
front end for test equipment like the 89441 vector signal analyzer (VSA), the PSA
series spectrum analyzers, the Infiniium oscilloscopes and 89600 VXI-based vector
signal analysis hardware.

In this example the 89601A software is controlling an 89441 VSA. There is a
lot involved in setting up this instrument for taking measurements on modulated
signals. Here we are describing only one VI used to set the frequency parameters
of the instrument. The first step is verifying there is no error input. An error input

will cause the code to exit without opening any references. In order to begin using the software you will need to open a reference to the 89601A. Once the reference to the application is open you will need to open the measurement property. Using the measurement property reference you can open a reference to the frequency property. Through this property all the necessary frequency properties such as center frequency, span, number of points and window are now available.

The center frequency and span properties require only numeric inputs. The number of points is also a numeric input, but for the ease of programming and configuration, the LabVIEW input for the number of points is an enumerated type control. The enumerated type contains the valid frequency point values: 51, 101, 201, 401, 801, 1601, and 3201. This enumerated type is converted to a string to pull out the selected value, and then is converted to a number for the property.

The window property requires an enumerated type input. The text of the enumerated type is not very user friendly. For this reason we created a separate enumerated type control for user input that will be converted to the appropriate 89601 enumerated type. The input options are Flat Top, Gaussian, Hanning, and Uniform. The index of the selected user enumerated type control is used as an input to an index array function. The other input for this function is an array constant containing the matching 89601 enumerated type values. The output is what is written to the Window property. Finally, the references for application are all closed before exiting. The code diagram of this VI is shown in Figure 8.59.

As you can see, there is no difference between automating functions with a Word document and automating the control of the 89441 VSA using the 89601A software. Obviously this was a simple example. To fully automate control of the instrument there are many settings such as the measurement type, sweep mode and averaging. All functions can be automated in the same way the frequency settings were written.

8.9.14 INSTRUMENT CONTROL USING .NET

In the last example we discussed a method for controlling a piece of test equipment using ActiveX. For newer equipment and software applications, the interface will likely be .NET. One such example of an instrument that can be controlled through a .NET interface is the Agilent ESG series vector signal generator. As always, standard communications to the instrument are available through GPIB and LAN connections. In order to extend the capabilities of the ESG to be able to capture waveforms, playback waveforms and fade signals, the processing for the generator was moved to the PC. Through the use of a PCI card installed on the host computer and a cable connecting the PCI card to the generator, the ability to manipulate the waveform data is now possible on the PC using Agilent's Baseband Studio software.

For this example we will be focusing on using the signal generator to fade a radio frequency (RF) signal. In RF receiver design and in faded signal analysis it is important to be able to simulate real-world scenarios without costly field testing. The results will also be more repeatable. You may be asking yourself what a faded signal is. One example is when a signal from a cellular base station is sent to a receiver (cell phone); the signal can take many paths to get there. This occurs when

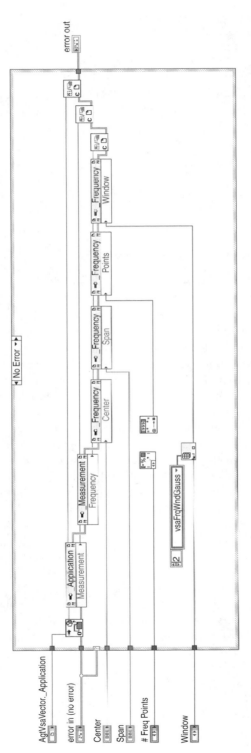

FIGURE 8.59

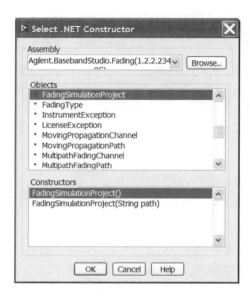

FIGURE 8.60

obstructions are encountered in the path between the tower and the receiver. The receiver will receive all these different paths causing the received signal amplitude and phase to fluctuate when the signals add constructively and destructively. There is also the matter of the receiver being in motion. These factors are all a part of fading. For this example, knowledge of fading is not needed, other than for the terminology used.

The first step in automating the faded signal generation is to create the Project. In .NET a constructor is used to open a reference to the application. A constructor node needs to be put on the code diagram. The constructor can be found in the .NET palette. When the constructor node is placed on the code diagram a dialog is opened for selecting the constructor. This dialog box is shown in Figure 8.60. Agilent.BasebandStudio.Fading is chosen from a list of registered .NET assemblies. The objects for this assembly are then populated below. The Fading Simulation Project is selected from the Objects section. When the object is selected, the available constructors show up in the bottom box in the dialog. Here we select the Fading Simulation Project.

Now that the constructor is linked we can select from a list of methods to call. We could open an existing project, save a project or apply the settings. To start we choose to open a new project. From here we are doing all of our operations on this new fading simulation project. The open project code is shown in Figure 8.61. An

FIGURE 8.61

issue that was observed when using this fader application with LabVIEW involved the reference timing out. If there is no activity between the LabVIEW application and the fading software for an extended period of time, the reference would close. In order to get around this problem, code was inserted at the main level of the application to query the fader enable state every 5 minutes. This effectively kept the connection open until the application was closed out.

From this point on, the programming required is similar to manually setting up the profile in the application window. To start we need to add a fading channel and define the paths. For this we must first add the channel. The Add Channel method is derived by first creating a channel property from the Simulation property. To add the channel we first need to use a constructor node to create a Multipath Fading Channel reference. This reference is used as an input to the Add Channel method. It is also an input to the Fading Channel paths property node. For this example there are 12 possible paths for the channel. We will need to use another constructor node to create the path reference. The code required to creating a new fading channel with 12 paths is shown in Figure 8.62.

Now that the paths have been created, we will need to enter information into each path (if used). To do this we will first need to obtain a reference to a specific path. This is done by opening a Fading Simulation Channel reference and selecting (get_Item) the desired channel. The resulting reference allows us to create a path reference. With the path reference we can select (get_Item) the specific path we want a reference to. The resulting reference is a generic path reference. In order to be able to select the needed properties and methods, the reference needs to be typecast to a more specific class. There is a function to do this in the .NET palette. The function takes the generic reference as an input. The other input is the type you are converting to. This is generated using a constructor node. The multipath fading path constructor is chosen. This will creates the correct reference for modifying the needed path parameters. The code to obtain the specific path is shown in Figure 8.63.

Now that we have a reference we need to configure the settings of the faded signal. Settings for speed and loss are entered as needed to create the desired profile. The remaining setup will not be discussed here as the implementation is fairly straightforward. The parameters for the faded path can be set using properties and methods just like any other .NET or ActiveX application.

8.9.15 CONTROLLING LABVIEW FROM OTHER APPLICATIONS

It is time to discuss using other languages to control LabVIEW. From time to time it may be desirable to reuse good code you wrote in LabVIEW to expand the abilities of an application that was written in another language. Control of LabVIEW VIs is performed through OLE automation. We will be using the same ActiveX interfaces that we have been working with for the last two chapters. As we mentioned earlier, OLE interfaces are a superset of ActiveX interfaces. Both technologies rely upon COM's specification.

We will be controlling LabVIEW through Microsoft Word. Visual Basic for Applications (VBA) is built into several Microsoft applications.

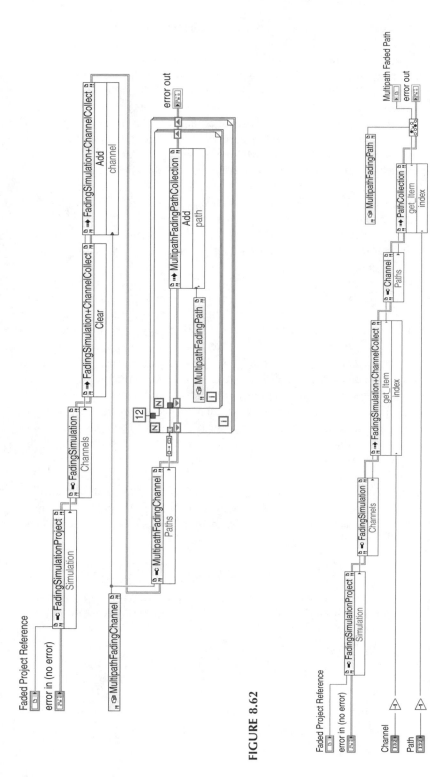

FIGURE 8.62

FIGURE 8.63

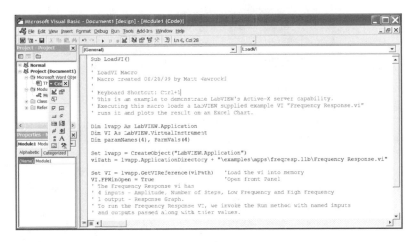

FIGURE 8.64

The first issue we learn about quickly is that Visual Basic and Visual Basic for Applications have no idea where LabVIEW's type library is. By default, you will not be able to use Intellisense to parse through the commands available. This is easily fixed; we need to import a reference to LabVIEW, and then we can get into the Intellisense feature that makes VBA a great language to write scripts with. Start by bringing up Microsoft Word. Make the Visual Basic toolbar active by right-clicking in an empty space next to the Word menu. Select Visual Basic, and a floating toolbar will make an appearance. The first available button is the Visual Basic Editor button. Click this button to start up the VBA editor. The VBA editor is shown in Figure 8.64. When performing this type of Word programming, we STRONGLY recommend that you do not append macros to the Normal template. The Normal template is the standard template that all Word documents begin with. Adding a macro to this template is likely to set virus detection software off every time you create a document and hand it over to a co-worker. This template is monitored for security reasons; it has been vulnerable to script viruses in the past. Right-click on Document 1 and add a code module. Code modules are the basic building blocks of executable code in the world of Visual Basic. Visual Basic forms are roughly equivalent to LabVIEW's VIs; both have a front panel (called a form in Visual Basic) and a code diagram (module in Visual Basic). This new module will be where we insert executable code to control LabVIEW.

Now that we have a module, we will enable Intellisense to locate LabVIEW's objects. Under the tools menu, select References. Scroll down the list of objects until you locate the LabVIEW type library (depends on which version of LabVIEW you are running). Click on this box, and hit OK. The reference box is shown in Figure 8.65.

LabVIEW exposes two main objects for other applications. The first is the Application object, which exposes methods and properties global to LabVIEW, and the Virtual Instrument object, which gives access to individual VIs. In order to have an object pointing to the LabVIEW application you need to create an object. Objects

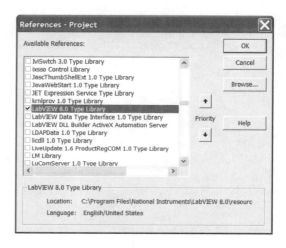

FIGURE 8.65

in VBA are created with the DIM statement, with the format DIM <name> as <object>. When you type "DIM view as…" you will get a floating menu of all the objects that VBA is currently aware of. Start by typing "L" and LabVIEW should appear. The word "LabVIEW" is simply a reference to the type library and not a meaningful object. When the word LabVIEW is highlighted in blue in the pull-down menu, hit the Tab key to enter this as your selection. You should now see DIM view as LabVIEW. Hitting the Period key will enable a second menu that you maneuver through. Select the word "Application" and we have created our first object of type LabVIEW application. Here is a helpful hint that gave the authors fits while trying to make a simple VBA script work: just because you have created a variable of type LabVIEW application does not mean the variable contains a LabVIEW application! Surprise! You need to inform VBA to set the contents of the variable, and you need the CreateObject method to do this. This will cause VBA to open LabVIEW and create the OLE link between the applications. Now we are in business to open a VI. Dimension a new variable of type LabVIEW.VirtualInstrument. This is, again, an empty variable that expects to receive a dispatch interface pointer to a real VI. We can do this with a simple Set command. The Set command needs to look like "Set vi = view.GetViReference("Test.VI")." This command will instruct LabVIEW to load the VI with this name. The line of code we gave did not include a path to the VI. Your application will need to include the path. The GetViReference returns a dispatch interface pointer to the variable. The interface pointer is what VBA uses to learn what methods and properties the Virtual Instrument object exposes.

Now that we have our channel to LabVIEW and a VI loaded into memory, we can set front panel controls and run the application. LabVIEW does not provide a mechanism to supply the names and types of front panel controls that exist on a particular VI. You need to know this information in advance.

There are a couple of methods that we can use to read and set the controls for a front panel. We can set them individually using a Set Control method. Set Control requires two properties; a string containing the name of the control to set and a variant with the control value. This is where variants can be a pleasure to deal with. Because variants can contain any type, we can pass integers, strings, arrays of any dimension, and clusters as well. Did we say you can pass clusters out of LabVIEW code? Yes, we did. Unlike dealing with CINs and DLLs, the OLE interface treats clusters like arrays of variants. Popping up on the cluster and selecting Cluster Order will tell you the order in which the cluster items appear.

Higher performance applications may take issue with all of the calls through the COM interface. To set all parameters and run the VI at the same time, the Call method can be invoked. Call takes two arguments; both are arrays of variants. The first array is the list of controls and indicators and the second array is the data values for controls and the location to store the indicators' values when the call is done. Call is a synchronous method, meaning that when you execute a Call method, the function will not return until the VI has completed execution.

To call VIs in an asynchronous method, use Run, instead. The Run method assumes that you have already set the control values in advance and requires a single Boolean argument. This argument determines if the call will be synchronous or asynchronous. If you do not want this function call to hold up execution of the VBA script, set this value to "true." The VI will start executing and your function call will return immediately.

LabVIEW exposes all kinds of nifty methods and properties for dealing with external applications. It is possible to show front panels and find out which subVIs a VI reference is going to call. You can also find out which VIs call your current VI reference as a subVI. Configuring the preferred execution subsystem is also possible if you are in a position to change the threading setup of LabVIEW. This is discussed in Chapter 9.

To find out more information on what methods and properties LabVIEW exposes, consult LabVIEW's online reference. There is complete documentation there for interfacing LabVIEW to the rest of the COM-capable community.

8.9.16 UNDERSTANDING ACTIVEX ERROR CODES

The ActiveX and the COM subsystems have their own sets of error codes. Virtually every ActiveX function has a return type that LabVIEW programmers cannot see. In C and C++, functions can have only one valid return type. COM functions generally use a defined integer called an HRESULT. The HRESULT is a 32-bit number that contains information about how well the function call went. Similar to LabVIEW error codes, there are possibilities for errors, warnings, informative messages, and no error values. The first two bits of an HRESULT contain the type of message. If the leading bit of the number is set, then an error condition is present. The lower 16 bits contain specific error information. In the event an ActiveX call goes south on you, you can get access to the HRESULT through the error cluster. The cluster will contain a positive error condition, the HRESULT will be the error code, and some descriptive information will be in the error description.

If you have access to Microsoft Visual C++, the error code can be located in winerror.h. Winerror.h lists the errors in a hexadecimal format. To locate the error in the file, you must convert the decimal error code to hex value and perform the search. The dispatch interface error (0x8000FFFF) is one of the more commonly returned error types, a "highly descriptive" error condition that is used when none of the other codes applies. Debugging ActiveX controls can be challenging because the error codes may be difficult for LabVIEW programmers to locate.

Some error codes are specific to the control. This will be made obvious when the string return value in the error cluster has information regarding where to find help. As a general troubleshooting rule, it helps when documentation for the control is available, and possibly the use of Visual Basic. LabVIEW has a custom implementation of the ActiveX specification and does not always behave the same as Visual Basic and Visual C++. We have seen cases in custom ActiveX controls where LabVIEW's interface was more accepting of a control than Visual Basic or the ActiveX test container supplied with Visual C++.

When troubleshooting an ActiveX problem, the obvious issue is that it is impossible to see inside the control. The first step is obvious: check the documentation! Verify all inputs are within spec, and that all needed properties are set. A number of controls have requirements that certain properties be initialized before a Method call can be successful. For example, in the Winsock driver, a call to connect must be made before a string can be written. Obvious, yes, but one author forgot to do this when testing the driver. After several reloads of LabVIEW and rebooting the system, the error was identified.

One thing to understand when looking at error codes is that Automation Open usually does not fail. The reason for this is that this VI will report problems only if the OCX could not be loaded. An ActiveX control is simply a fancy DLL. The act of calling Automation Open loads this DLL into LabVIEW's memory space (consult the multithreading chapter for more information on memory space). Close usually will not generate errors either. The Automation Close does not unload the DLL from memory, which came as a surprise to the authors. ActiveX controls keep a reference count; the Automation Close VI calls the Release function, but does not eliminate the control from LabVIEW's memory space. It is possible to get a control into an odd state, and exiting and restarting LabVIEW may resolve problems that occur only the second time you run a VI. Unfortunately, the only method we have found to force an ActiveX control out of LabVIEW's memory space is to destroy LabVIEW's memory space. The solution you need to look for is what properties/methods are putting the control into an odd state and prevent it from happening.

National Instruments had to supply its own implementation of the ActiveX specification. This is not a bad thing, but it does not always work identically to Microsoft's implementation in Visual Basic. The second recommended step of escalation is to try equivalent code in Visual Basic. When writing the MAPI driver, we discovered that example code written in LabVIEW just did not work. We tried the equivalent code in Visual Basic and, surprise! It also did not work. Reapplying the first rule, we discovered that we were not using the control correctly.

At another time, I was developing an ActiveX control in Visual C++. While hacking some properties out of the control, I did not pay attention to the modifications I made in the interface description file. The control compiled and registered correctly; LabVIEW could see all the controls and methods, but could not successfully set any properties. When I tested the control in Visual Basic and C++, only a few of the properties were available or the test application crashed. It appears that when National Instruments implemented the ActiveX standard it prioritized LabVIEW's stability. LabVIEW would not crash when I programmed it to use a bad control, but Visual Basic and C++ test applications crashed. The moral to this story is directed to ActiveX control developers: do not develop ActiveX controls and test them only in LabVIEW. LabVIEW processes interface queries differently than Visual Basic's and C++'s generated wrappers (.tlh files or class wizard generated files).

Programmers who do not "roll their own" controls can omit this paragraph. If you are using controls that you have programmed yourself, make sure the control supports a dual interface. Recall that LabVIEW only supports late binding. If the Idispatch interface is not implemented or behaves differently than the Iunknown interface, the control can work in Visual Basic, but will not work the same (if at all) in LabVIEW.

If a control does not work in either LabVIEW or Visual Basic, it is possible that the documentation for the control is missing a key piece of information. You will need to try various combinations of commands to see if any combinations work. This is not the greatest bit of programming advice, but at this point of debugging a control there are not many options left. Be aware of odd-looking error codes. If you get an error such as hex code 0x8000FFFF, varying the order of commands to the control is probably not going to help. This error code is the dispatch interface error and indicates that the control itself probably has a bug.

As a last resort, verify the control. If you cannot get a control to work properly in Visual Basic or LabVIEW, some sanity checks are called for. Have you recently upgraded the control? We experienced this problem when writing the Microsoft Agent driver. Apparently, Registry linkages were messed up when the control was upgraded from Version 1.5 to Version 2.0. Only the Automation Open VI would execute without generating an error. If possible, uninstall and reinstall the control, rebooting the machine between the uninstall and reinstall. It is possible that the Registry is not reflecting the control's location, interface, or current version correctly. Registry conflicts can be extremely difficult to resolve. We will not be discussing how to hack Registry keys out of the system because it is inherently dangerous and recommended as a desperate last resort (just before reinstalling the operating system).

If none of the above methods seem to help, it is time to contact the vendor (all options appear to now be exhausted). The reason for recommending contacting the vendor last is some vendors (like Microsoft, for example) will charge "an arm and a leg" for direct support. If the problem is not with the component itself, vendors such as Microsoft will ask for support charges of as much as $99/hr of phone support.

8.9.17 ADVANCED ACTIVEX DETAILS

This section is intended to provide additional information on ActiveX for programmers who are fairly familiar with this technology. LabVIEW does not handle interfaces in the same manner as Microsoft products such as Visual C++ and Visual Basic. Some of these differences are significant and we will mention them to make ActiveX development easier on programmers who support LabVIEW. Intensive instruction as to ActiveX control development is well beyond the scope of this book. Some information that applies strictly to LabVIEW is provided.

The three major techniques to develop ActiveX controls in Microsoft development tools are Visual Basic, Visual C++'s ATL library, and Visual C++'s MFC ActiveX Control Wizard. Each of the techniques has advantages and disadvantages.

Visual Basic is by far the easiest of the three tools in which to develop controls. Visual Basic controls will execute the slowest and have the largest footprint (code size). Visual Basic controls will only be OCX controls, meaning Visual Basic strictly supports ActiveX. Controls that do not have code size restrictions or strict execution speed requirements can be written in Visual Basic.

The ActiveX Template Library (ATL) is supported in Visual C++. This library is capable of writing the fastest and smallest controls. Learning to develop ATL COM objects can be quite a task. If a developer chooses to use custom-built components for a LabVIEW application, budgeting time in the development schedule is strongly recommended if ATL will be used as the control development tool. ATL is capable of developing simple COM objects that LabVIEW can use. LabVIEW documentation lists only support for ActiveX, but ATL's simple objects can be used in LabVIEW. Simple objects are the smallest and simplest components in the ATL arsenal of component types. Simple objects are DLLs, but we have used them in LabVIEW. The interesting aspect of simple controls is that they will appear in LabVIEW's list of available controls. They do not appear in Visual Basic's list of available controls; they must be created using Visual Basic's CreateObject command.

The MFC support for ActiveX controls will generate standalone servers or .ocx controls. When designing your own controls, MFC support may be desirable, but developing the object itself should be done with the ATL library. Going forward, ATL seems to be the weapon of choice that Microsoft is evolving for ActiveX control development. As mentioned, it is possible to get the support of the MFC library built into an ATL project.

Components that will be used by LabVIEW must be built with support for dual interfaces. Single interface controls will not work well in LabVIEW. Visual C++ will support the dual interface by default in all ActiveX/COM projects. A word of caution is to not change the default if your control will be used in LabVIEW. We have previously mentioned that LabVIEW only supports late binding. Controls that only have one interface will not completely support late binding.

ActiveX controls have two methods used to identify properties and methods: ID number and name. LabVIEW appears to address properties and methods by name, where Microsoft products use ID. This tidbit was discovered when one of the authors hacked a property out of a control he was working on. This property happened to have ID 3. LabVIEW could see all the properties in the control, but Visual Basic and C++ could see only the first two properties in the list. The Microsoft products were using ID numbers to identify properties and methods, but LabVIEW was using names. The lesson to learn here is that you need to exercise caution when hacking Interface Description Language (.idl) files.

It is possible to write components that accept LPDISPATCH arguments. LPDISPATCH is a Microsoft data type that means Dispatch Interface Pointer. LabVIEW can handle these types of arguments. When an LPDISPATCH argument appears in a method, it will have the coloring of an ActiveX refnum data type. All of the aggregated controls shown in this chapter use LPDISPATCH return types. You, as the programmer, will need to make it clear to users what data type is expected. The workaround for this type of issue is to have the control accept VARIANTS with LPDISPATCH contained inside. This will resolve the problem.

BIBLIOGRAPHY

.Net Framework Essentials, 3rd ed. Thuan Thai and Hoang Q. Lam. O'Reilly and Associates Inc., Sebastopol, 2003, ISBN 0596005059.

Understanding ActiveX and OLE: A guide for developers and managers. David Chapell. Microsoft Press, Redmond, 1996, ISBN 1572312165.

G Programming Reference, National Instruments, Austin, TX, 1999.

9 Multithreading in LabVIEW

This chapter discusses using multithreading to improve LabVIEW applications' performance. Multithreading is an advanced programming topic, and its effective use requires the programmer to possess a fundamental understanding of this technology. LabVIEW provides two significant advantages to the programmer when working with multitasking and multithreading. The first advantage is the complete abstraction of the threads themselves. LabVIEW programmers never create, destroy, or synchronize threads. The second advantage is the data flow model used by LabVIEW. This model provides G a distinct advantage over its textual language counterparts because it simplifies a programmer's perception of multitasking. The fundamental concept of multitasking can be difficult to visualize with text-based languages.

Multithreading, as a technology, has been around for quite awhile. Windows 95 was certainly not the first operating system to deploy multithreading, but it was a key driver to get this technology widely deployed.

Multithreading added a new dimension to software engineering. Applications can perform multiple tasks somewhat simultaneously. A classic example of an application that has added multithreading is a word processor such as Microsoft Word or Open Office. Both applications use multithreading to perform spell-checking while Word also offers background grammar validation. The threads added to perform this task allow the application to perform these tasks while the user is typing. Now archaic versions, such as Word 6.0 for Windows 3.1, or modern versions, such as the word processors commonly found on PDAs, cannot do this because they run only one task at a time; a user would have to stop typing and select Check Spelling. The first six sections of this chapter provide the basic knowledge of multithreading. This discussion focuses on definitions, multitasking mechanics, multithreading specific problems, and information on various thread-capable operating systems. Once the basic mechanics of multithreading have been explained, we will explore a new variant of this: Hyper-Threading (HT) technology.

A brief section on multithreading myths is presented. The impact of multithreading on applications is misunderstood by a number of programmers. Section 9.6 explains precisely what the benefits of multithreading are. Many readers will be surprised to learn that multithreading does little to increase the speed of an application. Multithreading does provide the illusion that sections of an application run faster.

The last three sections of this chapter are devoted to the effective use of multithreading in LabVIEW. A strategy to estimate the maximum number of useful threads will be presented. The focal point of this section is using subroutine VIs to maximize application performance. The use of threads adds a new dimension of benefits to both subroutine VIs and DLLs.

9.1 MULTITHREADING TERMINOLOGY

The following terminology will be used throughout this chapter. Programmers who require additional information on any of these topics should consult the chapter bibliography.

9.1.1 WIN32

Win32 is an Application Programming Interface (API) that is used by Microsoft's 32-bit operating systems: Windows XP home and office, Windows 2000, Windows NT, Windows 98, Windows 95, and Windows NT. The common API for programming also makes the operating systems look comparable to the programmer. Win32 replaced the Win16 API used in Windows 3.1, and in turn is being replaced by the Win64 API.

Windows XP and NT are designed to perform differently than Windows 95, 98, and ME. The primary focus of the Widows 9x line was to be as backward-compatible as possible with Windows 3.1. The Windows NT–derived product line, which includes XP, is designed to be as stable as possible. The differences in behavior of Windows NT line and Windows 9x are usually not obvious to most users. For example, many users do not know that Windows 9x will stop preemptive multithreading for certain Windows 3.1 applications whereas Windows XP will not.

9.1.2 UNIX

UNIX is an operating system that was conceived by AT&T Bell Labs. Like Win32, UNIX is a standard, but the Open Systems Foundation maintains the UNIX standard. Unlike Win32, UNIX is supported by a number of vendors who write and maintain operating systems to this standard. The most popular are Sun Microsystems' Solaris, IBM's AIX, Hewlett Packard's HP-UX, and the various Linux distributions.

Threads are specified in UNIX with the Portable Operating System Interface (POSIX) threads standard (pthread). At a low-level programming interface, pthreads are different than Win32 threads. This does not impact the LabVIEW programmer. Fundamentally, pthreads and Win32 threads operate in the same fashion. Their application-programming interfaces (API) are different, but conceptually, threads are threads.

9.1.3 MULTITASKING

Multitasking simply means to coordinate multiple tasks, which is a familiar concept to the LabVIEW programmer. As a dataflow-based language, LabVIEW has always supported multitasking — even the first edition of LabVIEW which ran on a single-threaded Macintosh OS. Operating systems, such as Windows 3.1 and MacOS, multitask operations such as multiple applications. The act of coordinating multiple

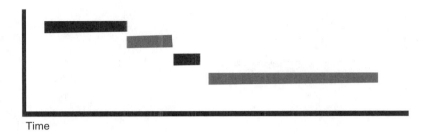

Time

FIGURE 9.1

tasks should not be confused with multithreading; multithreading is fundamentally different, and this section will help to explain why.

The best example of multitasking is you. At work, you have a priority list of tasks that need to get accomplished. You will work on tasks one at a time. You may accomplish only some parts of a task, but not all before working briefly on another task. This is simple multitasking; you work on one thing for a while and then work on another topic. The process of switching between tasks is not multithreading; it is simply multitasking. The act of deciding how long to work on a task before working on a different one is your scheduling algorithm. This is the principle that Windows 3.1, MacOS, and some real-time operating systems were built on. In "the bad old days," the applications in Windows 3.1 and MacOS would decide how long they would run before surrendering the CPU to another process. These environments often had failures when an application would never decide to return the CPU to the system. Figure 9.1 demonstrates time utilization of a CPU when cooperative multitasking is used.

A simple demonstration of multitasking in LabVIEW is independent While loops. It is important for the reader to clearly understand that multitasking has always been available, and multithreading does not add or subtract from LabVIEW's ability to multitask operations. One of the key questions this chapter hopes to answer for the LabVIEW programmer is when multithreading will be of benefit.

9.1.3.1 Preemptive Multithreading

Taking the previous example of multitasking, you at work, we will explain the fundamental concept of multithreading. Imagine that your hands were capable of independently working. Your right hand could be typing a status report while the left dialed a phone number. Once the left hand completed dialing the number, it began to solder components on a circuit board. If you were capable of talking on the phone, typing your status report, and soldering components at the same time you would be multithreading. Your body is effectively a process, and your hands and mouth are threads of execution. They belong to the same process, but are functioning completely independent of each other.

This is fundamentally what multithreading is doing in a computer program. Each thread has a task it works on regardless of what the rest of the program is doing. This has been a difficult concept for many programmers to grasp. When programming with text-based languages such as C/C++, programmers associate lines of code

Time

FIGURE 9.2

as operating sequentially. The concept that is difficult to grasp is that threads behave as their own little program running inside a larger one. LabVIEW's graphical code allows programmers to visualize execution paths much easier.

Preemptive multithreading uses CPU hardware and an operating system capable of supporting threads of execution. Preemption occurs at the hardware level; a hardware timer interrupt occurs and the CPU takes a thread of execution off the CPU and brings in another one. A lot of interesting things happen with CPUs that support multithreading operating systems. First, each program has a memory map. This memory map is unique to each application. The memory maps are translated into real memory addresses by hardware in the CPU. When preemption occurs, the timer in the CPU informs the CPU to change maps to the operating system's management. The operating system will then determine which thread is next to run and inform the CPU to load that thread's process memory map. Since the first edition of this book, all operating systems that LabVIEW supports are now capable of supporting multithreading. The act of scheduling threads and processes will be explained in Section 9.2. Figure 9.2 shows the timelines for multiple processes.

9.1.4 KERNEL OBJECTS

Kernel objects are small blocks of memory, often C structures, that are owned by the operating system. They are created at the request of programs and are used to protect sections of memory. Later in this chapter the need for protection will be clearly explained, but a short definition of this term is provided now.

9.1.5 THREAD

Before we begin describing a thread, a few terms used for programs must be quickly defined. A program that has one thread of execution is a single-threaded program and must have a call stack. The call stack retains items like variables and what the next instruction is to be executed. C programmers will quickly associate variables placed on the stack as local variables. In some operating systems, the program will also have a copy of CPU registers. CPU registers are special memory locations inside the CPU. The advantage of using CPU registers is that the CPU can use these memory locations significantly faster than standard memory locations. The disadvantage is that there are relatively few memory locations in the registers. Each thread has its own call stack and copy of CPU registers.

Effectively, a thread of execution is a miniature program running in a large, shared memory space. Threads are the smallest units that may be scheduled time for execution on the CPU and possess a call stack and set of CPU registers. The call stack is a first-in-first-out (FIFO) stack that is used to contain things like function calls and temporary variables. A thread is aware of only its own call stack. The registers are loaded into the CPU when the thread starts its execution cycle, and pulled out and loaded back into memory when the thread completes its execution time.

9.1.6 PROCESS

The exact definition of a process depends on the operating system, but a basic definition includes a block of memory and a thread of execution. When a process is started by the operating system, it is assigned a region of memory to operate in and has a list of instructions to execute. The list of instructions to begin processing is the "thread of execution." All applications begin with a single thread and are allowed to create additional threads during execution.

The process's memory does not correlate directly to physical memory. For example, a Win32 process is defined as four gigabytes of linear address space and at least one thread of execution. The average computer has significantly less memory. Banished are the bad old days of DOS programming where you had to worry about concepts such as conventional memory, extended memory, and high memory. These memory types still exist, but are managed by the operating system. The region created for a process is a flat, or linear, range of memory addresses. This is referred to as a memory map, or protected memory. The operating system is responsible for mapping the addresses the program has into the physical memory. The concept behind protected memory is that a process cannot access memory of other processes because it has no idea where other memory is. The operating system and CPU switch memory in and out of physical memory and hard disk space. Memory mapped to the hard disk is referred to as "virtual memory" in Windows and "swap space" in UNIX.

A process also has security information. This information identifies what the process has authorization to do in the system. Security is used in Windows XP professional and UNIX, but is generally not used in Windows XP Home or older Win32 operating systems such as Windows ME or 98.

9.1.7 APPLICATION

An application (or program) is a collection of processes. With concepts like Distributed Computing Environments (DCE) and the .NET environment, clustered applications, and even basic database applications, are not required to execute in one process, or even on one computer. With the lower cost of computers and faster network speeds, distributed computing is feasible and desirable in many applications. Applications that use processes on multiple machines are "distributed applications."

As an example of a distributed application, consider LabVIEW. With .NET support and VI Server functionality, VIs can control other VIs that are not resident and executing on the same computer. .NET is discussed in Chapter 8.

9.1.8 PRIORITY

Every process and thread has a priority associated with it. The priority of the process or thread determines how important it is when compared to other processes or threads. Priorities are integer numbers, and the higher the number, the more important the process. Process priorities are relative to other processes while thread priorities are relative only to other threads in the same process. In other words, two running programs such as a LabVIEW executable running at priority 16 is less important than an operating system task operating a priority level 32. LabVIEW 8 by default has four threads. If you were to set one of these thread's personal thread priorities to 32 it would outrank the other LabVIEW process threads, but since the LabVIEW process still is out ranked by the operating system task the high priority LabVIEW thread will still be out ranked by all threads in the operating system process. LabVIEW programmers have access to priority levels used by LabVIEW. Configuring LabVIEW's thread usage will be discussed in Section 9.7.4.

9.1.8.1 How Operating Systems Determine which Threads

Both Win32 and POSIX have 32-integer values that are used to identify the priority of a process. The implementation of priority and scheduling is different between Win32 and pthreads. Additional information on both specifications appears in Sections 9.3 and 9.4.

9.1.9 SECURITY

Windows XP and UNIX systems have security attributes that need to be verified by the operating system. Threads may operate only within the security restrictions the system places on them. When using .NET objects security permissions can become an issue. The default security attributes associated with an application are the level of permissions the user who started the application has. System security can limit access to files, hardware devices, and network components.

Windows .NET support adds additional security in the form of "managed code." It needs to be understood that .NET overlays onto the core Windows operating system. If the operating system does not support security features (Windows 9X and ME), then it does not fully support .NET's security features.

9.1.10 THREAD SAFE

Programmers often misunderstand the term "thread safe." The concept of thread-safe code implies that data access is atomic. "Atomic" means that the CPU will execute the entire instruction, regardless of what external events occur, such as interrupts. Assembly-level instructions require more than one clock cycle to execute, but are atomic. When writing higher-level code, such as LabVIEW VIs or C/C++ code, the concept of executing blocks of code in an atomic fashion is critical when multiple threads of execution are involved. Threads of execution are often required to have access to shared data and variables. Atomic access allows for threads to have complete access to data the next time they are scheduled.

The problem with preemption is that a thread is removed from the CPU after completion of its current machine instruction. Commands in C can be comprised of dozens of machine-level instructions. It is important to make sure data access is started and completed without interference from other threads. Threads are not informed by the operating system that they were preemptively removed from the CPU. Threads cannot know shared data was altered when it was removed from the CPU. It is also not possible to determine preemption occurred with code; it is a hardware operation that is hidden from threads. Several Kernel objects can be used to guarantee that data is not altered by another thread of execution.

Both UNIX Pthreads and Win32 threads support semaphores and mutexes. A Mutual Exclusion (mutex) object is a Kernel object that allows one thread to take possession of a data item. When a thread requires access to a data item that is protected by a mutex, it requests ownership from the operating system. If no other threads currently own the mutex, ownership is granted. When preemption occurs, the owning thread is still shifted off the CPU, but when another thread requests ownership of the mutex it will be blocked. A thread that takes possession of a mutex is required to release the mutex. The operating system will never force a thread to relinquish resources it has taken possession of. It is impossible for the operating system to determine if data protected by the mutex is in a transient state and would cause problems if another thread were given control of the data.

A semaphore is similar to a mutex, but ownership is permitted by a specified number of threads. An analogy of a semaphore is a crowded nightclub. If capacity of the club is limited to 500 people, and 600 people want to enter the club, 100 are forced to wait outside. The doorman is the semaphore, and restricts access to the first 500 people. When a person exits, another individual from outside is allowed to enter. A semaphore works the same way.

Mutexes and semaphores must be used in DLLs and code libraries if they are to be considered thread-safe. LabVIEW can be configured to call DLLs from the user interface subsystem, its primary thread, if it is unclear that the DLL is thread safe. A programmer should never assume that code is thread safe; this can lead to very difficult issues to resolve.

9.2 THREAD MECHANICS

All activities that threads perform are documented in an operating system's specification. The actual behavior of the threads is dependent on a vendor's implementation. In the case of Windows, there is only one vendor, Microsoft. On the other hand, Linux and UNIX have a number of vendors who implement the standard in slightly different ways. Providing detailed information on operating system-specific details for all the UNIX flavors is beyond the scope of this book.

Regardless of operating system, all threads have a few things in common. First, threads must be given time on the CPU to execute. The operating system scheduler determines which threads get time on the CPU. Second, all threads have a concept of state; the state of a thread determines its eligibility to be given time on the CPU.

9.2.1 THREAD STATES

A thread can be in one of three states: active, blocked, or suspended. Active threads will be arranged according to their priority and allocated time on the CPU. An active thread may become blocked during its execution. In the event an executing thread becomes blocked, it will be moved into the inactive queue, which will be explained shortly.

Threads that are blocked are currently waiting on a resource from the operating system (a Kernel object or message). For example, when a thread tries to access the hard drive of the system, there will be a delay on the order of 10 ms for the hard drive to respond. The operating system blocks this thread because it is now waiting for a resource and would otherwise waste time on the CPU. When the hard drive triggers an interrupt, it informs the operating system it is ready. The operating system will signal the blocked thread and the thread will be put back into a run queue.

Suspended threads are "sleeping." For example, in C, using the Sleep statement will suspend the thread that executed the command. The operating system effectively treats blocked and suspended threads in the same fashion; they are not allowed time on the CPU. Both suspended and blocked threads are allowed to resume execution when they have the ability to, when they are signaled.

9.2.2 SCHEDULING THREADS

The operation of a scheduling algorithm is not public knowledge for most operating systems, or at least a vendor's implementation of the scheduling algorithm. The basic operation of scheduling algorithms is detailed in this section.

The operating system will maintain two lists of threads. The first list contains the active threads and the second list contains blocked and suspended threads. The active thread list is time-ordered and weighted by the priority of the thread in the system. The highest priority thread will be allowed to run on the CPU for a specified amount of time, and will then be switched off the CPU. This is referred to as a "round robin" scheduling policy.

When there are multiple CPUs available to the system, the scheduler will determine which threads get to run on which CPU. Symmetric Multiprocessing (SMP) used in Windows XP Professional allows for threads of the same process to run on different CPUs. This is not always the case. A dual-CPU machine may have threads of different processes running on the pair of CPUs, which is determined by the scheduling algorithm. Some UNIX implementations allow only a process's threads on a single CPU.

The blocked/suspended queue is not time-ordered. It is impossible for the operating system to know when a signal will be generated to unblock a thread. The scheduling algorithm polls this list to determine if any threads have become available to execute. When a thread becomes unblocked and has higher priority than the currently running thread, it will be granted control of the CPU.

9.2.3 CONTEXT SWITCHING

The process of changing which thread is executing on the CPU is called "context switching." Context switching between threads that are in the same process is relatively fast. This is referred to as a "thread context switch." The CPU needs to offload the current thread's instruction pointer and its copy of the CPU registers into memory. The CPU will then load the next thread's information and begin executing the next thread. Since threads in the same process execute in the same protected memory, there is no need to remap physical memory into the memory map used by the process.

When context switching occurs between threads of different processes, it is called a "process context switch." There is a lot more work that is required in a process context switch. In addition to swapping out instruction pointers and CPU registers, the memory mapping must also be changed for the new thread.

As mentioned in Section 9.2.2, when a thread becomes signaled (eligible to run) and has a higher priority than the currently running thread, it will be given control of the CPU. This is an involuntary context switch. This is a potential problem for LabVIEW programmers. Section 9.5 will discuss multithreading problems such as starvation and priority inversion that can be caused by poorly-designed configurations of LabVIEW's thread pools. Configuring the threads that LabVIEW uses is discussed in Section 9.7.4.

9.3 WIN32 MULTITHREADING

The Win32 model expands the capabilities of the Win16 model. Threading and security are two of the features that were added to Windows. Win32 systems branched with the release of Windows NT 3.51. Windows 2000 and XP leveraged the threading model used in Windows NT. Windows NT and Windows 95 operate differently when threads are considered, as stated-modern Windows operating systems use the Windows NT model for thread handling. The API for these operating systems are the same. Windows 98 and ME ignore several attributes given to a thread. For example, Windows 98 and ME ignore security attributes. Windows 98 and ME were designed for home usage so thread/process security is not used in these operating systems. Issues involving security and permissions are usually not a problem in LabVIEW applications. Some issues may surface when using the VI server or .NET objects.

Windows XP will always operate using preemptive multithreading. The primary design goal of Windows NT branch was stability. Windows 95x platforms and 98 were designed with backward compatibility in mind. Windows 9x will run as cooperative multithreaded environments, similar to Windows 3.1. This is a consideration for LabVIEW programmers to remember when they anticipate a need to support users running legacy applications such as Microsoft Office Version 4 or interfacing with very old test applications. Older versions of Microsoft Word (Version 6.0) may be given cooperative multitasking support so they feel at home in a Windows 3.1 environment. Legacy DOS applications may behave the same way. Either way, when Windows 9x systems drop into a cooperative multitasking mode it will not inform the user that this decision was made. Performance degradation may be seen in Windows 9x environments when legacy applications are run. If users raise issues related to performance of LabVIEW applications on Windows 9x systems, ask if any other applications are being used concurrently with LabVIEW.

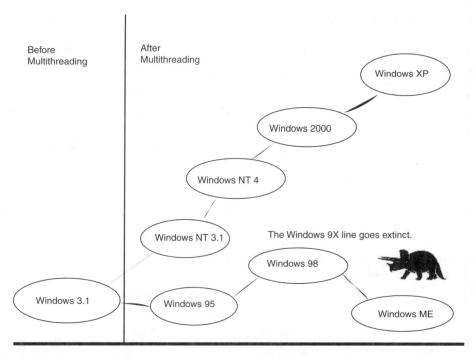

FIGURE 9.3

Another difference between the Windows 9x and NT branch is hardware accessibility. Windows 9x allows applications to access hardware directly. This allows drivers to have faster access to devices such as GPIB cards and DAQ boards. Windows NT's model for stability forbids direct hardware access. Drivers must be written specifically to act as operating system components.

A useful chart showing the progression of Microsoft's operating systems is shown in Figure 9.3 Windows operating systems did a "split" with Windows 3.1 and Windows NT 3.1. The Windows 95 branch has essentially died out. All operating systems since Windows ME have been on the Windows NT branch.

9.4 PTHREADS

Many topics on UNIX multithreading are well beyond the scope of this book. The chapter bibliography lists a number of sources that contain additional information. One of the difficulties in writing about UNIX is the number of vendors writing UNIX. The POSIX standard is intended to provide a uniform list of design requirements for vendors to support. This does not translate directly to uniform behavior of UNIX operating systems. Vendors write an operating system that conforms to their interpretation of the specification.

Priority and scheduling are different for Pthreads; Pthreads have defined scheduling policies: round robin; first-in, first-out; and others. The FIFO policy lets a thread execute until it completes its execution or becomes blocked. This policy is multitasking by any other name, because there is no preemption involved. The round-robin policy is preemptive multithreading. Each thread is allowed to execute for a maximum amount of time, a unit referred to as a "quantum." The time of a quantum is defined by the vendor's implementation. The "other" policy has no formal definition in the POSIX standard. This is an option left up to individual vendors. Pthreads expand on a concept used in UNIX called "forking." A UNIX process may duplicate itself using a fork command. Many UNIX daemons such as Telnet use forking. Forking is not available to the Win32 programmer. A process that generates the fork is called the Parent process, while the process that is created as a result of the fork command is referred to as the Child process. The Child process is used to handle a specific task, and the Parent process typically does nothing but wait for another job request to arrive. This type of multitasking has been used for years in UNIX systems.

As an example of forking, consider a Telnet daemon. When a connection request is received, the process executes a fork. The Telnet daemon is replicated by the system and two copies of the daemon are now running in memory; the Parent process and Child process. The Parent continues to listen on the well-known Telnet port. The Child process takes over for the current Telnet connection and executes as an independent process. If another user requests another Telnet session, a second Child process will be spawned by another fork. The Parent process, the original Telnet daemon, will continue to listen to the well-known Telnet port. The two Child processes will handle the individual Telnet sessions that users requested.

Forking has both advantages and disadvantages. The major disadvantage to forking is that Parent and Child processes are independent processes, and both Parent and Child have their own protected memory space. The advantage to having independent memory spaces is robustness; if the Child process crashes, the Parent process will continue to execute. The disadvantage is that since Parent and Child are independent processes, the operating system must perform process context switches, and this requires additional overhead.

9.5   MULTITHREADING PROBLEMS

This section will outline problems that multithreading can cause. This is an important section and will be referenced in Section 9.6, "Multithreading Myths." Some of these problems do not occur in LabVIEW, but are included for completeness. Many of these problems can be difficult to diagnose because they occur at the operating system level. The OS does not report to applications that cause these problems. Some issues, such as race conditions and deadlock, cannot be observed by the operating system. The operating system will not be able to determine if this is normal operation or a problem for the code.

It is important to understand that any multitasking system can suffer from these problems, including LabVIEW. This section is intended to discuss these problems relative to LabVIEW's multithreading abilities. As a programmer, you can create race conditions, priority inversion, starvation, and deadlock in LabVIEW's VIs when working with VI priorities. We will identify which problems you can cause by poor configuration of LabVIEW's thread counts.

9.5.1 RACE CONDITIONS

Many LabVIEW programmers are familiar with the concept of a "race condition." Multithreading in general is susceptible to this problem. Fortunately, when writing LabVIEW code, a programmer will not create a race condition with LabVIEW's threads. A race condition in multithreaded code happens when one thread requires data that should have been modified by a previous thread. Additional information on LabVIEW race conditions can be found in the LabVIEW documentation or training course materials. LabVIEW execution systems are not susceptible to this problem. The dedicated folks at National Instruments built the threading model used by LabVIEW's execution engine to properly synchronize and protect data. LabVIEW programmers cannot cause thread-based race conditions; however, there is still plenty of room for LabVIEW programmers to create race conditions in their own code.

Thread-based race conditions can be extremely hazardous. If pointer variables are involved, crashes and exception errors are fairly likely. This type of problem can be extremely difficult to diagnose because it is never made clear to the programmer the order in which the threads were executing. Scheduling algorithms make no guarantees when threads get to operate.

9.5.2 PRIORITY INVERSION

A problem occurs when two threads of different priority require a resource. If the lower-priority thread acquires the resource, the higher-priority process is locked out from using the resource until it is released. Effectively, the higher-priority process has its priority reduced because it must now wait for the lower-priority process to finish. This type of blocking is referred to as Priority Inversion. The resource could be in the application, or the resource could be external to the application, such as accessing a shared file on the hard drive. Internal resources include things like variables. External resources include accessing the hard drive, waiting on .NET components, and waiting on Kernel-level operating system components such as mutexes and semaphores.

Priority inversion will degrade an application's performance because high-priority threads do not execute as such. However, the program might still execute properly. Inversion is a problem that can be caused by a LabVIEW programmer who errantly alters the priority level of LabVIEW's thread pools. Priority levels of "Normal" should be used, and this will prevent priority inversion problems. When threads have no active work to do, they will be blocked.

Most LabVIEW applications should not require modification of priority levels of threads. Errant priority levels can cause a number of problems for a LabVIEW application. An example of when a programmer would consider adjusting the priority level of a subsystem is a high-speed data acquisition program. If the DAQ subsystem required a majority of the CPU time, then a programmer may need to raise priority levels for the DAQ subsystem. If a common queue were used to store data brought in from the DAQ card, priority inversion would occur. VIs performing numerical processing or data display that are resident in the user subsystem will execute with lower priority than the DAQ threads. This becomes a problem when the user interface threads have access to the queue and spend a lot of time waiting for permission to execute. The DAQ threads trying to put data into the queue become blocked until lower-priority interface threads complete. The lesson to learn here is important: when priority inversion occurs, the high-priority threads end up suffering. This type of problem can seriously impact application performance and accuracy. If the DAQ card's buffer overflows because the DAQ subsystem was blocked, application accuracy would become questionable.

9.5.3 STARVATION

Starvation is essentially the opposite of priority inversion. If access to a resource that two threads need is for very short periods of time, then the higher-priority thread will almost always acquire it before the lower priority thread gets the opportunity to do so. This happens because a higher-priority thread receives more execution time. Statistically, it will get the resource far more often than the lower-priority thread.

Like, priority inversion, "starvation" is resolved differently by the Windows NT and 9x branches — recall Windows XP is on the NT branch. Again, like priority inversion, a LabVIEW programmer can cause this multithreading problem. We will discuss prevention of both priority inversion and starvation issues later in this chapter.

If Windows will actively seek to resolve starvation and priority inversion, then why bother to prevent them from happening? The reason you do not want to cause either problem is that both reduce the efficiency of an application. It is poor programming practice to design an application that requires the operating system to resolve its deficiencies.

If there is a valid reason for a thread to execute at a higher priority than others, then the program design should make sure that its execution time is not limited by priority inversion. Lower-priority threads can suffer from starvation. This is highly undesirable because execution time of the higher-priority threads will become limited while Windows allows the lower threads to catch up. A balanced design in thread priorities is required. In multithreaded code, it is often best just to leave all threads at the same priority.

9.5.4 DEADLOCKING

Deadlock is the most difficult multithreading problem that is encountered. Deadlock can occur only when two or more threads are using several resources. For example, there are two threads, A and B. There are also two resources, C and D. If both threads need to take ownership of both resources to accomplish their task, deadlocking occurs when each thread has one resource and is waiting for the other. Thread A acquires resource C, and thread B acquires resource D. Both threads are blocked until they acquire the other resource.

The bad news as far as deadlocking is concerned: Windows and Linux have no mechanism to resolve this type of problem. The operating system will never force a thread to release its resources. Fortunately, deadlocking is highly unlikely to be caused by a LabVIEW programmer. This is a thread-level problem and would be caused by the execution engine of LabVIEW. Once again, the dedicated folks at

National Instruments have thoroughly tested LabVIEW's engine to verify that this problem does not exist. This eliminates the possibility that you, the programmer, can cause this problem.

9.5.5 OPERATING SYSTEM SOLUTIONS

Operating systems try to compensate for starvation and priority inversion. Requiring the operating system to compensate is poor programming practice, but here is how UNIX and Win32 try to resolve them.

Priority inversion is resolved differently by Windows 9x and NT branches. Windows NT-derived systems (Windows XP) will add a random number to the priority of every thread when it orders the active queue. This obviously is not a complete solution, but a well-structured program does not require the operating system to address thread problems.

Windows 9x will actually increase the priority of the entire process. Increasing the priority of the process gives the entire application more of the CPU time. This is not as effective a solution as the one used by NT. The reason for this type of handling in Windows 95 is for backward compatibility. Legacy Win16 programs are aware of only one thread of execution. Elevating the entire process's priority makes Win16 applications feel at home in a Windows 3.1 environment. You may be very hard pressed to find systems out there still using Win16 applications, but the support for these types of applications still exist in the Win9x operating systems.

Since the first edition of this book, the entire Win9x line of operating systems has largely fallen out of use and Microsoft has ended support for the operating systems in this line.

9.6 MULTITHREADING MYTHS

This section discusses some common myths about multithreading. We have heard many of these myths from other LabVIEW programmers. Multithreading is one of the exciting new additions to the language that first appeared in LabVIEW 5.0; unfortunately, it is also one of the most dangerous. The following myths can lead to performance degradation for either a LabVIEW application or the entire operating system. The single biggest myth surrounding multithreading is that it makes applications run faster. This is entirely false, and the case is outlined below.

9.6.1 THE MORE THREADS, THE MERRIER

The first myth that needs to be addressed is "More threads, better performance." This is just not in line with reality; application speed is not a function of the number of running threads. When writing code in languages like C++, this is an extremely dangerous position to take. Having a thread of execution for every action is more likely to slow an application down than speed it up. If many of the threads are kept suspended or blocked, then the program is more likely to use memory inefficiently. Either way, with languages like C++ the programmer has a lot of room to cause significant problems.

LabVIEW abstracts the threading model from the programmer. This threading model is a double-edged sword. In cases like the number of threads available, LabVIEW will not always use every thread it has available, and the programmer will just waste memory. When applications are running on smaller computers, such as laptops where memory might be scarce, this could be a problem.

The rule of thumb for how many threads to have is rather vague: do not use too many. It is often better to use fewer threads. A large number of threads will introduce a lot of overhead for the operating system to track, and performance degradation will eventually set in. If your computer has one CPU, then no matter what threading model you use, only one thread will run at a time. We'll go into customizing the LabVIEW threading model later in this chapter. We will give guidelines as to how many threads and executions systems programs might want to have.

9.6.2 MORE THREADS, MORE SPEED

This myth is not as dangerous as "The more threads the merrier" but still needs to be brought back to reality. The basic problem here is that threads make sections of a program appear to run at the same time. This illusion is not executing the program faster. When the user interface is running in its own thread of execution, the application's GUI will respond more fluidly. Other tasks will have to stop running when the user interface is updating. When high performance is required, a nicely updated GUI will degrade the speed that other subsystems operate.

The only true way to have a program run faster with multiple threads is to have more than one CPU in the computer. This is not the case for most computers, and, therefore, threads will not always boost application performance. Most modern Intel-based machines for both Windows and Linux have Hyper-Threading capable processors. We will discuss Hyper-Threading in the next section. Oddly enough, Hyper-Threading does not always improve the performance of a multithreaded application. Windows 9x and XP home do not support Symmetric Multiprocessing, and will make use of only one CPU in the system.

Later in this chapter, we will show how performance gains can be made without making significant changes to the thread configuration. As a rule of thumb, threads are always a difficult business. Whenever possible, try to tweak performance without working with threads. We will explain the mechanics of multithreading in the next sections, and it will become clear to the reader that multithreading is not a "silver bullet" for slow applications.

9.6.3 MAKES APPLICATIONS MORE ROBUST

There is nothing mentioned in any specification regarding threads that lends stability to an application. If anything, writing multithreaded code from a low level is far more difficult than writing single-threaded code. This is not a concern for the LabVIEW programmer because the dedicated folks at National Instruments wrote the low-level details for us. Writing thread-safe code requires detailed design and an intimate understanding of the data structures of a program. The presence of multiple threads does not add to the stability of an application. When a single thread in a multithreaded application throws an exception, or encounters a severe error, it will crash the entire process. In a distributed application, it could potentially tear down the entire application. The only way to ensure the stability of a multithreaded application is a significant amount of testing. National Instruments has done this testing, so there is no need to spend hours doing so to verify your threading models.

9.6.4 CONCLUSION ON MYTHS

Multithreading gives new abilities to LabVIEW and other programming languages, but there is always a price to be paid for gains. It is important to clearly understand what multithreading does and does not provide. Performance gains will not be realized with threads running on a single-processor system. Large numbers of threads will slow an application down because of the overhead the system must support for the threads to run. Applications' graphical interfaces will respond more fluidly because the graphics subsystems' threads will get periodic time to execute. Intensive computational routines will not block other sections of code from executing.

9.7 HYPER-THREADING

Before we can begin a discussion of how Hyper-Threading works, we will need to review the "instruction pipeline" that a processor uses. Modern processors use a variety of buffers, queues, caches of memory, architecture states, stack points, and registers. These items tend to place limits on the performance of a given processor. For example, cache memory is always a performance tradeoff. It is always desirable for a larger memory cache near the processor, but the cache's speed performance is inversely related to its size; make the cache small and it runs faster; make it larger and it runs more slowly.

Intel has added a new technology to their processors; marketing literature refers to it as Hyper-Threading. Hyper-Threading's technical name is *simultaneous multi-threading*. Hyper-Threading adds a second set of CPU resources to a processor to make it appear in most respects to have two processors.

Hyper-Threading is not the same as Symmetric Multiprocessing. SMP is two completely independent processors; a Hyper-Threading processor has two cores that have to share some key resources of the processor. In order to take advantage of Hyper-Threading, the processor, BIOS, chip set, and operating system all need to be capable of supporting Hyper-Threading. Windows XP was the first Microsoft operating system with Hyper-Threading support.

Hyper-Threading implements the concepts of logical and physical processors. An HT processor is one physical processor that implements two logical processors. The operating system will present the logical processors as two processors to the system. If you look at Window's task manager on an HT-enabled machine, the performance tab will show two CPUs. The two logical processors are not two completely independent processors — elements of a CPU core are either replicated, shared, or partitioned. The HT processor will start up and shut down the second logical core as needed to improve performance of the entire chip.

Items the CPU core has to share between the two cores are the cache memory and the out of order execution engine. The out of order execution engine is available on P6 or later Intel processors. This engine reviews running code for dependencies. If two instructions do not conflict on resources running on the CPU, it will run them at the same time in parallel. The instructions and their results have to be re-sorted into order after the parallel operation. The memory cache is a set of high speed memory near the processor that allows for the CPU to have quick access to relevant memory. These two items are shared by the CPU cores in HT processors. As the two cores have to share, the performance of these segments can be reduced per thread running.

The buffers to reorder instructions from the out of order execution engine and queues in the core are partitioned. There is one set of each, and parts are allocated specifically to one of the cores. When the performance of a single core would be greatly improved by allocating all these resources to the single core, the HT processor can shut down the second virtual processor which partitions all of the queues and recording buffers to the single core.

Each core will have its own instruction pointer and thread state. In the event the second core is shut down, this hardware will not be used by the processor.

A Hyper-Threading microprocessor has a common process context between the cores. In other words, a HT-enabled system can run two threads belonging to the same process together. Multiple processes cannot be run simultaneously because every process has an independent memory mapping — the hardware to do the map to real memory translation is shared by both cores. In general, there will always be operating system specific tasks running. Of importance to the LabVIEW programmer will be processes that handle I/O such as Internet-related traffic. Windows XP and Linux do not permit direct hardware access. Anytime I/O is being used, it can be assumed that a driver running in a different mode of operation is in use.

LabVIEW 8 by default uses four threads of execution. Hyper-Threading will not always run two threads at a time. If three of the threads are idle or blocked, the processor will shut down the second core and provide all processor resources to the active thread. When two threads are running, they will run a bit slower at the same time compared to the execution speed of one thread running with all CPU resources. It is possible to combine SMP and HT. In the event you have a dual processor machine with two HT-enabled processors the system will operate with four logical processors.

The average thread of execution uses roughly 35% of a processor's capabilities. Hyper-Threading adds enough logic circuitry to the processor to allow for a second thread to run and take advantage of the remaining resources. The relatively small increase in chip hardware for generally better performance makes Hyper-Threading a very useful technology. HT-enabled processors do not significantly increase the power consumption of a processor, which gives a significant improvement in performance relative to power consumption. In environments with a lot of hardware running, there are realizable savings in electrical usage for the machines and the HVAC systems that support them.

9.8 MULTITHREADED LABVIEW

Fundamentally, the dataflow operation of LabVIEW is not impacted by multithreading. The real differences are abstracted from the programmer. In this section we

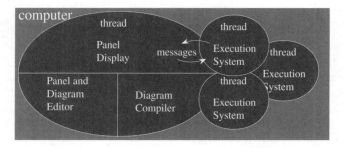

FIGURE 9.4

discuss the architecture of multithreaded LabVIEW, the main run queue, and how to configure the thread systems. Understanding how threads interact with the Lab-VIEW subsystems will allow a programmer to use threading effectively. Threads in one subsystem may access VIs you thought were operating in another subsystem. The topic is somewhat abstract, and this section intends to clarify the interactions of VIs and threads. Once the reader understands the subsystem architecture and main run queue, an introduction to thread configuration will be presented.

9.8.1 EXECUTION SUBSYSTEMS

The various activities that LabVIEW performs are handled by six subsystems. The LabVIEW subsystems are User, Standard, I/O, DAQ, Other 1, and Other 2. The original design of LabVIEW 5.0 used these subsystems to perform tasks related to the system's name. This rigid partitioning did not make it into the release version of LabVIEW 5.0, and tasks could run in any LabVIEW subsystem. This subsystem architecture still exists in LabVIEW 8. Figure 9.4 depicts LabVIEW broken up into its constituent subsystems.

Each subsystem has a pool of threads and task queue associated with it. Lab-VIEW also maintains a main run queue. The run queue stores a priority-sorted list of tasks that are assigned to the threads in the subsystem. A LabVIEW subsystem has an "array" of threads and priorities. The maximum number of threads that can be created for a subsystem is 40; this is the maximum of 8 threads per priority and 5 priority levels. Section 9.8.4 discusses thread configuration for subsystems. Figure 9.5 shows a subsystem and its array of threads.

The User subsystem is the only subsystem that is required for LabVIEW to run, because all other subsystems are optional to running LabVIEW. Configuring thread counts and priorities for subsystems is covered in 9.8.4. The User subsystem maintains the user interface, compiles VIs, and holds the primary thread of execution for LabVIEW. When a DLL or code fragment with questionable thread safety is run, the User subsystem is where it should always be called. LabVIEW can be configured to run in as a single threaded application. The single thread runs and it rides on the User subsystem.

The Standard subsystem was intended to be the default subsystem for LabVIEW executable code. If you are interested in keeping dedicated execution time to the user interface, assign the main level VIs to this subsystem. This will guarantee that

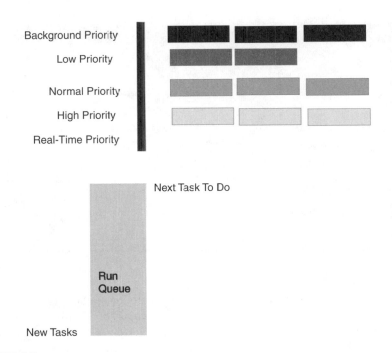

Background Priority

Low Priority

Normal Priority

High Priority

Real-Time Priority

Next Task To Do

Run
Queue

New Tasks

FIGURE 9.5

the User subsystem threads have plenty of time to keep the display updated. Like
the User subsystem, the Standard subsystem can be configured to run with an array
of threads.

The DAQ subsystem was originally intended to run data acquisition-specific
tasks. It is currently available to any VI. The I/O subsystem was intended for VXI,
GPIB, Serial, and IP communications (TCP and UDP). This design is interesting
when the programmer considers using the VISA communication suite. VISA is a
communications subsystem, and a dedicated thread pool is certainly a good idea.
Remember that having dedicated threads guarantees some amount of execution time.
On single CPU systems without Hyper-Threading, there is still a "borrow from Peter
to pay Paul" issue, but communications is fundamental to many LabVIEW applications and justification for guaranteed execution time is sometimes appropriate. The
priority levels of a subsystem's threads relative to the other subsystem threads would
serve as a rough gauge of the amount of execution time available. Other 1 and Other
2 were intended for user-specified subsystems. Again, these two subsystems can be
used for any purpose desired. Most applications do not need to assign VIs to specific
subsystems. A simple litmus test to decide if a VI should be dedicated to a subsystem
is to write a one-paragraph description of why the VI should only be executed in a
specific subsystem. This is a simple Description Of Logic (DOL) statement that
should be included in any application design. Description of Logic was discussed
in Chapter 4. If the DOL cannot describe what the benefits of the assignment are,
then the assignment is not justified. Valid reasons for dedicating a set of VIs to a
specific subsystem include the need to guarantee some amount of execution time.

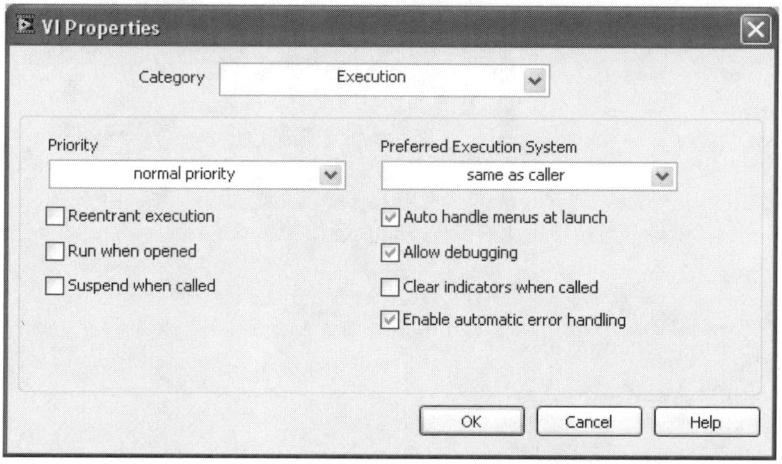

FIGURE 9.6

If a programmer decides to modularize the execution of an application, then assign-ment of a VI to a subsystem can be done in VI Setup as shown in Figure 9.6.

When a new VI is created, its default priority is normal and the system is "same as caller." If all VIs in a call chain are listed as run under "same as caller," then any of the subsystems could potentially call the VI. The default thread configuration for LabVIEW is to have four threads per subsystem with normal priority. The subsystem that calls a VI will be highly variable; during execution it is impossible to determine which threads were selected by the scheduler to execute. When a VI is assigned to a particular subsystem, only threads belonging to the specified subsystem will execute it.

Now that a simple definition of the LabVIEW subsystems has been presented, let's consider threads. Subsystem threads execute in a round-robin list and are scheduled by the operating system. When a VI is listed to execute in a specific subsystem, only threads assigned to that subsystem can execute it. As an example, a VI, test_other1.vi is permitted to be executed only by the Other 1 subsystem. When test2_other2.vi is executed and told to call test_other1.vi, the thread that is executing test2_other2.vi will block. The data flow at the call is blocked until a thread from the Other 1 subsystem is available to execute it. This is up to the scheduler of the operating system, and also one of the points where errant priorities can cause priority inversion or starvation. LabVIEW cannot directly switch to a thread in Other 1 to execute the thread. Only the operating system can decide which thread gets to execute when. Other 1 threads will not be considered special to the operating system, and they must wait in the thread queue until it is their turn to execute.

When assigning VIs to particular subsystems, use caution when assigning thread priorities. If VIs are called by two subsystems, there should not be large differences in the priorities of the threads belonging to these subsystems. An example is if the subsystem Other 1 has 2 threads at "Above Normal" priority and Other 2 has a "Background" priority thread. If a VI called in subsystem Other 1 calls a subVI in Other 2, not only does the thread in Other 1 block, it has to wait for a background priority thread in Other 2 to get scheduled time to execute! This is a simple example of priority inversion that can be caused by a LabVIEW programmer. If Other 1 and Other 2 were both executing with "Normal" priority, the scheduling algorithm would have mixed the scheduling of the threads and no priority inversion would occur. The thread in Other 1 would have needed to wait on the thread context switches, but those delays are relatively minor.

Another threading problem that can be caused by errant priority settings is starvation. Consider the following case: VIs other1.vi, other2.vi, and io.vi. The Other 1 VI is listed with "time-critical" priority level, Other 2 VI is listed with "background" priority, and io.vi is listed "same as caller." Same as caller allows a VI to execute in any subsystem with the same priority thread that executed the calling VI. Both other1.vi and other2.vi need to call io.vi. Since other1.vi is running in a subsystem with a time-critical priority thread, it is going to get a significant amount of execution time compared with other2.vi. Access to io.vi will be granted to other1.vi far more often than other2.vi. Other2.vi will become starved because it does not get enough access to io.vi. Scheduling algorithms are notoriously unforgiving when they schedule threads. If a thread is available to execute, it will get put in the active list based entirely on its priority. If a thread with low priority is always available, it will still make the active list, but will always be near the bottom.

9.8.2 THE RUN QUEUE

LabVIEW maintains several run queues consisting of a main run queue and a run queue for each subsystem. A run queue is simply a priority-ordered list of tasks that are executed. When a VI is executed, the LabVIEW execution engine determines which elements in the block diagram have the needed inputs to be executed. The engine then orders inputs by priority into the run queues. The run queue is not strictly a first-in first-out (FIFO) stack. VIs have priorities associated with them (the default priority is "normal"). After execution of each element, the run queue is updated to reflect elements (subVIs or built-in LabVIEW functions, such as addition, subtraction, or string concatenation) that still need to be executed. It is possible for VIs to take precedence over other VIs because they have higher priority. Wildly changing VI priorities will likely result in performance issues with LabVIEW. One key point to understand is that VI priorities are in no way associated with thread priorities. The thread that pulls it off the run queue will execute a VI with high priority. If the thread has background priority, a slow thread will execute the high-importance VI. The execution engine will not take a task away from one thread and reassign it to a thread with a more suitable thread priority.

To help illustrate the use of run queues, consider the in box on your desk. With LabVIEW 4.1 and earlier, there was a single in box, and each time a VI was able to run it would be put into the in box and you would be able to grab the task and perform it. LabVIEW 5.0 and later have multiple run queues which equate to one in box for each of your hands. As tasks become available they get put into an appropriate in box and the hand that corresponds to that in box can grab the task. Another comparison for Windows programmers is the message pump. Windows 3.1 had a single message loop. Each time an event occurred, such as a mouse click, the

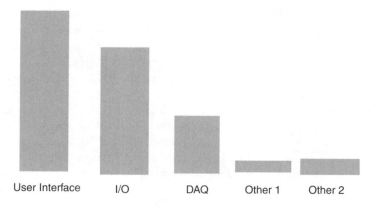

User Interface   I/O   DAQ   Other 1   Other 2

FIGURE 9.7

event would be put into the system-wide message loop. All applications shared the same message loop, and numerous problems were caused because some applications would not return from the message loop and would lock up windows. Windows 95 and later have message loops for each application. Every application can continue to run regardless of what other applications are doing. We have the same benefit in LabVIEW. VIs assigned to different subsystems can now operate with their own threads and their own run queues.

A thread will go to the run queue associated with its subsystem and pull the top task off the list. It will then execute this task. Other threads in the subsystem will go to the run queue and take tasks. Again, this is not a FIFO stack — the highest-priority VI will be handed to a thread. This leaves a lot of room for both performance tweaking and performance degradation. Priorities other than "normal" should be the exception, and not status quo.

When a VI is configured to run only in a particular subsystem, it will be put onto the run queue of that particular subsystem, and then the VI must wait for a thread belonging to this system to be assigned to execute it. This can cause performance degradation when thread priorities are different between subsystems. Section 9.8.2 discusses thread configurations in multisubsystem LabVIEW applications.

Figure 9.7 shows the run queues that are developed when a LabVIEW code diagram is run. When VIs are scheduled to run in the "same as caller subsystem," a thread belonging to the subsystem will end up executing the VI. A subtle point is that if there are multiple threads belonging to the subsystem, there are no guarantees which thread will execute which VI.

9.8.3 DLLs in Multithreaded LabVIEW

Special care must be taken when working with DLLs in multithreaded LabVIEW. DLLs can potentially be called from several different threads in LabVIEW. If the DLL has not been written to handle access by multiple threads, it will likely cause problems during execution. Recall thread safe in Section 9.1.10. If mutexes, semaphores, or critical sections are not explicitly designed into the DLL, then it is not guaranteed to be thread safe.

Threading problems are not always obvious. Several million calls may need to be made to the DLL before a problem surfaces. This makes troubleshooting thread problems extremely difficult. Bizarre program operation can be extremely difficult to troubleshoot, especially when the code can execute for days at a time without failure. When working with DLLs, and crashes occur only occasionally, suspect a thread problem.

When writing C/C++ code to be called by LabVIEW, you need to know if it will possibly be called by multiple threads of execution. If so, then you need to include appropriate protection for the code. It is fairly simple to provide complete coverage for small functions. The Include file that is needed in Visual C++ is process.h. This file contains the definitions for Critical Sections, Mutexes, and Semaphores. This example is fairly simple and will use Critical Sections for data protection. A Critical Section is used to prevent multiple threads from running a defined block of code. Internal data items are protected because their access is within these defined blocks of code. Critical Sections are the easiest thread protection mechanism available to the Windows programmer, and their use should be considered first.

```
#include <process.h>

//Sample code fragment for CriticalSections to be
used by a //LabVIEW function.

CRITICAL_SECTION Protect_Foo

Void Initialize_Protection(void)

{

  INITIALIZE_CRITICAL_SECTION(&Protect_Foo);

}

Void Destroy_Protection(void)

{

  DELETE_CRITICAL_SECTION(&Protect_Foo);

}

int foo (int test)

{

  int special_Value;

  ENTER_CRITICAL_SECTION(&Protect_Foo); //Block
      other threads from accessing

  Special_Value = Use_Values_That_Need_
      Protection(void);

  LEAVE_CRITICAL_SECTION(&Protect_Foo);//Let other
      threads access Special Value, I'm finished.
```

```
    Return special_Value;

}
```

The fragment above does not do a lot of useful work as far as most programmers are concerned, but it does illustrate how easy thread protection can be added. When working with Critical Sections, they must be initialized prior to use. The INITIALIZE_CRITICAL_SECTION must be called. The argument to this function is a reference to the Critical Section being initialized. Compile errors will result if the Critical Section itself is passed. The Critical Section must also be destroyed when it will no longer be used. Initialization and destruction should be done at the beginning and end of the application, not during normal execution.

Using a Critical Section requires that you call the Enter and Leave functions. The functions are going to make the assumption that the Critical Section that is being passed was previously initialized. It is important to know that once a thread has entered a Critical Section, no other threads can access this block until the first thread calls the Leave function.

If the functions being protected include time-consuming tasks, then perhaps the location of the Critical Section boundaries should be moved to areas that access things like data members. Local variables do not require thread protection. Local variables exist on the call stack of the thread and each thread has a private call stack, which makes local variables completely invisible to other threads.

C++ programmers must also remember that LabVIEW uses C naming conventions. The keyword extern C must be used when object methods are being exposed to LabVIEW. LabVIEW does not guarantee that C++ DLLs will work with Lab-VIEW. In the event you encounter severe problems getting C++ DLLs to operate with LabVIEW, you may have hit a problem that cannot be resolved.

A few additional notes on DLLs being called from LabVIEW before we complete this section.

LabVIEW uses color-coding to identify DLLs that are executed by the user interface thread from DLLs that are listed as "re-entrant." If a DLL call is shown in an orange icon, this identifies a DLL call that will be made from the User Interface subsystem. If the standard off-yellow color is shown, it will be considered re-entrant by LabVIEW and will allow multiple threads to call the DLL. Library call functions default to User Interface subsystem calls. If a DLL was written to be thread safe, then changing this option to reentrant will help improve performance. When a User Interface Only DLL call is made, execution of the DLL will wait until the user interface thread is available to execute the call. If the DLL has time-consuming operations to perform, the user interface's performance will degrade.

When working with DLLs of questionable thread safety, always call them from the user interface. When it is known that threading protection has been built into a DLL, make the library call re-entrant. This will allow multiple threads to call the DLL and not cause performance limitations. If you are stuck with a DLL that is not known to be thread safe, be careful when calling the DLL from a loop. The number of thread context switches will be increased, and performance degradation may set in. We did get a tip from Steve Rogers, one of the LabVIEW development team members, on how to minimize the number of context switches for DLLs. This tip works when you are repeatedly calling a DLL from a loop that is not assigned to the user subsystem. Wrap the DLL call in a VI that is assigned to the user subsystem. The thread context switches have been moved to the VI call and not the DLL call. Effectively, this means that the entire loop will execute in the user subsystem, and the number of needed context switches will drop dramatically.

Execution of a DLL still blocks LabVIEW's multitasking. The thread that begins executing the DLL will not perform other operations until the DLL call has completed. Unlike LabVIEW 4.0 and earlier, other threads in the subsystem will continue to perform tasks.

9.8.4 CUSTOMIZING THE THREAD CONFIGURATION

The number of threads in each LabVIEW execution subsystem is specified in the labview.ini file. This information should not be routinely altered for development workstations. The default configuration will work just fine for most applications. The default configuration for a multithreaded platform is four threads of execution per subsystem of normal priority.

In situations where a LabVIEW application is running on a dedicated machine, tweaking the thread configuration can be considered. National Instruments provides a useful VI for configuring the ini file used by LabVIEW: threadconf.vi. The Thread Configuration VI should be used to alter LabVIEW's thread configuration. Vendors are expected to supply tools to easily modify application configurations. The most difficult aspect of using this VI is locating it! The VI can be found in vi.lib\utilities\sysinfo.llb. Figure 9.8 shows the front panel of this VI.

To change the configuration of the thread engine, select Configure. A second dialog box will appear, as shown in Figure 9.8. Each execution subsystem may be configured for up to eight threads per priority. Some readers may feel compelled to review Section 9.5. A plethora of options are available, and the advice is not to alter most of them. In the next section, information on estimating the maximum number of useful threads is presented.

A few words about thread priorities: Time-critical priorities are, in general, hazardous. Consider Windows XP and time-critical threads. Mouse clicks and keypad activity are reported to the operating system; the operating system then sends a message to the application that should receive the user input. In the event that time-critical threads are running, they may take priority over operating system threads. The commands you enter to Quit may never arrive, and the application will have to be killed (end task from task manager).

Background priorities may be useful but, in general, keeping all threads at normal priority is best. If all threads are running on normal priority, none of them will suffer from starvation or priority inversion. The easiest way to avoid complex threading issues is to avoid creating them. This is the biggest caveat in this section; review Section 9.5 for descriptions on priority inversion and starvation. If more threads are available than can be used, LabVIEW's execution engine will allow the thread to continue checking the queue for VIs that belong to its subsystem. This will require a minimal amount of CPU time and avoids thread problems.

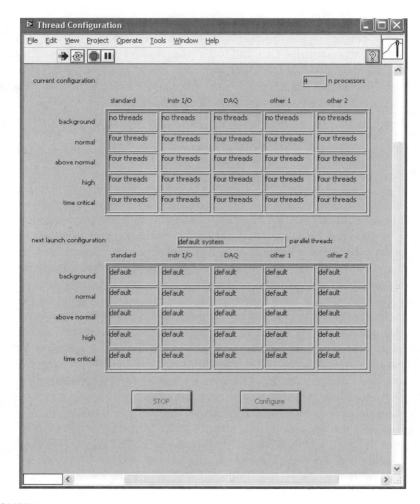

FIGURE 9.8

In a Hyper-Threading environment, the CPU will shut down the second CPU core if only one thread is actively running. Having excessive threads will potentially hinder the CPU from optimizing its performance. LabVIEW will start only the threads specified for the user subsystem when it starts up. The reason is to minimize system resource usage. We do not want to create more threads than necessary because that would cause extra overhead for the operating system to track. LabVIEW defers creation for other subsystems' threads when it loads a VI that is assigned to a subsystem into memory. Performance tuned applications may want to run an empty VI for each subsystem in use during application startup. This allows for the application to create the needed threads in advance of their usage. The LabVIEW INI file specifies the number of threads created for any particular subsystem.

Windows users may see an additional thread created when dialog boxes are used. The dialog boxes for Printers, Save File, Save File As, and Color Dialog are actually

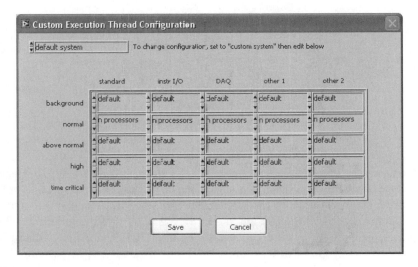

FIGURE 9.9

the same dialog box. This dialog box is called the common dialog and is responsible
for creating that extra thread. This is normal operation for Windows. The common
dialog thread is created by the dialog box, and is also destroyed by the dialog box.
High-performance applications should refrain from displaying dialog boxes during
execution except when absolutely necessary.

When working with the Application Builder, the thread information will be
stored in a file that is located with the executable generated by LabVIEW. The same
rules apply to application builder programs as mentioned above. Normal priority
threads are all that a programmer should ever need; do not create more threads than
an application can use. Recall that the existence of threads translates to performance
gains only when multiple processors are available on the machine.

Another word of caution when configuring the threads LabVIEW uses: since
LabVIEW is an application like any other running on the system, its threads are
scheduled like any other, including many operating system components. When
LabVIEW threads are all high priority, they just might steal too much time from the
operating system. This could cause system-wide inefficiency for the operating system, and the performance of the entire computer will become degraded or possibly
unstable. The fundamental lesson when working with thread priorities is that your
application is not the only one running on the system, even if it is the only application
you started.

9.9 THREAD COUNT ESTIMATION FOR LABVIEW

When discussing thread count, this section will refer exclusively to the maximum
number of useful threads, which is the number of threads that will be of benefit to
the execution systems without any threads being left idle. The minimum number of
useful threads to the LabVIEW execution engine is always one.

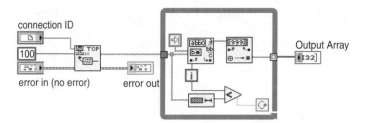

connection ID

100

error in (no error)   error out

Output Array

FIGURE 9.10

Consider the VI code diagram presented in Figure 9.10. The number of useful threads for this VI is one. This VI obviously does not consider events such as errors, fewer than 100 bytes read, or the spreadsheet String-to-Array function. In spite of the fact that it is not well-thought-out, it is an excellent dataflow example. Everything will happen in the VI in a well-established order: the TCP Read function will execute, and then the While loop will execute. There is only one path of execution, and multiple threads will do nothing to optimize the execution of this VI. The fact that the While loop will execute multiple times does not suggest multiple threads can help. It does not matter which thread in which subsystem is currently processing this VI; execution will still happen in a defined order. Thread context switches will do nothing to help here. The lesson learned in this simple example: if order of execution is maintained, multithreading is not going to improve application performance.

Alert readers would have noticed the location of the String Length function. In the While loop it is possible for two threads to perform work, but one will only need to return a string length, which is a very simple function. This is not a significant amount of work for a thread to do. Also, it would be far more efficient to locate the String Length function outside the While loop and feed the result into the loop. When optimizing code for threading, look for all performance enhancements, not just the ones that impact threading potential. Having an additional thread for this example will not improve performance as much as moving the string length outside the While loop.

The VI code diagram presented in Figure 9.11 presents a different story. The multiple loops shown provide different paths of execution to follow. Threads can execute each loop and help maintain the illusion that all loops seem to execute at the same time. If multiple CPUs are involved, application speed will improve. The number of useful threads for this example is three. If the internal operations of the loops are time-intensive operations, then a fourth thread may be desirable. This fourth thread will be added to help support the front panel. If there are graphs or other intensive operations, consider an additional thread for supporting the display. Recall that additional threads will take some time away from other threads.

Now consider the VI code diagram shown in Figure 9.12. There appears to be a single path for this VI to take, but considering that several of the VIs are not waiting on inputs, they can be scheduled to run right away. There are four paths of execution that merge into one at the end of the code diagram. Threads can help here, and the maximum number of useful threads will be equal to the number of paths of execution. In this case, the maximum number is four. If several of these subVIs are

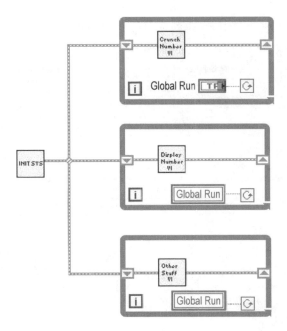

FIGURE 9.11

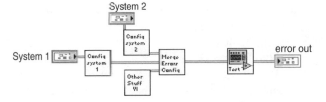

FIGURE 9.12

expected to execute quickly, then they do not require a thread to exist on their behalf, and you should consider reducing the number of threads. Recall that each thread is going to be scheduled and requires some time on the CPU. The lower the thread count, the more execution time per thread.

The code diagram presented in Figure 9.13 is basically a mess. Multiple threads could potentially be beneficial here, but if the operations splattered about the display were modularly grouped into subVIs, then the benefit seen in Figure 9.12 would still exist. You can consider prioritizing the subVIs as subroutines; the benefit of reduced overhead would make the VI run nearly as fast, and a lot of readability will be gained. Section 9.9 describes criteria for using subroutine VIs in multi-threaded LabVIEW.

We have gone through a few simple examples concerning only a single VI. When a large-scale application is going in development, the maximum number of useful threads will probably not skyrocket, but the determination can be much more difficult. An application consisting of 250 subVIs will be time-intensive for this

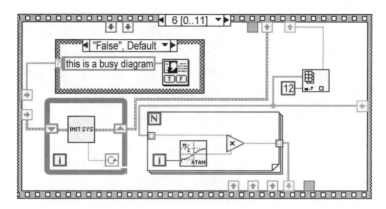

FIGURE 9.13

type of analysis. The programmer's intuition will come into play for application-wide analysis. Also, never forget that at some point, adding threads is not going to make an improvement. Unless you are working with a Hyper-Threading quad-CPU system, having hundreds of threads of execution is not going to buy much in terms of performance.

9.9.1 Same as Caller or Single Subsystem Applications

When attempting to determine the maximum number of useful threads for an application, the maximum number of execution paths for the application must be determined. This can be difficult to accomplish. For example, look at the hierarchy window of a medium to large-scale application. Each branch of a hierarchy window does not equate to one branch of execution. A programmer who is familiar with the functionality of each subVI will have an understanding of the tasks performed in the subVI. Look at subVIs that have descriptions that suggest parallel operations. This is a difficult piece of advice to generalize, but programmers should be familiar with their application. Order of execution may be forced in a number of branches, and this will limit the number of useful threads. Having more threads than necessary will cause minor performance hits. The number of thread context switches that will be incurred when threads are given time will be increased. If the thread configuration includes threads of differing priority, then lower-priority threads may receive little execution time and not be of much help to the application.

The simplest case to analyze is when all threads are running in a single LabVIEW subsystem or all VIs are assigned to a "same as caller" subsystem. Then there are no needed considerations to be made regarding which subsystems require threads. On a system dedicated to running an application of this type, consider modifying the thread configuration so that only one subsystem has threads — the User subsystem.

The following VI code diagram simply demonstrates a main level VI and its three independent loops. Obviously, three threads may support this VI. If the three loops require heavy numerical processing, then a fourth thread may be desired if a lot of display updates are also desired. Since the three subVIs are going to keep busy running numerical calculations, a fourth thread could be brought in for GUI updates. Understand that LabVIEW is not going to allocate a thread to each loop and the fourth to the VI, but there will always be four threads looking into the run queue for a new task. If threads take a lot of time to complete one iteration of a loop, then three threads may periodically become bogged down in a loop. The fourth thread exists to help out when circumstances like this arise. When no intensive GUI updates are required, the fourth thread is not desirable. Additional thread context switches can be avoided to improve performance.

Reconsidering the above example, if one of the loops performs a very simple operation, then reducing the number of threads to two may also be beneficial. Having fewer threads means less work for the operating system to do. This is a judgment call the programmer is going to have to consider. The fundamental trade-off is going to be parallel operation versus operating system overhead. The general guideline is to have fewer threads and minimize overhead. When looking to estimate the number of threads, look for operations that are time consuming. Examples of time-consuming operations are large array manipulation, DLL calls, and slow data communications, such as serial ports.

9.9.2 MULTIPLE SUBSYSTEM APPLICATIONS

Determining how many threads can support a LabVIEW application with VIs running in dedicated subsystems requires additional work. The number of useful threads per subsystem must now be considered. The solution to this problem is to analyze the number of paths of execution per subsystem. Considerations must be made that threads may become blocked while waiting for VIs to be assigned to other subsystems. Again, an additional thread may be considered to help out with display updates. Do not forget that an additional thread will take some time away from other threads in LabVIEW. High-performance applications may still need to refrain from displaying graphs during run-time.

It is still possible to write many multithreading optimized applications without resorting to using multiple subsystems. LabVIEW's configuration allows for a maximum of eight threads per subsystem. When you conclude that the maximum number of useful threads is well beyond eight, then forcing some VIs to execute in different subsystems should be considered. If fewer than nine threads can handle a VI, do not force multiple subsystem execution. Performance limitations could arise with the extra contact switching.

A special case of the multiple subsystem application is a distributed LabVIEW application. Optimization of this application should be handled in two distinct parts. As LabVIEW is executing independently on two different machines, you have two independent applications to optimize. Each machine running LabVIEW has two separate processes and each will have their own versions of subsystems. When the threading model is being customized, each machine should have its own threading configuration. One machine may be used solely for a user interface, and the other machine may be executing test or control code. Consider using the standard configuration for the user interface machine. It is unlikely that sophisticated analysis of the user interface is required. Consider investing engineering time in the more important task of the control code. In situations where each instance of LabVIEW

is performing some hard-core, mission-critical control code, both instances of Lab-
VIEW may have their threading configurations customized.

Your group should deploy a coding standard to indicate information regarding
the subsystem a VI is assigned to. When trying to identify problems in an application,
the subsystem a VI is assigned to is not obvious. A programmer must actively look
for information regarding that. A note above or below the VI should clearly indicate
that the VI has been forced into a subsystem. An alternative to using notes is to color-
code the icon, or a portion of it, to clearly indicate that the VI has been forced into
a nonstandard mode of execution. This will simplify debugging and maintenance.
Multiple subsystem applications will almost always be very large-scale applications;
these types of techniques will simplify maintenance of such large applications.

9.9.3 Optimizing VIs for Threading

When you are writing code for which you would like to have the maximum benefit
of the threading engine, avoid forcing the order of execution whenever possible.
When a VI is coded for tasks to happen in a single-file fashion, the tasks assigned
to the run queue must also be assigned in a single-file fashion. This limits the
ability of the threads to handle tasks because they will always be waiting for a
task to become available. If possible, avoid the use of sequences; they are going
to force an order of execution. Let the error clusters force an order of execution
for things like read and write operations. Operations that are handled, such as
loading strings into a VI and determining the value of some inputs, can be done
in a very parallel fashion. This will maximize the ability of the threads to handle
their jobs. All simple operations will have their data available and will be scheduled
to run.

As an example of maximizing data flow, consider the code diagram in Figures
9.14, 9.15, and 9.16. These three diagrams describe three sequences for a simple

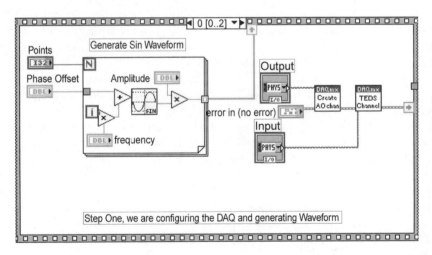

FIGURE 9.14

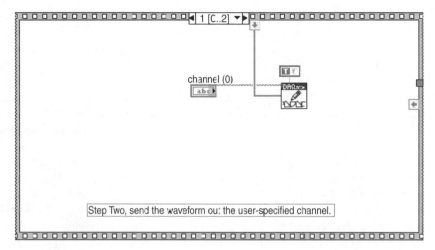

FIGURE 9.15

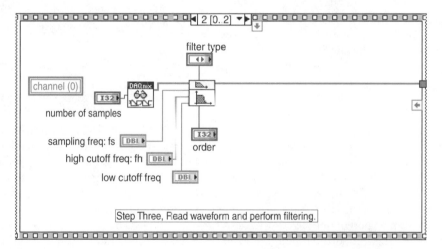

FIGURE 9.16

data acquisition program. The items in the first sequence must be handled and completed before the second can be executed. The second sequence is fairly simple, and the waveform is shipped out. The third sequence reads in a signal and filters it. The DAQ experts may criticize the appearance of this VI, but it serves as an example of how sequences limit the thread's ability to operate.

In the first sequence there are two paths of execution to follow. The first is the generation of the sine waveform to be used. The second path to follow is the Analog Output and Analog Input VIs. Please note that the error cluster forces an order of execution; the Output VI must be executed, then the Input VI. There is some initial loading of values on the wire table that needs to be done. The threads will also handle this.

The second sequence diagram simply sends out the waveform. The inputs here cannot be processed and moved on the wire table until this sequence starts executing. Had this VI been in the first sequence, the constants could have already been shifted in LabVIEW's wire table.

The third sequence reads an input waveform and runs it through a Butterworth filter. Many DAQ experts will argue about the timing delays and choice of a Butterworth filter, but we are putting emphasis on the threading issues. The constants in this sequence also may not be loaded into new sections of the wire diagram until this sequence begins execution.

Let us quickly rethink our position on the number of paths that could be followed in the first sequence. Two was the decided number, one for the signal generation, and one for the Configuration VIs. Recall the Setup VIs have multiple subVIs with the possibility of dozens of internal paths. We are unable to maximize the number of executable paths because the order of execution is strongly forced. The "thread friendly" version is shown in Figure 9.17. Wrapping the Output Generation VI in a sequence was all that was needed to force the Configuration, Generation, and Read functions. The one-step sequence cannot execute until the error cluster output becomes available.

The Configuration VIs are set in parallel with a little VI inserted to add any errors seen in the clusters. This is a handy little VI that is included on the companion CD to this book. The multiple execution paths internal to these VIs are now available to the threading engine.

All constants on the block diagram can be loaded into appropriate slots on the wire table without waiting for any sequences to start. Any of these functions can be encapsulated into subVIs to make readability easier. VIs that easily fit on a 17-in. monitor should not require 46-in. flat-panel displays for viewing after modification.

The lesson of this example is fairly simple: do not force order of execution in multithreaded LabVIEW. If you want to take full advantages of the threading engine, you need to leave the engine a little room to have execution paths. Obviously, some order must exist in a VI, but leave as many execution paths as possible.

This next part of optimization has less to do with threads than the above example, but will stress good programming practice. Polling loops should be minimized or eliminated whenever possible. Polling loops involve some kind of While loop continuously checking for an event to happen. Every time this loop is executed, CPU cycles are burned while looking for an event. In LabVIEW 4.1 and earlier versions, you may have noticed that the CPU usage of your machine ran up to 100%. That is because the polling loop was "tight." Tight loops do very little in a cycle. This allows the loop to complete its execution quickly and take more time. Because there is always something in LabVIEW's run queue to do (run the loop again), LabVIEW appears to be a very busy application to the system. LabVIEW will get all kinds of time from the scheduling algorithms, and the performance of the rest of the system may suffer. In LabVIEW 5.0 the threads that are assigned tasks for the loop will be just as busy, and therefore make LabVIEW again look like a very busy application. Once more, LabVIEW is going to get all kinds of time from the operating system which will degrade performance of the system.

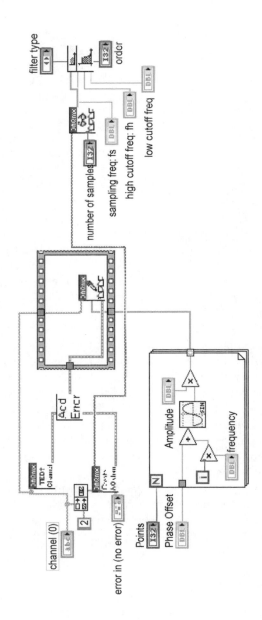

FIGURE 9.17

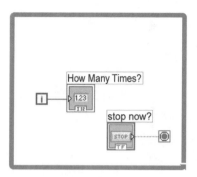

FIGURE 9.18

Figure 9.18 shows a simple loop that increments a counter. This is a short example of a tight loop. There is very little activity going on inside the loop, and millions of iterations happen in very little time. If the value of the loop iterator is wired to a terminal, the execution speed will slow down because of the volume of graphics updates that need to happen. The System Monitor (Windows 95), Task Manager (Windows XP), or an application such as Top or Monitor (UNIX) will show a significant amount of CPU usage. The problem is there isn't much useful happening, but the amount of CPU usage will top out the processor. Other applications will still get time to run on the CPU, but they will not receive as much time because the tight loop will appear to always need time to run when viewed by the operating system scheduler.

Tight loops and polling loops cannot always be avoided. When an application is looking for an external event, polling may be the only way to go. When this is the case, use a Wait Milliseconds command if possible. It is unlikely that every polling loop needs to check a value at CPU speeds. If this is the case, the selection of LabVIEW may be appropriate only on Concurrent PowerMAX systems, which are real-time operating systems. If the event does not absolutely need to be detected, make the loop sleep for a millisecond. This will drastically reduce the CPU utilization. The thread executing the wait will effectively be useless to LabVIEW while sleeping, but LabVIEW's other threads do not need to fight for CPU time while the thread is executing the tight loop.

When waiting on events that will be generated from within LabVIEW, using polling loops is an inefficient practice. Occurrence programming should be used for this. This will prevent any of LabVIEW's threads from sitting in polling loops. The code that is waiting on an occurrence will not execute until the occurrence is triggered. No CPU cycles are used on nonexecuting code.

9.9.4 USING VI PRIORITIES

The priority at which the VI executes may also be considered when configuring VIs. The VI execution priority is not directly related to the threads that execute the VI, or the priority of the threads that execute the VI. Figure 9.19 shows the configuration of a VI. The priority levels assigned to a VI are used when LabVIEW schedules

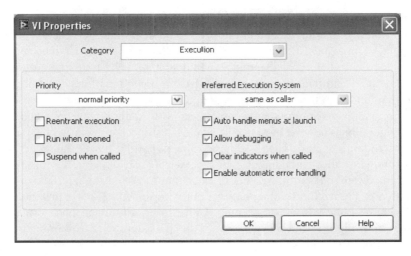

FIGURE 9.19

tasks in the run queue. The priority of the VI has nothing to do with the priority of the thread that is executing it. When a thread with high priority is assigned a VI with background priority, the thread will not reduce its priority to accommodate the VI. The background importance VI will be executed with blazing speed. The reverse is also true.

When working with VI priorities, recall multithreading problem definitions; several of them can be caused in LabVIEW's scheduling routines. Starvation is the easiest problem to cause. When a VI is listed as background priority, VIs with higher priorities will be put into the run queue ahead of the low-priority VI. This will cause the execution of the low-priority VI to be delayed. This could impact the performance of the code diagram. What will end up happening is that all other eligible tasks will be run until the low-priority VI is the only available task to be executed. This would form a bottleneck in the code diagram, potentially degrading performance. The use of VI priorities should not be used to force the order of execution. Techniques using error clusters should be used instead. LabVIEW's engine makes no promises regarding execution time, much like a multithreaded operating system's scheduling algorithm. In the event that parallel executing loops are involved, it is possible for the background priority VI never to be executed.

Priority inversion can also be caused by VI priorities. Recall that the priority of a VI does not impact or change the priority of the thread(s) executing it. If a VI with high priority depends on the outputs of a VI with lower priority, execution of the high-priority VI will be delayed until the low-priority VI has completed execution. This potential for performance limitations should be avoided.

Race conditions can also be induced with VI priorities. The threading model used does not induce these race conditions. These would be race conditions caused by the code diagram itself.

The best logic to use to prevent VI priority problems is similar to preventing problems with the threading engine. A priority other than "normal" should be an exception, not the norm. If a convincing case cannot be put into the Description of Logic of the VI, then its execution priority should be normal. In general, we avoid changing VI priorities. Forcing the order of execution is a better mechanism to accomplish control of a code diagram. In addition, it is much easier for a programmer to look at a VI and understand that the order of execution is forced. VI priority is somewhat hidden in the VI's configuration; a programmer must actively search for this information. Assuming that programmers will examine your code and search the configuration is unwise; most people would not suspect problems with VI priority.

As a coding standard, when a VI has an altered priority, a note should be located above or below to clearly indicate to others who may use the VI that there is something different about it. Another flag that may be used is to color-code the icon or portion of the icon indicating that its priority is something other than normal.

If you absolutely insist on keeping a VI as a priority other than normal, then use the following tip from Steve Rogers (LabVIEW developer extraordinaire): VIs of high priority should never be executing continuously. High-priority VIs should be kept in a suspended mode, waiting on something such as an occurrence, before they are allowed to execute. Once the VI completes executing, it should be suspended again and wait for the next occurrence. This allows for the high-priority VI to execute as the most important VI when it has valid data to process, and to not execute at all when it is not needed. This will prevent programmers from creating priority inversion or starvation issues with LabVIEW's run queue management.

9.10 SUBROUTINES IN LABVIEW

As hinted throughout the chapter, subroutine VIs have strong advantages when using multithreading. First, we need to review the rules on subroutine priority VIs:

1. Subroutine VIs may not have a user interface.
2. Subroutine VIs may call only other subroutine-priority VIs.
3. Subroutines may not call asynchronous nodes (dialog boxes, for example; nodes that do not have a guaranteed return time).

It is important to understand that subroutine classification is not a true priority. "Subroutine" denotes that this VI is no longer a standard VI and that its execution and compilation are radically different from other VIs. Subroutine priority VIs do not have a priority associated with them, and they are never placed into the run queues of LabVIEW. Once all inputs for the subroutine are available, the subroutine will execute immediately, bypassing all run queues. The subsystem associated with the subroutine will stop processing tasks until the subroutine has completed execution. This might sound like a bad idea, but it is not. Having a routine complete execution ASAP is going to get its operation over as quickly as possible and allow LabVIEW to do other things fairly quickly. Subroutines are a bad idea when very time-intensive tasks need to be done because you will block the run queue for a subsystem for an extended amount of time.

Subroutines execute faster than standard VIs because they use less overhead to represent instructions. You may not have a user interface on subroutine priority VIs because, technically, a subroutine does not have a user interface. This is part of the reduced overhead that subroutines have.

Subroutine VIs may call only other subroutines because they are executed in an atomic fashion. Once execution of a subroutine VI starts, single-threaded LabVIEW execution engines will not do anything else until this subroutine has finished. Multitasking becomes blocked in single-threaded LabVIEW environments. Multithreaded LabVIEW environments will continue multitasking when one thread enters a subroutine. The thread assigned to work on the subroutine may do nothing else until the subroutine is executed. Other threads in the system are free to pull jobs off the run queue. In the next section, we will discuss the data types that LabVIEW supports; this is relevant material when subroutine VIs are considered.

9.10.1 EXPRESS VIs

With LabVIEW 7's release a new type of programming construct became available, namely express VIs. These VIs contain precompiled code that accepts inputs. The question with respect to this chapter is how do express VIs interact with LabVIEW's thread model.

As far as LabVIEW's execution engine is concerned, an express VI is an atomic function, meaning it is equivalent to a subroutine. Once an express VI has been put into a run queue, the thread that begins to execute it will finish execution before it can process other tasks.

In the event an express VI is broken out into editable code, then it is handled as any other VI with respect to thread count estimation.

9.10.2 LABVIEW DATA TYPES

Every LabVIEW programmer is familiar with the basic data types LabVIEW supports. This section introduces the low-level details on variables and data storage. Table 9.1 shows the LabVIEW data types. Of concern for application performance is how fast LabVIEW can process the various data types. Most numerical processing can always be assumed to be relatively fast.

As stated in Table 9.1, Booleans are simply 16-bit integers in LabVIEW 4.0 and earlier, and 8-bit integers in LabVIEW 5.0 and later. Their storage and creation is fairly quick; arrays of Booleans can be used with minimal memory requirements. It must be noted that computers are minimum 32-bit machines, and are transitioning to 64 bit access. Four bytes is the minimum amount of memory that can be addressed at a time. One- and two-byte storage is still addressed as four-byte blocks, and the upper blocks are ignored.

Integer sizes obviously depend on byte, word, or long word selections. Integer arithmetic is the fastest numerical processing possible in modern hardware. We will show in the next section that it is advantageous to perform integer processing in one thread of execution.

Floating-point numbers also support three precision formats. Single- and double-precision numbers are represented with 32- or 64-bit numbers internal to LabVIEW. Extended precision floating-point numbers have sizes dependent on the platform you are using. Execution speed will vary with the types of operations

TABLE 9.1
LabVIEW Data Types

Data Type	Size	Processing Speed	Notes
Boolean	16 bits (LabVIEW 4) 8 bits (LabVIEW 5), high bit determines true/false	Fast	High bit determines true or false.
Integers	8, 16, 32, 64 bits	Fast	Signed and Unsigned
Floating Point	Depends on type and platform	Fast	Extended precision size is machine-dependent; single and double are 32- and 64-bit numbers
Complex	Depends on type and platform	Medium?	Slower than floating points.
String	4 bytes + length of string	Slow	First 4 bytes identify length of string
Array	Variable on type	Slow	Faster than strings but can be slow, especially when the array is dimensioned often.
Cluster	Depends on contents	Slow	Processing speed depends heavily on contents of cluster.

performed. Extended-precision numbers are slower than double-precision, which are slower than single-precision numbers. In very high performance computing, such as Software Defined Radio (SDR) where large volumes of floating point data need to be processed in fairly short order, a tradeoff of numerical precision versus processing time is made quite often. The lower precision numbers contribute to quantization noise, but in some cases make or break the performance of an application. Floating-point calculations are always slower than integer arithmetic. Each floating point stores sections of a number in various parts of the memory allocated. For example, one bit is used to store the sign of the number, several bytes will be used to store the mantissa, one byte will store the sign of the exponent, and the rest will store the integer exponent of the number. The format for single- and double-precision numbers is determined by National Instruments, and they are represented internally in LabVIEW. Extended-precision number formats depend on the hardware supporting your system.

Complex numbers use a pair of floating-point numbers for representation. Complex numbers use the same precision as floating-point numbers, but they are slower for processing. Each complex multiplication involves four floating-point calculations. Additions and subtractions involve two floating-point calculations. When necessary, complex calculations need to be done, but their execution speed must be considered in performance-critical applications.

String processing can be very slow. LabVIEW uses four bytes to indicate the length of the string internally, and the contents of the string following the length preamble. This is an advantage LabVIEW programmers have over their C counterparts. C style strings must end with an ASCII 0 (NULL); these NULL-terminated strings assume that there are no NULL values occurring in the middle of the string.

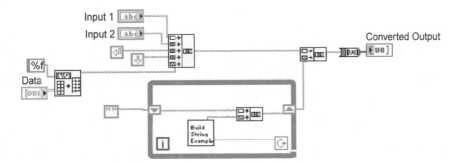

FIGURE 9.20

Microsoft compilers also use a string construct called the "BSTR" which looks a lot like LabVIEW's string, 4 bytes of length information preceding the character data. Oddly, most of the BSTR support functions still are not capable of supporting embedded NULLs. LabVIEW strings do not have this no-embedded NULLs requirement. This is advantageous when working with many devices and communications protocols.

Any time you perform an operation on a string, a duplication of the string will be performed. In terms of C programming, this will involve a "memcopy." BSTR types or OO string classes still use a memory copy; the memcopy is abstracted from the programmer. Memory copies involve requesting an allocation of memory from the memory manager and then duplicating the memory used. This is a performance hit, and, although it cannot be entirely avoided, performance hits can be minimized. Whenever possible, major string manipulation should be avoided when application performance is required. Examine Figure 9.20 for an illustration for where memory copies are made. Memory copies will be made for other variable types, but sizes for integers are 4 bytes, floating points are a maximum of 8 bytes, and Booleans require a minimum of 32 bits for storage. The shortest string representation in LabVIEW is an empty string, which requires five bytes, the four-byte preamble, and one blank byte. Most strings contain information, and longer strings require more time to copy internally.

Array processing can be significantly faster than string processing, but can also be hazardous to application performance. When using arrays in performance-critical applications, pre-dimension the array and then insert values into it. When pre-dimensioning arrays, an initial memory allocation will be performed. This prevents LabVIEW from needing to perform additional allocations, which will cause performance degradation. Figure 9.21 illustrates two array-handling routines. Array copying can be as CPU-intensive as string manipulation. Array variables have four bytes for storage of the array dimensions, and a number of bytes equivalent to the size of the dimensions times the storage size of the type.

9.10.3 WHEN TO USE SUBROUTINES

Now that we know the benefits and penalties of using threads and are familiar with implications of data type choices, it is time to determine when subroutines should

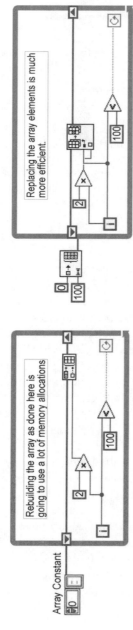

FIGURE 9.21

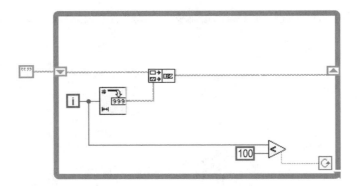

FIGURE 9.22

be used. Knowing that numerical operations are fast and that string and dynamic arrays are slow, the obvious conclusion is that numerical-intensive VIs are prime candidates for subroutine priority.

When handling large strings or variable-dimension arrays, do not use subroutines. When LabVIEW tries to copy strings, the thread performing the copying will incur overhead while the memory allocation is being performed. If this thread is running in a subroutine, other threads may become blocked waiting for one thread to cycle through the subroutine. It is preferable in these situations to have multiple threads working in a VI to minimize blocking points.

Numerical processing is fairly fast; it is possible for threads to completely execute most subroutines in a single timeslice. With modern processors running multiple cores, handling processor instructions out of order, and doing it all with clock rates in the gigahertz, millions of numerical calculations can be accomplished in a single timeslice on a single thread.

As an example of this, we will consider the following problem: the VI depicted in Figure 9.22 shows a simple string-generation routine. The counter is converted to a string and then concatenated to another string and fed through the shift register. Consider the copying and memory allocations that need to be done. Every time the integer is converted to a string, five to six bytes are allocated from heap memory to store the number; five bytes are used when the number is between zero and nine, and six bytes when the number has two digits. Recall that four bytes of length information are stored in a string preamble. The string concatenation requires an additional allocation of length: four bytes + length of old string + length of new string. These allocations are relatively small but add overhead. Figure 9.23 shows the VI profile from numerous runs of this VI.

The timing profile demonstrates that at least one execution required 15 ms to complete. In terms of CPU time, this is significant.

A thread may not have enough time in a quantum to always complete this operation. The thread may then be preempted for higher-priority threads and take some time before it can resume execution. During at least one execution, the thread was preempted and took an order of magnitude more time to complete. Larger string manipulation routines will even take longer. If a single thread is dedicated to per-

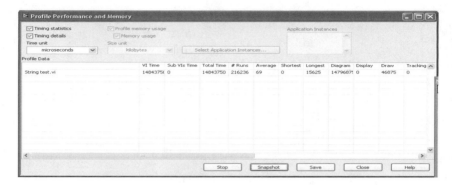

FIGURE 9.23

forming all the manipulations, this could reduce performance of the application. Outputs of this VI will probably be required by other VIs. These other VIs would be blocked from execution while this current VI is completed. The conclusion this example demonstrates is that intensive string manipulations should be performed in VIs that are not subroutines. This will allow multiple threads to perform tasks contained inside the VI. Other threads will not become blocked waiting on a single thread to perform large amounts of memory allocations.

Integer operations require significantly less time to complete, and are often good candidates for subroutine priority. The VI shown in Figure 9.24 shows a simple 100-element manipulation. This type of manipulation may not be common in everyday computing, but it serves as a similar example to the one mentioned above. Figure 9.25 shows the profile window for a large number of runs. Notice the significantly lower timing requirements. It is much less likely that a thread will become blocked during execution of this subVI; therefore, it is desirable to give this subVI subroutine priority because subroutine VIs have less overhead and will execute faster than standard VIs.

When working with arrays in LabVIEW, try to use fixed-length or pre-dimensioned arrays as much as possible. When using the Initialize Array function, one block of memory will be taken from the heap memory. Replacing individual elements in the array will not require the array to be reallocated. Once a fixed array size is defined, then a VI manipulating this array can be a candidate for subroutine priority.

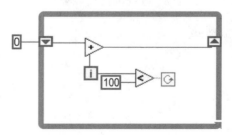

FIGURE 9.24

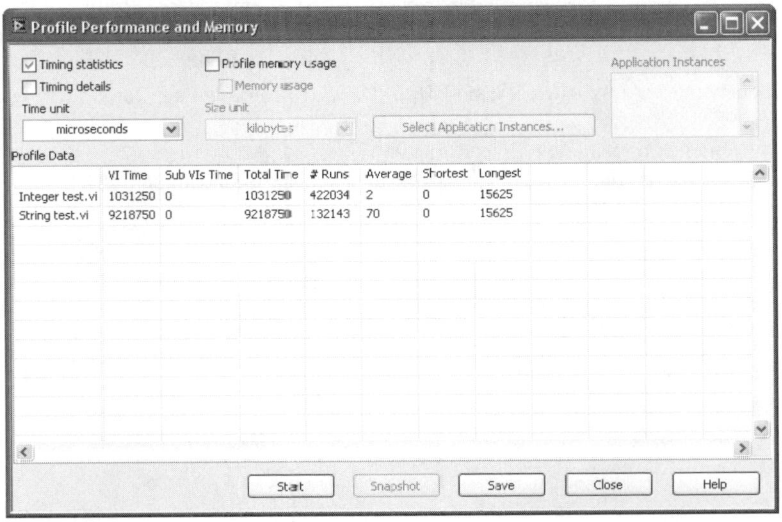
Profile Performance and Memory
Timing statistics
Timing details
Time unit
microseconds
Profile memory usage
Memory usage
Size unit
kilobytes
Select Application Instances...
Application Instances
Profile Data
VI Time
Sub VIs Time
Total Time
Runs
Average
Shortest
Longest
Integer test.vi 1031250 0 1031250 422034 2 0 15625
String test.vi 9218750 0 9218750 132143 70 0 15625
Start
Snapshot
Save
Close
Help

FIGURE 9.25

There will not be significant memory allocations that need to be performed. The overhead on the VI will be reduced, improving the performance of the VI.

9.11 SUMMARY

This chapter began with core multithreading terminology. Threads, processes, applications, and several operating systems were explained. The basics of multithreading — scheduling, priorities, processes, and thread basics — were discussed and defined.

Major myths involving multithreading were addressed. It should never be assumed that threads will improve application performance. Many data flow applications force a serial order of execution; this is precisely where multithreading will be of the least benefit. Another common misunderstanding regarding threads is the idea that the application performance is proportional to the number of threads running. This is only true if multiple processors are available. Rotating threads of execution in and out of the CPU will cause more performance problems than solutions.

Estimating the optimum number of threads is challenging, but not entirely impossible. The programmer must identify where the maximum number of executable elements is generated in the code. Using this number as the maximum number of useful threads will prevent performance limitations.

Subroutine priority VIs can lead to performance gains.

BIBLIOGRAPHY

Microsoft Press, *Windows Architecture for Developers Training*. Redmond: Microsoft Press, 1998.

Aeleen Frisch. *Essential System Administration*, 2nd ed. Cambridge: O'Reilly, 1995.

Bradford Nichols, Dick Buttler, and Jacqueline Prowly Farrel. *Pthreads*. Cambridge: O'Reilly, 1996.

Aaron Cohen and Mike Woodring, *Win32 Multithreaded Programming*. Cambridge: O'Reilly, 1998.

http://www.intel.com/technology/hyperthread/.

http://msdn.microsoft.com/library/default.asp?url=/library/en-us/dnucmg/html/UCMGch02.asp.

10 Object-Oriented Programming in LabVIEW

This chapter applies a different programming paradigm to G: Object-Oriented Programming (OOP). New languages like Java and its use on the Internet have created a lot of interest in this programming paradigm. This chapter explains the concepts that make object-oriented programming work, and applies them to programming in LabVIEW.

This chapter begins with definitions of objects and classes. These are the fundamental building blocks of OOP. Key definitions that define OOP are then presented which give a foundation for programmers to view applications in terms of their constituent objects.

Once the basics of OOP are described, the first stage of objects is presented — object analysis. Fundamentally, the beginning of the design is to identify the objects of the system. Section 10.4 discusses object design, the process by which methods and properties are specified. The interaction of objects is also defined in the design phase. The third and last phase is the object programming phase. This is where the code to implement the methods and properties is performed.

This type of structuring seems foreign or even backward to many programmers with experience in structured languages such as LabVIEW. Object-oriented is how programming is currently being taught to computer science and engineering students around the world. A significant amount of effort has been put into the design of a process to produce high-quality software. This section introduces this type of philosophy to LabVIEW graphical programming.

Object-oriented design is supported by a number of languages, including C++ and Java. This book tries to refrain from using rules used specifically by any particular language. The concept of object-oriented coding brings some powerful new design tools, which will be of use to the LabVIEW developer. The concept of the VI has already taught LabVIEW programmers to develop applications modularly. This chapter will expand on modular software development.

This chapter discusses the basic methodology of object coding, and also discusses a development process to use. Many LabVIEW programmers have backgrounds in science and engineering disciplines other than software engineering. The world of software engineering has placed significant emphasis into developing basic design processes for large software projects. The intent of the process is to improve software quality and reduce the amount of time it takes to produce the final product. Team development environments are also addressed in this methodology.

As stated in the previous paragraph, this chapter provides only a primer on object design methodology. There are numerous books on this topic, and readers who decide to use this methodology may want to consult additional resources.

10.1 WHAT IS OBJECT-ORIENTED?

Object-oriented is a design methodology. In short, object-oriented programming revolves around a simple perspective: divide the elements of a programming problem into components. This section defines the three key properties of object-oriented: encapsulation, inheritance, and polymorphism. These three properties are used to resolve a number of problems that have been experienced with structured languages such as C.

It will be shown that LabVIEW is not an object-oriented language. This is a limitation to how much object-oriented programming can be done in LabVIEW, but the paradigm is highly useful and it will be demonstrated that many benefits of object-oriented design can be used successfully in LabVIEW. This chapter will develop a simple representation for classes and objects that can be used in LabVIEW application development.

10.1.1 THE CLASS

Before we can explain the properties of an object-oriented environment, the basic definition of an object must be explained. The core of object-oriented environments is the "class." Many programmers not familiar with object-oriented programming might think the terms "class" and "object" are interchangeable. They are not. A "class" is the core definition of some entity in a program. Classes that might exist in LabVIEW applications include test instrument classes, signal classes, or even digital filters. When performing object programming, the class is a definition or template for the objects. You create objects when programming; the objects are created from their class template. A simple example of a class/object relationship is that a book is a class; similarly, *LabVIEW Advanced Programming Techniques* is an object of the type "book." Your library does not have any book classes on its shelves; rather, it has many instances of book classes. An object is often referred to as an instance of the class. We will provide much more information on classes and objects later in this chapter. For now, a simple definition of classes and objects is required to properly define the principles of object-oriented languages.

A class object has a list of actions or tasks it performs. The tasks objects perform are referred to as "methods." A method is basically a function that is owned by the class object. Generally speaking, a method for a class can be called only by an instance of the class, an object. Methods will be discussed in more detail in Section 10.2.1.

The object must also have internal data to manipulate. Data that are specified in the class template are referred to as "properties." Methods and properties should be familiar terms now; we heard about both of those items in Chapter 7, ActiveX. Active X is built on object-oriented principals and uses the terminology extensively.

Experienced C++ programmers know the static keyword can be used to work around the restriction that objects must exist to use methods and properties. The implementation of objects and classes in this chapter will not strictly follow any particular implementations in languages. We will follow the basic guidelines spelled out in many object-oriented books. Rules regarding objects and classes in languages like C++ and Java are implementations of object-oriented theory. When developing objects for non-object-oriented languages, it will be helpful to not strictly model the objects after any particular implementation.

LabVIEW does not have a built-in class object. Some programmers might suspect that a cluster would be a class template. A cluster is similar to a structure in C. It does not directly support methods or properties, and is therefore not a class object. We will use clusters in the development of class objects in this chapter. One major problem with clusters is that data is not protected from access, which leads us to our next object-oriented principal, encapsulation.

10.1.2 ENCAPSULATION

Encapsulation, or data hiding, is the ability for an object to prevent manipulation of its data by external agents in unknown ways. Global variables in languages like C and LabVIEW have caused numerous problems in very large-scale applications. Troubleshooting applications with many global variables that are altered and used by many different functions is difficult, at best. Object-programming prevents and resolves this problem by encapsulating data. Data that is encapsulated and otherwise inaccessible to outside functions is referred to as "private data." Data that is accessible to external functions is referred to as "public data."

The object-oriented solution to the problem of excessive access to data is to make most data private to objects. The object itself may alter only private data. To modify data private to an object, you must call a function, referred to as a method, that the object has declared public (available to other objects). The solution that is provided is that private data may be altered only by known methods. The object that owns the data is "aware" that the data is being altered. The public function may change other internal data in response to the function call. Figure 10.1 demonstrates the concept of encapsulated data.

Any object may alter data that is declared public. This is potentially dangerous programming and is generally avoided by many programmers. As public data may be altered at any time by any object, the variable is nearly as unprotected as a global variable. It cannot be stressed enough that defensive programming is a valuable technique when larger scale applications are being written. One goal of this section is to convince programmers that global data is dangerous. If you choose not to pursue object-oriented techniques, you should at least gather a few ideas on how to limit access to and understand the danger of global data.

A language that does not support some method for encapsulation is not object-oriented. Although LabVIEW itself is not object-oriented, objects can be developed to support encapsulation. Encapsulation is extremely useful in large-scale LabVIEW

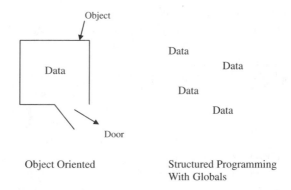

FIGURE 10.1

applications, particularly when an application is being developed in a team environment. Global data should be considered hazardous in team environments. It is often difficult to know which team member's code has accessed global variables. In addition to having multiple points where the data is altered, it can be difficult to know the reason for altering the data. Using good descriptions of logic (DOL) has minimized many problems associated with globals. Using encapsulation, programmers would have to change the variable through a subVI; this subVI would alter variables in a known fashion, every time. For debugging purposes, the subVI could also be programmed to remember the call chain of subVIs that called it.

Encapsulation encourages defensive programming. This is an important mindset when developing large-scale applications, or when a team develops applications. Application variables should be divided into groups that own and control the objects. A small degree of paranoia is applied, and the result is usually an easier to maintain, higher quality application. Global variables have been the bane of many college professors for years. This mindset is important in languages like C and C++; LabVIEW is another environment that should approach globals with a healthy degree of paranoia.

10.1.3 AGGREGATION

Objects can be related to each other in one of two relationships: "is a" and "has a." A "has a" relationship is called *aggregation*. For example, "a computer has a CD-ROM drive" is an aggregated relationship. The computer is not specialized by the CD-ROM, and the CD-ROM is not interchangeable with the computer itself. Aggregation is a fundamental relationship in object design. We will see later in the chapter that an aggregated object is a property of the owning object.

FIGURE 10.2

Aggregation is a useful mechanism to develop complex objects. In an object diagram, boxes represent classes, and aggregation is shown as an arrow connecting the two objects. The relationship between the computer and CD-ROM is shown in Figure 10.2.

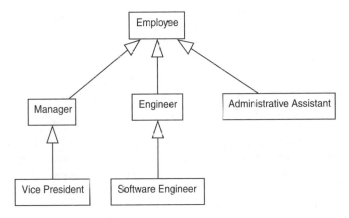

FIGURE 10.3

10.1.4 INHERITANCE

Inheritance is the ability for one class to specialize another. A simple example of inheritance is that a software engineer is a specialization of an engineer. An engineer is a specialization of an employee. Figure 10.3 shows a diagram that demonstrates the hierarchy of classes that are derived from an employee class. Common in object-oriented introductions is the "is a" relationship. A class inherits from another if it is a specialization or is a type of the superclass. This is a fundamental question that needs to be asked when considering if one class is a specialization of another. Examples of this relationship are engineer "is a" employee, and power supply "is a" GPIB instrument.

When one class inherits from another, the definition of the class is transferred to the lower class. The class that is inherited from is the "superclass" and the inheriting class is the "subclass." For example, consider a class employee. An engineer "is a" employee (please excuse the bad grammar, but it's difficult to be grammatically correct and illustrate an "is a" relationship!). This means that the definition of an employee is used by and expanded by the engineer class. An engineer is a specialization of employee. Other specializations may be vice-president, human resources personnel, and manager. Everything that defines an employee will be used by the subclasses. All employees have a salary; therefore, engineers have salaries. The salary is a property of employee and is used by all subclasses of employee.

All employees leave at the end of the day, some later than others. The function of leaving is common, and is a function that all employee subclasses must use. This method is directly inherited; the same leave function may be used by engineers, vice presidents, and marketing subclasses.

All employees work. Engineers perform different jobs than human resource employees. A definition in the employee class should exist because all employees do some kind of work, but the specifics of the function vary by class. This type of function is part of the employee specification of the employee class but must be done differently in each of the subclasses. In C++, this is referred to as a "pure virtual function." When a class has a pure virtual function, it is referred to as an "abstract class." Abstract classes cannot be created; only their subclasses may be created. This is not a limitation. In this example, you do not hire employees; you hire specific types of employees.

There is another manner in which classes acquire functions. If employee has a method for taking breaks, and normal breaks are 15 minutes, then most subclasses will inherit a function from employee that lets them take a 15-minute break. Vice presidents take half-hour breaks. The solution to implementing this method is to have a pure virtual method in employee, and have each subclass implement the break function. Object programming has virtual functions. The employee class will have a 15-minute break function declared virtual. When using subclasses such as engineer and the break function is called, it will go to its superclass and execute the break function. The vice president class will have its own version of the break function. When the vice president class calls the break function, it will use its own version. This allows for you to write a single function that many of the subclasses will use in one place. The few functions that need to use a customized version can do so without forcing a rewrite of the same code in multiple places.

Inheritance is one of the most important aspects of object-oriented programming. If a language cannot support inheritance, it is not object-oriented. LabVIEW is not an object-oriented language, but we will explore how many of the benefits of this programming paradigm can be supported in LabVIEW.

10.1.5 POLYMORPHISM

Polymorphism is the ability for objects to behave appropriately. This stems from the use of pointers and references in languages like C++ and Java (Java does not support pointers). It is possible in C++ to have a pointer to an employee class and have the object pointed to be an engineer class. When the work method of the pointer is called, the engineer's work method is used. This is polymorphism; this property is useful in large systems where collections of objects of different type are used.

LabVIEW does not support inheritance, and cannot support polymorphism. We will show later in this chapter how many of the benefits of object-oriented programming can be used in LabVIEW, despite its lack of support for object-oriented programming. Polymorphism will not be used in our object implementation in this chapter. It is possible to develop an object implementation that would support inheritance and polymorphism, but we will not pursue it in this chapter.

10.2 OBJECTS AND CLASSES

The concept of OOP revolves around the idea of looking at a programming problem in terms of the components that make up the system. This is a natural perspective in applications involving simulation, test instruments, and data acquisition (DAQ). When writing a test application, each instrument is an object in the system along with the device under test (DUT). When performing simulations, each element being simulated can be considered one or more classes. Recall from Section 10.1.1 that an instance of a class is an object. Each instrument in a test rack is an instance of a test instrument class or subclass.

10.2.1 Methods

Methods are functions; these functions belong to the class. In LabVIEW, methods will be written as subVIs in our implementation. The term "method" is not indigenous to object-oriented software, but recall from Chapter 7, ActiveX, that ActiveX controls use methods. Methods may be encapsulated into a class. Methods are considered private when only an instance of the class may use the method.

Methods that are public are available for any other object to call. Public methods allow the rest of the program to instruct an object to perform an action. Examples of public methods that objects should support are Get and Set functions. Get and Set functions allow an external object to get a copy of an internal variable, or ask the object to change one of its internal variables. The Get functions will return a copy of the internal data; this would prevent an external object from accidentally altering the variable, causing problems for the owning object later. Public methods define the interface that an object exposes to other elements of the program. The use of defensive programming is taken to individual objects in object-oriented programming. Only public methods may be invoked, which allows objects to protect internal data and methods.

Only the object that owns itself may call private methods. These types of functions are used to manipulate internal data in a manner that could be dangerous to software quality if any object could alter the internal data. As an example of using objects in a simulation system, consider a LabVIEW application used to simulate a cellular phone network. A class phone has methods to register to the system, make call, and hang up. These methods are public so the program can tell phone objects to perform those actions. Each method, in turn, calls a Transmit method that sends data specific to registration, call setup, or call teardown. The information for each type of message is stored in the specific methods and is passed to the Transmit function. The Transmit function is private to the object; it is undesirable for any other part of the program to tell a phone to transmit arbitrary information. Only specific message types will be sent by the phones. The transmit method may be a common use function internal to the class

10.2.1.1 Special Method — Constructor

Every class requires two special methods. The first is the Constructor. The Constructor is called whenever an instance of the class is created. The purpose of the Constructor is to properly initialize a new object. Constructors can effectively do nothing, or can be very elaborate functions. As an example, a test instrument class for GPIB instruments would have to know their GPIB address. The application may also need to know which GPIB board they are being used on. When a test instrument object is instantiated, this information is passed to the function in the Constructor. This allows for the test instrument object to be initialized when it is created, requiring no additional configuration on the part of the programmer. Constructors are useful when uninitialized objects can cause problems. For example, if a test instrument object ends up with default GPIB address of 0 and you send a message to this instrument, it goes back to the system controller. In Section 10.7.1 we will implement Constructor functions in LabVIEW.

The Constructor method is something that cannot be done with simple clusters. Clusters can have default values, but a VI to wrap around the cluster to provide initialization will be necessary. The Constructor function in LabVIEW will be discussed in Section 10.7. Initialization will allow an object to put internal data into a known state before the object becomes used. Default values could be used for primitive data types such as integers and strings, but what if the object contains data that is not a primitive type, such as a VISA handle, TCP handle, or another object? Constructors allow us to set all internal data into a known state.

10.2.1.2   Special Method — Destructor

The purpose of the Destructor is the opposite of the Constructor. This is the second special method of all classes. When an object is deleted, this function gets called to perform cleanup operations such as freeing heap memory. LabVIEW programmers are not concerned with heap memory, but there are cases when LabVIEW objects will want to have a Destructor function. For instance, if when an object is destroyed it is desirable to write information on this object to a log file. If a TCP conversation were encapsulated into a class object, the class Destructor may be responsible for closing the TCP connection and destroying the handle.

In languages such as C++, it is possible to have an object that does not have a defined Constructor or Destructor. The compiler actually provides default functions for objects that do not define their own Constructor and Destructor. Our implementation does not have a compiler that will graciously provide functions that we are too lazy to write on our own. The object implementation presented later in this chapter requires Constructors for all classes, but Destructors will be optional. This is not usually considered good programming practice in object-oriented programming, but our implementation will not support the full features of OOP.

10.2.2   PROPERTIES

Properties are the object-oriented name for variables. The variables that are part of a class belong to that class. Properties can be primitive types such as Booleans, or can be complex types such as other classes. Encapsulating a class inside of another class is aggregation. We will discuss aggregation again later in this chapter. An example of a class with class properties is a *bookshelf*. The bookshelf itself is a class with an integer property representing the number of shelves. If the shelf were defined to have a single book on the "A" shelf, then a property to describe the book would be necessary. The description of the book is defined as a class with its own properties, such as number of pages.

Properties defined for a class need to have some relevance to the problem to be solved. If your shelf class had a color constant to represent the color of the shelf, this information should be used somewhere in the program. Future considerations are acceptable; for instance, we do not use the color information now, but the next revision will definitely need it. If extra properties are primitive types, such as Booleans, there will not be any significant problems. When extra properties are complex types or use resources such as TCP conversations, performance issues could be created because of the extra resources the classes use in the system.

The values of an object's properties make the object unique. All objects of a class have the same methods and property types. Differentiation between objects can be done only with the property values. An example of this would be a generic GPIB instrument class. This class may have properties such as GPIB board and GPIB address. The values of board and address make different GPIB instruments unique. All GPIB objects would have the same methods and property types (address and board number). The value of the address and board make the particular GPIB object unique.

Most properties are private to the class. This means that only the class itself may modify the member variables (properties). This is another measure of defensive programming. Global data has caused countless headaches for C programmers, and encapsulation is one solution to preventing this problem in object-oriented applications. The implementation for objects in this chapter will effectively make all properties private. This means that we will have to supply methods for modifying data from outside the class.

10.3 OBJECT ANALYSIS

Object analysis is the first stage in an object-oriented design process. The objects that comprise the system are identified. The object analysis phase is the shortest phase, but is the most important. Practical experience has shown us that when the object analysis is done well, many mistakes made in the design and program phases have reduced impacts. Good selection of objects will make the design phase easier. Your ability to visualize how objects interact will help you define the needed properties and methods.

When selecting objects, every significant aspect of an application must be represented in one of the objects. Caution must be exercised to not make too many or negligible-purpose objects. For example, when attempting to perform an object analysis on a test application using GPIB instruments, an object to represent the GPIB cables will not be very useful. As none of the GPIB interfaces need to know about the cables, encapsulating a cable description into an object will not be of any benefit to the program. No effort is being made at this point to implement the objects; a basic understanding of which objects are necessary is all that should be needed. The design and interactions should be specified well in advance of coding. Having the design finalized allows for easier tracking of scheduling and software metric collection. This also eliminates the possibility of "feature creep," when the definition of what a function is supposed to do keeps getting changed and expanded. Feature creep will drag out schedules, increase the probability of software defects, and can lead to spaghetti code. Unfortunately, spaghetti code written in LabVIEW does have a resemblance to noodles.

TABLE 10.1
Object Analysis Example #1

Equipment	Purpose
Multimeter	Measure DC bias on output signal line to Device under Test(DUT).
Arbitrary waveform generator	Generate test signal stimuli. The signal is generated on the input signal line to the DUT.
Device under test	The Meaningless Object in Example (MOIE).

Example 1:

This example attempts to clarify the purpose of the object analysis on the design of an application to control an Automated Test Equipment (ATE) rack. This example will start out with a relatively simple object analysis. We will be testing Meaningless Objects in Example (MOIE). Table 10.1 identifies the equipment used in the rack. The MOIE has two signal lines, one input line and one output line.

Solution 1:

We will attempt to make everything available an object. Figure 10.4 shows a model of the objects that exist in the rack. This model is not very useful; there are far too many objects from which to build a solution.

The objects for the GPIB cables are not necessary as the test application will not perform any operations directly with the cables. GPIB read and write operations will be performed, but the application does not need to have internal representations for each and every cable in the system. The objects for the input voltages to the test instruments also make for thorough accounting for everything in the system. Nevertheless, if the software is not going to use an object directly, there is no need to account for it in the design.

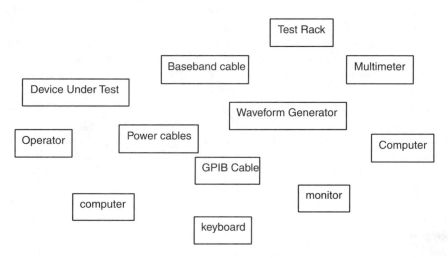

FIGURE 10.4

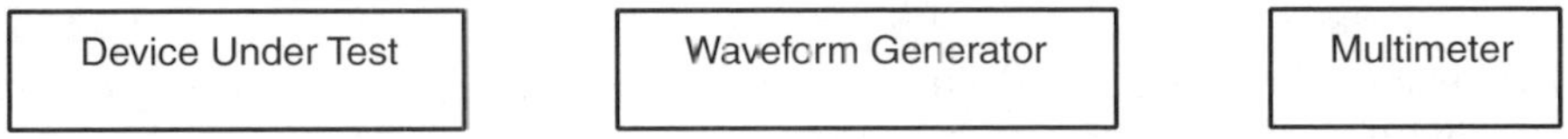

| Device Under Test | | Waveform Generator | | Multimeter |

FIGURE 10.5

We are not using objects to represent GPIB cables, but there are times when a software representation of a cable is desirable. If you are working with cables for signal transmission or RF/microwave use, calibration factors may be necessary. An object would be useful because you can encapsulate the losses as a function of cable length, current supplied, or any other relevant factors.

Solution 2:
The easiest object analysis to perform is to declare the three items in the test rack to be their own objects. The simple diagram in Figure 10.5 shows the object relation. This could be an acceptable design, but it is clearly not sophisticated. When performing object analysis, look for common ground between objects in the system. If two or more objects have common functions, then look to make a superclass that has these methods or properties, and have the objects derive from them. This is code reuse in its best form; you need to write the code only once.

Solution 3:
Working from Solution 1 with the three objects (the DUT, waveform generator, and multimeter objects), consider the two GPIB instruments. Both obviously read and write on the GPIB bus; there is common ground between the instruments. A GPIB instrument class could be put into the model, and the meter and waveform generator could inherit the read and write functions of the superclass. Figure 10.6 shows the model with the new superclass.

Which is the best solution, 2 or 3? The answer to that question is: It depends. Personal preferences will come into play. In this example, GPIB control is not necessarily a function that needs to be encapsulated in an object. LabVIEW functions easily supply GPIB functions, and there may be little benefit to abstracting the control to a superclass. The GPIB instrument object needs to be implemented, and this may become more work for the programmer.

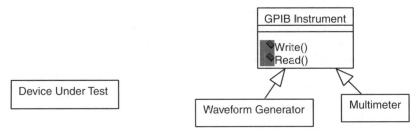

FIGURE 10.6

Example 2:

This example will perform an object analysis on generic test instruments. We will be using the results of this example throughout the chapter. The analysis here will be the focus of a design example in the next section. We have a need to design a hierarchy of classes to support the development of test instrument drivers. An object analysis starting this vaguely needs to start with a base class named "instrument." This base class defines the basic properties and methods that all test instruments would have. Base classes such as this typically have a few properties and methods, and that is about it. Making common ground can be difficult.

A common property would be an address. This address could be used for the GPIB primary address, or a COM port number for a serial instrument. If we define an address to be a 32-bit number, then it could also be the IP address for Ethernet instruments (should they come into existence soon). This is really the only required property in the abstract base class because it is the only common variable to the major communications protocols. For example, we would not want a baud rate property because TCP/IP or GPIB instruments would not use it. Information on the physical cable lengths is not necessary because only high-speed GPIB has any need for this type of information.

All instruments will likely need some type of read and write methods. These methods would be "pure virtual." Pure virtual means that every subclass must support these methods, and the base class will supply no implementation, just the requirement that subclasses have these methods. Pure virtual methods allow each of the base classes to supply custom functionality to the read and write method, which is inherently different for a serial-, GPIB-, or TCP/IP-based instrument. By defining read and write methods, we have effectively standardized the interface for the objects. This is essentially what the VISA library did for LabVIEW; we are repeating this effort in an object-oriented fashion.

The subclasses of instruments for this example will be serial, GPIB, and IP instruments. Obviously, IP is not a common communication protocol for the test industry, but its representation in the object diagram allows for easy future expansion. The GPIB class will cover only the IEEE 488 standard (we will make 488.2 instruments a subclass of GPIB). The following paragraphs will identify the properties and methods that are required for each of the subclasses.

Serial instruments need to implement the read and write methods of the instrument class. Their COM port information will be stored in the address property, which is inherited from instrument. Specific to serial instruments are baud rates and flow control information. These properties will be made private to serial instruments. A Connect method will be supplied for serial instruments. This method will allow for the object to initialize the serial port settings and send a string message if desired by the programmer. The Connect method will not be required in the base class instrument because some instrument subclasses do not require a connection or initialization routine to begin operation — namely, GPIB instruments.

GPIB instruments require a GPIB board and have an optional second address. These two properties are really the only additional items for a GPIB instrument. GPIB instruments do not require a Connect method to configure their communications port. This object must supply Read and Write methods because they derive from "instrument." Other than read, write, board number, and secondary address, there is little work that needs to be done for GPIB instruments.

As we mentioned previously, we intend to have an IEEE 488.2 class derived from the GPIB instrument class. Functionally, this class will add the ability to send the required commands (such as *RST). In addition to read and write, the required commands are the only members that need to be added to the class. Alert readers will have noticed that we have not added support for high-speed GPIB instruments. Not to worry, this class makes an appearance in the exercises at the end of this chapter.

IP instruments are going to be another abstract base class. This consideration is being made because there are two possible protocols that could be used beneath IP, namely UDP and TCP. We know that both UDP- and TCP-based instruments would require a port number in addition to the address. This is a property that is common to both subclasses, and it should be defined at the higher-level base class. Again, IP will require that UDP and TCP instruments support read and write methods. An additional pair of properties would also be helpful: destination port and address. These properties can again be added to the IP instrument class.

To wrap up the currently fictitious IP instruments branch of the object tree, UDP and TCP instruments need an initialization function. We will call UDP's initialization method "initialize," and TCP's method will be called "connect." We are making a differentiation here because UDP does not maintain a connection and TCP does. TCP instruments must also support a disconnect method. We did not include one in the serial instrument class because, in general, a serial port can effectively just go away. TCP sockets, on the other hand, should be cleaned up when finished with because they will tie up resources on a machine. The object diagram of this example can be seen in Figure 10.7. Resulting from this object analysis is a hierarchy describing the instrument types. This hierarchy can be used as the base for deriving classes for specific types of instruments. For example, a Bitter-2970 power supply may be a particular serial instrument. Serial Instrument would be the base class for this power supply, and its class could be put into the hierarchy beneath the serial instrument. All properties — COM port, methods, read and write — would be supported by the Bitter-2970 power supply, and you would not need to do a significant amount of work to implement the functionality.

Example 3:

This example is going to be a bit more detailed than the previous one. We are going to perform the object analysis for the testing of a Communications Analyzer. This is a compound test instrument that has the functionality of an RF analyzer, RF generator, Audio Analyzer, and Audio Generator. The instrument has a list of other functions that it is capable of doing, but for the purposes of this example we will consider only the functions mentioned.

The first step in the object analysis is to determine what the significant objects are. The RF generator, AF generator, RF analyzer, and AF analyzer are fairly obvious. The HP8920 is compliant with IEEE 488.2, so it has GPIB control with a couple of standard instrument commands. As the instrument is GPIB-based, we will start with abstract classes for instrument and GPIB instrument. A GPIB instrument is a specialization of an instrument. Further specification results in the 488.2 GPIB

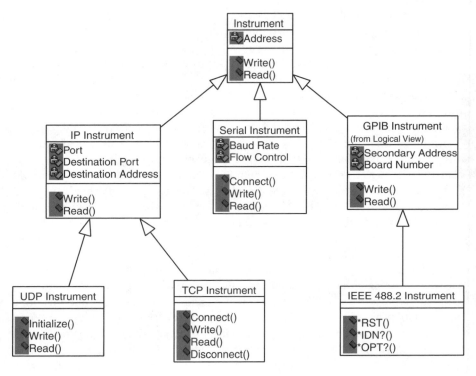

FIGURE 10.7

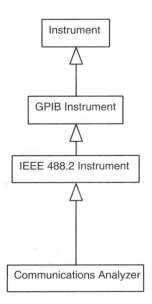

FIGURE 10.8

Instrument subclass. Figure 10.8 shows how our hierarchy is progressing. Some communication analyzers have additional capabilities, including spectrum analyzers, oscilloscopes, and power supplies for devices under test. These objects can also be added to the hierarchy of the design.

An HP-8920 is a communications test set. There are a number of communications analyzers available on the market; for example, the Anritzu 8920 and Motorola 2600 are competitive communications analyzers. All of the instruments have common subsystems, namely the RF analyzers and generators, and the AF analyzers and generators. The preceding sentences suggest that having an HP-8920 as a subclass of a 488.2 instrument is not the best placement. There is a communications analyzer sub-

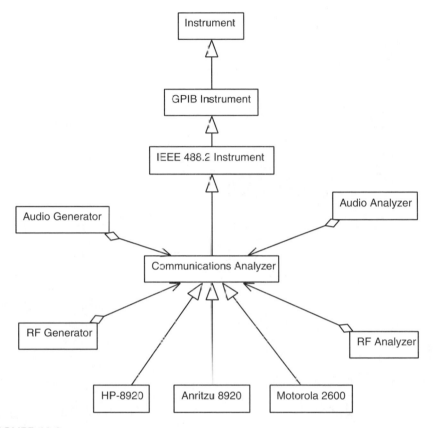

FIGURE 10.9

class of 488.2 instrument. There may also be other subclasses of the 488.2 instrument, such as power supplies. Figure 10.9 shows the expanding hierarchy.

All of the frequency generators have common elements, such as the frequency at which the generator is operating. The RF generator is going to use a much higher frequency, but it is still generating a signal at a specified frequency. This suggests that a generator superclass may be desirable. Our communications analyzer is going to have several embedded classes. Recall that embedded classes are referred to as aggregated classes in object-oriented terminology. We have a new base class to start working with, namely Generator. The AF and RF analyzers have similar starting points, and their analysis is left as an exercise for the reader. Figure 10.10 shows a breakdown of the Analyzer, Generator, and Instrument classes. The dotted line connecting the communications analyzer and the generator/analyzers symbolizes aggregation, encapsulating one class into another. We have a basic design on the core objects. Each class should have a paragraph or so description of the purpose for their design. This would complete an object analysis on this project.

The last objects, oscilloscope, spectrum analyzer, and power supply, are independent objects. There are no base objects necessary to simplify their description. The power supply properties and methods should be fairly easy to decide. The

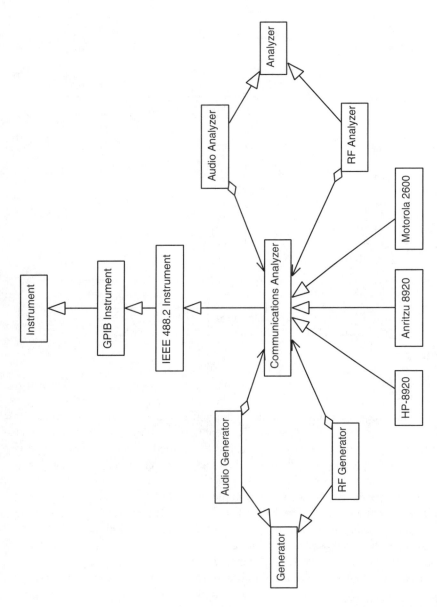

FIGURE 10.10

oscilloscope and spectrum analyzer are significantly more complex and we are going to leave their analysis and design to the reader.

Object analysis is the first step in object programming, but it is arguably the most important. Good selection of objects makes understanding their interrelations much easier. Now that the basic objects in this system are defined, it is time to focus our attention on how the objects interrelate. The process of analysis and design is iterative; the spiral design model allows you to select your objects, design the objects, and program them. Once you complete one cycle, you can start over and tweak the object list, the function list, and the resulting code.

10.4 OBJECT DESIGN

Object design is where the properties, methods, and interactions of classes are defined. Like object analysis, this phase is pure paperwork. The idea is to have a clearly designed application before code is written. Schedules for coding can be developed with better precision because definitions of functions are well defined. It is much more difficult to estimate code schedules when VIs are defined ad hoc. Coordination of medium to large-scale applications requires project management in addition to a solid game plan. Object designs are well suited for project management processes. Chapter 4, Application Architecture, goes through the details of the waterfall and spiral design models.

A full application design forces architects and programmers to put a significant amount of thought into how the application will be built in the startup phase of a project. Mistakes commonly made during the analysis phase are to neglect or miss application details such as exception handling. Focus at this point of an application should revolve around the interaction of the objects. Implementing individual functions should not be performed unless there are concerns about feasibility.

It is not acceptable to write half the application at this phase of the design unless you are concerned about whether or not a concept is possible. For example, if you were designing a distributed application and had concerns about the speed of DCOM, by all means build a prototype to validate the speed of DCOM. Prototyping is acceptable at this phase because it is much easier to change design decisions when you have not written 100 VIs.

In this section we will explain identifying methods and properties for a class. Once the methods and properties are defined, their interactions can then be considered. The actual implementation of the specified methods and properties will not be considered until the next section. It is important to go through the three distinct phases of development. Once this phase of design is completed, an application's "innards" will be very well defined. The programmers will know what each object is for, how it should function, and what its interactions are with other objects in the system. The idea is to make programming as trivial an exercise as possible. Conceptually, programmers will not have to spend time thinking about interactions of objects because we have already done this. The last phase of the design, programming, should revolve around implementing objects on a method-by-method basis. This allows programmers to focus on the small tasks at hand.

Before we begin a full discussion of object design techniques, we need to expand our class understanding: Sections 10.4.1 and 10.4.2 introduce two concepts to classes. The Container class is useful for storing information that will not be used often. Section 10.4.2 discusses the Abstract class. Abstract classes are useful for defining classes and methods that are used to define interfaces that subclasses must support.

10.4.1 CONTAINER CLASSES

Most, but not all, objects in a system perform actions. In languages such as C++ it is often desirable to encapsulate a group of properties in a class to improve code readability. These container classes are similar in purpose to C structures (also available in C++) and clusters in LabVIEW. Container classes have an advantage over structures and clusters, known as "constructors." The constructor can provide a guaranteed initialization of the container object's properties. Knowing from Chapter 6, "Exception Handling," that the most common problems seen are configuration errors, the object constructor of a container class helps prevent many configuration errors.

Container classes need to support Set and Get functions in addition to any constructors that are needed. If a class performs other actions, then it is not a container class. Container classes should be considered when large amounts of configuration information are needed for objects. Putting this information into an embedded class will improve the readability of the code because the property list will contain a nested class instead of the 20 properties it contains.

10.4.2 ABSTRACT CLASSES

Some classes exist to force a structure on subclasses. Abstract classes define methods that all subclasses use, but cannot be implemented the same way. An example of an abstract class is an automobile class. A sport utility has a different implementation for Drive than a subcompact car. The automobile class requires that all subclasses define a Drive method. The actual implementation is left up to the subclass. This type of method is referred to as "pure virtual." Pure virtual methods are useful for verifying that a group of subclasses must support common interfaces. When a class defines one or more pure virtual methods, it is an abstract class. Abstract classes may not be instantiated by themselves. An abstract class has at least one method that has no code supporting it.

An abstract class available to the LabVIEW programmer is the Instrument abstract class we designed in Section 10.3. This class requires several methods be supported. Methods to read and write should be required of all instrument derivatives, but serial instruments have different write implementations than GPIB instruments. Therefore, the Write method should be specified as a pure virtual function in the instrument base class. A pure virtual function is defined, but not implemented. This means that we have said this class has a function, but we will not tell you how it works. When a pure virtual function is defined in a class, the class cannot be instantiated into an object. A subclass may be instantiated, but the subclass must also provide implementation for the method.

Continuing on the instrument base class, consider adding a Connect method. Future TCP-based instruments would require a Connect method. Considering future expansions of a class is a good design practice, but a Connect method is not required by all subclasses. Some serial devices may have connect routines, but some will not. GPIB instruments do not require connections to be established. Connections to the VISA subsystem could be done in the constructor. The conclusion is that a Connect method should not be defined in the instrument base class. This method may become a required interface in a subclass; for example, the Connect method can be defined in a TCP Instrument subclass.

The highest class defined in many hierarchies is often an abstract class. Again, the main purpose of the top class in a hierarchy is to define the methods that subclasses must support. Using abstract classes also defines the scope of a class hierarchy. A common mistake made by many object designers is to have too many classes. The base class in a hierarchy defines the scope of that particular class tree. This reduces the possibility of introducing too many classes into a design.

To expand on the object analysis begun in the previous section, consider Example 2. We have the base class "Instrument." All instruments have a primary address, regardless of the communications protocol they use. For a GPIB instrument, the address is the primary address. For a serial instrument, the COM port number can be the primary address. Leaving room for the future possibility of Ethernet and USB instruments, the address property will be a 32-bit number and will be used by all instruments. A 32-bit number was chosen because IP addresses are actually four 8-bit numbers. These 8-bit numbers can be stored in a single 32-bit number. This is actually what the String to IP function does in LabVIEW. Because the address is common to all major subsystems, we will define it as a property of the Instrument base class.

We have identified one property that is common to all instruments; now on to common methods. We know that all instruments must read and write. The Read and Write functions will be different for each type of instrument; therefore, the Instrument class must have two pure virtual methods, Read and Write. Languages like C++ use strong type checking, which means you must also define the arguments to the function and the return types. These arguments and return types must match in the subclass. The good news for us is that we are not required to follow this rule. All instruments so far must read and write strings and possess an address. This seems like a good starting point to the instrument architecture. Figure 10.11 shows a Rational Rose drawing of the Instrument class.

The subclass of Instrument that we will design now is the GPIB instrument subtype. Here we are forced with a decision: which properties does a GPIB instru-

FIGURE 10.11

ment require? Earlier, we decided to use the VISA subsystem for communications. VISA will handle the communications for our GPIB instruments in this example also. The property that a GPIB instrument class requires is a VISA handle. To generate the handle, the primary address, secondary address, and GPIB board must be given in the constructor. The GPIB Instrument class now has one required constructor. Support for read and write must be provided. Read and Write functions are abstract in the Instrument base class, so we must provide the functionality in this class. These methods and properties pretty much encapsulate the functionality of most GPIB (IEEE 488) instruments.

Some instruments, namely IEEE 488.2-compliant instruments have standard commands, such as Reset (*RST), Identity (*IDN?), and Options (*OPT?). Literally, IEEE 488.2 instruments are a subclass of GPIB instruments, and they will be in this object design. The 13 standard commands will be encapsulated as functions in the 488.2 Instrument subclass. The problem that we are developing is that we are in the second phase of the programming, object design. This class should have been thought of during the object analysis. When following a waterfall design methodology, we should stop performing the design and return to the analysis phase. After a moment of thought, several new features for the 488.2 subclass could become useful. For example, the Identification (*IDN?) command could be used in the constructor. This would allow the application to validate that the instrument on the GPIB bus is, in fact, a 488.2 instrument. If the instrument did not respond, or responds incorrectly to the *IDN? command, an error could be generated by the constructor. This type of functionality could be very useful in an application — better user validation without adding a lot to the program.

Now that methods required of Instrument, GPIB Instrument, and IEEE 488.2 Instrument have been defined, it is time to make use of them. Any instruments built using objects will be subtypes of either IEEE 488.2 or GPIB Instrument. If the physical instrument complies to the 488.2 standard, it will be a subclass of the 488.2 Instrument class. In the event we are working with an older, noncompliant instrument, then it will descend directly from GPIB Instrument. This will allow us to make sure that 488.2 commands will never be sent to a non compliant instrument.

As far as defensive programming goes, we have made excellent progress in defending the communications ports. Each instrument class encapsulates the communications port information and does not directly give access to any of the external code. This will make it impossible for arbitrary commands to be sent to any of the instruments. We will limit the commands that are sent to instruments to invoke the objects issue. Assuming the objects are written correctly, correct commands can only be sent on the communications lines.

Another set of methods that you can define for classes is operators. If you had an application that used vector objects, how would you perform addition? An Addition operator that accepts two vector objects can be written. It is possible to specify operators for all the arithmetic operators such as Addition, Subtraction, Multiplication, and Division. Also, Equality operators can be defined as methods that a class can support. It may be possible to use LabVIEW's Equality operator to directly compare the string handles to our objects, but there are some instances where comparing the flattened strings may not yield the results desired.

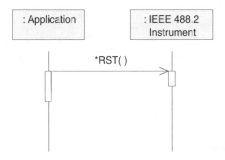

FIGURE 10.12

We have just identified a number of methods that will exist in the Instrument class hierarchy. To help visualize the interaction among these objects, we will use an interaction diagram. Software design tools such as Rational Rose, Microsoft Visual Modeler, and Software through Pictures provide tools to graphically depict the interaction among classes.

As the implementation of classes that we will be using later in this chapter does not support inheritance, we will aggregate the superclasses. The interaction diagram will capture the list of VI calls necessary to accomplish communication between the objects. In an interaction diagram, a class is a box at the top of the diagram. A vertical line going down the page denotes use of that class. Arrows between class objects descend down the vertical lines indicating the order in which the calls are made. Interaction diagrams allow programmers to understand the intended use of classes, methods, and their interactions. As part of an object design, interaction diagrams provide a lot of information that class hierarchies cannot.

If we had an application that created an IEEE 488.2 Instrument object and wanted to send an *RST to the physical instrument, the call chain would be fairly simple. At the upper left of the interaction diagram, an object we have not defined appears — Application. This "class" is a placeholder we are using to indicate that a method call is happening from outside an object. Next in line appears IEEE 488.2 Instrument. The arrow connecting the two classes indicates that we are invoking the *RST method. When the *RST method is called, a string *RST will be created by the object and sent to the GPIB Write function. This encapsulation allows us to control what strings will appear on the GPIB bus. In other words, what we have done is defined a set of function calls that will only allow commands to be sent that are valid. This is effectively an Application Programming Interface (API) that we are defining through this method. The diagram for this interaction appears in Figure 10.12.

This diagram is just one of possibly hundreds for a large-scale application. Interaction diagrams do not clearly define what the methods are expected to do, however; a description of logic (DOL) is required. In addition to interaction diagrams for every possible sequence of function calls, a written description of each call is required. In the DOL should appear function inputs and outputs and a reasonable description of what the function is expected to accomplish. DOLs can be simple, as in the case of the *RST command:

```
*RST-
```

```
inputs: error cluster, string handle for object
```

```
outputs: error cluster
```

This function sends a string, *RST on the GPIB bus to the instrument defined in the string handle.

For complex functions, the description may become a bit more complicated. Remember that the interaction diagrams are there to help. In a DOL you do not need to spell out an entire call sequence; that is what the interaction diagrams are there for. The DOL and interaction diagrams should leave no ambiguity for the programmers. Together, both pieces of information should minimize the thought process needed to program this particular sequence of events.

The class hierarchy, interaction diagrams, and DOL provide a complete picture for programmers to follow. The interaction diagrams provide graphical direction for programmers to follow in the next phase of development, object programming. Thus far we have managed to define what objects exist in the system, and what methods and properties the objects have. Object interactions are now defined, and each method and property has a paragraph or so describing what it is expected to do. This leaves the small matter of programming the methods and objects. This is not a phase of development for taking shortcuts. Leaving design details out of this phase will cause ambiguities in the programming phase. Any issues or possible interactions that are not covered in the design phase will also cause problems in the programming phase. Software defects become a strong possibility when the design is not complete. Some programmers may neglect any interactions they do not see, and others may resolve the issue on their own. This could well cause "undocumented features," which can cause well-documented complaints from customers. When the software model is complete and up-to-date, it is an excellent resource for how an application behaves. In languages like C++ when an application can have over 100 source code files, it is often easier to look at interaction diagrams and get an idea of where a problem is. Surfing through thousands of lines of source code can be tedious and cause programmers to forget the big picture of the application.

If you do not choose to follow an object-oriented software design methodology, a number of concepts in this chapter still directly apply to your code development. Start with a VI hierarchy and determine what pile of VIs will be necessary to accomplish these tasks. Then write out the interaction diagrams for the possible sequences of events. When writing out interaction diagrams, be sure to include paths that occur when exception handling is in process. Once the interactions are defined, write out descriptions of logic for each of the VIs, and voila! All that is left is to code the individual VIs. Writing the code may not be as easy as just described, but much of the thought process as to how the application should work has been decided. All you have to do is code to the plan.

10.5 OBJECT PROGRAMMING

This section concludes our discussion of the basic process of developing object-oriented code. This is the last phase of the programming, and should be fairly easy to do. You already know what every needed object is, and you know what all the methods are supposed to do. The challenge is to keep committed to the waterfall model design process. Waterfall design was discussed in Chapter 4, Application Structure. Object programming works well in large-scale applications, and programming technique should be coordinated with a process model for control of development and scheduling.

Once the object analysis and design are done, effort should be made to stick to the design and schedule. In the event that a defect is identified in the design of an application, work on the particular function should be halted. The design should be reviewed with a list of possible solutions. It is important in a large-scale application to understand the effects a design change can have on the entire application. Making design changes at will in the programming phase can cause issues with integration of the application's objects or subsystems. Software quality can become degraded if the impact of design changes is not well understood. If objects' instances are used throughout a program, it should be clearly understood what impact a design change can have on all areas of the application.

In a general sense, there are two techniques that can be followed in assembling an object-oriented application. The first technique is to write one object at a time. The advantage of this technique is that programmers can be dispersed to write and test individual objects. Once each object is written and tested, the objects are integrated into an application. Assuming that all the objects were written and properly tested, the integration should be fairly simple. If this technique is being used, programmers must follow the object design definitions precisely; failure to do so will cause problems during integration.

The second technique in writing an object-oriented application is to write enough code to define the interfaces for each of the objects. This minimally functional set of objects is then integrated into the application. The advantage of this technique is to have the skeleton of the entire application together, which minimizes integration problems. Once the skeleton is together, programmers can fill in the methods and internal functionality of each of the objects. Before embarking on this path, define which need to be partially functional and which need to be fully functional. External interfaces such as GPIB handles may need to be fully functional, whereas report generation code can only be functional enough to compile and generate empty reports.

10.6 DEVELOPING OBJECTS IN LABVIEW

This section begins to apply the previous information to programming in LabVIEW. Our object representations will use clusters as storage containers for the properties of an object. SubVIs will be used as methods. We will develop VIs to function as constructors and destructors, and to perform operations such as comparison. In addition, methods identified in the object design will be implemented.

Cluster type definitions will be used as containers, but we will present only strings to the programmer. Strings will be used as a handle to the object. Clusters will be used for methods because this prevents programmers from "cheating." The idea here is to encapsulate the data and prevent access as a global variable. Programmers will need to use the Get/Set or other defined methods for access to object "innards." This satisfies the requirement that data be encapsulated. This may seem like a long-winded approach, but it is possible to add functionality to the Set methods that log the VI call chain every time this method is invoked. Knowing which VI modified which variable at which time can be a tremendous benefit when you need to perform some emergency debugging. This is generally not possible if you have a set of global variables that are modified in dozens of locations in the application.

This implementation will not provide direct support for inheritance. Inheritance would require that the flattened string be recognizable as one of a list of possible clusters. Having a pile of clusters to support this functionality is possible, but the parent class should not need to maintain a list of possible child classes. In languages like C++, a virtual function table is used "under the hood" to recognize which methods should be called. Unfortunately, we do not have this luxury. Identifying which method should be called when a virtual method is invoked would be a significant undertaking. This book has a page count limit, and we would certainly exceed it by explaining the logistics of object identification. We will simulate inheritance through aggregation. Aggregation is the ability for one object to contain another as a property. Having the parent class be a property of the child class will simulate inheritance. This technique was used with Visual Basic 4.0 and current versions; Visual Basic does not directly support inheritance. Polymorphism cannot be supported directly because inheritance cannot be directly supported. This is a limitation on how extensive our support for objects can be.

The object analysis has shown us which classes are necessary for an application, and the object design has identified the properties and methods each object has. The first step in implementing the object design is to define the classes. Because Lab-VIEW is not an object-oriented language, it does not have an internal object representation. The implementation we are developing in this section is not unique. Once you understand the underlying principles behind object-oriented design, you are free to design your own object representations.

10.6.1 Properties

All objects have the same properties and methods. What makes objects of the same type unique are the values of the properties. We will implement our objects with separated properties and methods. The methods must be subVIs, and the class templates for properties will be type definitions using clusters. The cluster is the only primitive data type that can encapsulate a variety of other primitive types, such as integers. Class definitions will be encapsulated inside clusters. The internal representation for properties is clusters, and the external representation will actually be strings. As it is impossible for a programmer to access a member variable without invoking a Get or Set method, our implementation's properties are always private members of the class.

When you program a class definition, the cluster should be saved as a type definition. We will use this type definition internally for all the object's methods. The type definition makes it convenient for us to place the definition in class methods. The cluster definition will be used only in internal methods to the class; programmers may not see this cluster in code that is not in direct control of the class. External to the object, the cluster type definition will be flattened to a string. There are a number of reasons why this representation is desirable, several of which will be presented in the next section.

Some readers may argue that we should just be passing the cluster itself around rather than flattening it into a string. The problem with passing the class data around as a cluster is that it provides temptation to other programmers to not use the class methods when altering internal data. Flattening this data to a string makes it difficult, although not impossible, for programmers to cheat. Granted, a programmer can always unflatten the cluster from a string and cheat anyway, but at some point we are going to have to make a decision to be reasonable. Flattening the cluster to a string provides a reasonable amount of protection for the internal data of the object.

10.6.2 CONSTRUCTORS

An object design may determine that several constructors will be necessary. The purpose of the constructor is to provide object initialization, and it may be desirable to perform initialization in several different manners. Our class implementation will require that each object have at least one available constructor. The constructor will be responsible for flattening the Typedef cluster into a string as an external handle to the object. Object-oriented languages such as C++ do not require absolutely that each class have a constructor. Like Miranda rights, in C++, if you do not have a constructor, the compiler will appoint one for you. We are supplying an object implementation for LabVIEW, and our first rule is that all objects must have at least one constructor.

A simple example for a class and its constructor is a point. A point has two properties, an x and y coordinate. Figure 10.13 shows the front panel for a constructor function for a point class. The required two inputs, the x and y coordinates, are supplied, and a flattened string representing the points is returned. The code diagram is shown in Figure 10.14. A simple Build Cluster function was used. For more

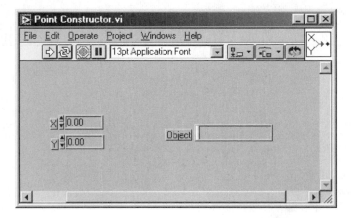

FIGURE 10.13

FIGURE 10.14

complicated functions, you want to build a cluster and save it as a control. This control can be placed on the front panel and not assigned as a connector. This would facilitate the cluster sizing without giving up the internals to external entities.

Another object we will develop using the point is the circle. A circle has a radius and an origin. The circle object will have the properties of a point for origin and a double-precision number for its radius. The cluster control for a circle is shown in Figure 10.15. The circle object constructor, like the point constructor, uses floating-point numbers for radius, x, and y. The front panel is shown in Figure 10.16, and the code diagram is shown in Figure 10.17.

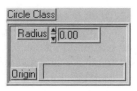

FIGURE 10.15

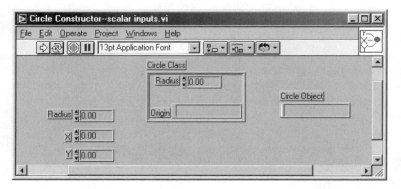

FIGURE 10.16

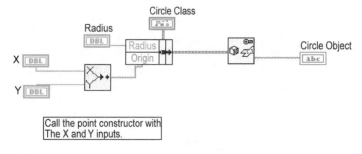

FIGURE 10.17

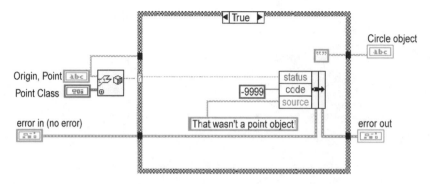

FIGURE 10.18

The front panel uses type definitions for the contents of the clusters. Point coordinates are fed into the point constructor and the string input is fed into the circle's cluster definition. The clusters are used only as inputs to the Bundle function. Next, the cluster is then flattened and returned as a string. The circle uses a nested class as one of its properties.

Now, say that a programmer has the radius and the point itself in a string. Another constructor can be built using the string point object and the radius as inputs. The block diagram for this VI is shown in Figure 10.18. Note that the first thing the constructor does is validate the point object. The error clusters are included in this VI; it is a trivial matter to include them in the other constructors. The circle object demonstrates where multiple constructors are desirable. Functions that have the same purpose, but use different arguments can be called "overloaded." In C++ and Java, a function name can be used many different times as long as the argument list is different for each instance of the function. The function name and argument list is the function's signature in C++; this does not include the return type. As long as different signatures are used, function name reuse is valid. LabVIEW does not have this restriction if desired multiple constructors with the same input list can be used.

Example 10.6.1

Develop a constructor for a GPIB instrument class. All GPIB instruments have primary addresses, secondary addresses, and GPIB boards they are assigned to. All instrument communications will be performed through the VISA subsystem.

Solution:

First, we need to consider the design of the object. The problem statement made it obvious that addresses and GPIB boards are to be used by the object. The statement also says that the communications will be performed via VISA calls. The cluster really needs just one item, a VISA handle. The inputs of addresses and board number can be formatted into a VISA descriptor. The VISA Open VI will open the connection and return the needed VISA handle. Figure 10.19 shows the constructor. Not shown is the front panel, in which the only notable features are the limitations placed on the GPIB address: the primary and secondary addresses must be between 1 and 31. If the secondary address is zero, then we will use the primary address as the secondary address. In addition, the GPIB board number must be between 0 and 7.

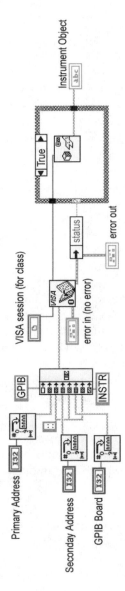

FIGURE 10.19

As this VI will perform communications outside LabVIEW and into the VISA subsystem, error clusters are used to indicate the success or failure of the operation. If VISA Open returns an error, then an empty string is returned as the object in addition to setting the error cluster. Our implementation of objects will return empty strings when objects cannot be created.

10.6.3 DESTRUCTORS

Destructors in our object implementation need to exist only when some activity for closing the object needs to be done. Any objects used to encapsulate TCP, UDP, VISA, ActiveX (automation), or synchronization objects should destroy those connections or conversations when the object is destroyed. If the object's string is allowed to go out of scope without calling the destructor, LabVIEW's engine will free up the memory from the no-longer-used string. The information holding the references to the open handles will not be freed up. Over long periods of time, this will cause a memory leak and degrade LabVIEW's performance.

Classes such as the point and circle do not require a destructor. The information they hold will be freed up by the system when they are eliminated; no additional handles need to be closed. This is actually consistent with other programming languages such as C++ and Java. Programmers need to implement destructors only when some functionally needs to be added.

Example 10.6.2
Implement a destructor for the GPIB class object created in Example 10.6.1.

Solution:
All that needs to be done for this example is to close the VISA handle. However, when we recover the cluster from the object string, we will verify that it is a legitimate instance of the object. The destructor code diagram is shown in Figure 10.20.

The significance of destructors is important for objects that communicate with the outside world. For internally used objects, such as our point and circle, or for objects simulating things like signals, no destructor is necessary. Unlike constructors, there is only one destructor for an object. In our implementation it is possible to construct multiple destructors, but this should be avoided. It will be obvious in the object design what items need to be closed out; this should all be done in one point. The act of destroying an object will require the use of only one destructor. Having multiple destructors will serve to confuse programmers more than it will help clarify the code.

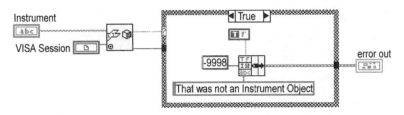

FIGURE 10.20

10.6.4 METHODS

In this section we will begin implementing methods that objects will use. The two distinct classifications of methods will be Public and Private methods. Protected methods will not be supported; their implementation will be much more difficult. A stronger class design in the future may allow for protected interfaces. Private and public methods will be handled separately because their interfaces will be different.

Our object implementation will use methods to interface with the outside application exclusively. Setting or retrieving the values of properties will be handled through methods called Set and Get. This is how ActiveX is implemented. Each property that you have read and/or write access to is handled through a function call.

10.6.4.1 Public Methods

Public methods take a reference to the object in the form of a string. This prevents outside code from modifying private properties without a defined interface. A special example of a Public function is the constructor. The constructor takes inputs and returns a string reference to the object. This is a public method because it was called from outside the object and internal properties were configured by a method the object supports.

Class designs may have many public methods, or just a few. Object design tools like Rational Rose generate code that, by default, will include Get and Set functions for access to object properties. We will be using Get and Set functions to alter properties of our objects. The Get/Set functions will form the protective interface for member variables. Set methods will allow us to control or force rules on setting internal data members. This gives us the ability to perform a lot of intelligent coding for the object users. From our previous example, a For GPIB Instrument, we could have a Set GPIB Address method as public. This method allows programmers to allow the object to determine if its GPIB address has been changed during execution. If this is the case, the object could be made capable of refusing to make the change and generate an error, or of closing its current VISA session and creating a new one. Public methods enable the object to make intelligent decisions regarding its actions and how its private data is handled. This is one of the strengths of encapsulation. When designing objects, consideration can be made as to how the Get and Set interfaces will operate. Intelligent objects require less work of programmers who use them because many sanity-checking details can be put into the code, freeing object users from trivial tasks.

10.6.4.2 Private Methods

The major distinction between public and private methods in our LabVIEW implementation is that private methods may use the type definition cluster as input. Private methods are considered internal to the class and may directly impact the private properties of the class. The extra step of passing a string reference is unnecessary; basically, a private method is considered trustable to the class. When using this object implementation in a LabVIEW project, a private method should always appear as a subVI to a public method VI. This will enable you to verify that defensive programming techniques are being followed. As a quick check, the VI hierarchy can be examined to verify that private methods are only called as subVIs of public methods. Public methods serve as a gateway to the private methods; private methods are generally used to simplify reading the code stored in public methods.

By definition, private methods may be invoked only by the object itself. When can an object call an internal method? Public methods may call private methods. When a public method is executing, it may execute private methods. This may seem to be a waste, but it really is not. This is a good defensive programming practice. The public method is an interface to the external program, and the private methods will be used to accomplish tasks that need to be performed.

As an example of a private method that an object would use, consider implementing a collection of VIs to send e-mail using Simple Mail Transfer Protocol (SMTP). To send mail using SMTP, a TCP connection must be established to a server. The Read and Write TCP functions are not something you would want a user of the SMTP mail object to directly have access to. Implementing the SMTP protocol is done internally to the SMTP object. Again, this is defensive programming. Not allowing generic access to the TCP connection means that the SMTP object has complete control over the connection, and that no other elements of code can write data to the mail server that has not been properly formatted.

Simple objects may not require private methods to accomplish their jobs. A simple object such as a vector does not require a private method to determine its magnitude. In our implementation, private methods are necessary only if they simplify code readability of a public method. There are a number of places where this is desirable. In our GPIB Instrument class, public methods would store the strings needed to send a given command. The common ground between all the public methods would be the need to send a string through the GPIB bus. A Write VI can be made as a private method so you do not need to place GPIB write commands and addressing information in each of the VIs. This can be done in the single private method and dropped into public methods.

10.7 EXAMPLES IN DEVELOPING INSTRUMENT DRIVERS

This section will develop several instrument drivers to illustrate the benefits of object modeling in LabVIEW. This technique works very well with SCPI instruments. SCPI command sets are modularized and easily broken down into class templates. Object-based instruments can be considered alternatives to standard instrument drivers and IVI Instruments. Our first example will concentrate on power supplies, specifically one of the IVI-based instruments.

A simple object model for a power supply would have the core definition of a power supply. It would be fair to assume that GPIB controllers will control a vast majority of power supplies. It would make sense to reuse the core GPIB Instrument class to control GPIB behavior. This design will use a one-class-fits-all approach. In a true object-oriented implementation, it would make sense to have a power supply abstract base class that defines voltage and current limit properties. In addition to the current limit property, we will be supplying a read only property: current. This will allow users to read the current draw on the supply as a property. As we are adding object implementations to a non-object-oriented language, many of the advantages of abstract base classes do not apply. Instead, we will define the methods and properties that are used by the vast majority of power supplies and implement them with the addition of a model property. This technique will work well for simple instruments, but complex instruments such as oscilloscopes would be extremely difficult. Multiple combinations of commands that individual manufacturer's scopes would require to perform common tasks would be an involving project, which it is for the IVI Foundation.

The purpose of the Model property is to allow users to select from an enumerated list of power supplies that this class supports. Internally, each time a voltage or current is set, a Select VI will be used to choose the appropriate GPIB command for the particular instrument model. From a coding standpoint, this is not the most efficient method to use to implement an instrument driver, but the concepts of the objects are made clear.

The first step in this object design is to determine which properties go inside the cluster Typedef for this class. Obvious choices are values for current limit and voltage. This will allow the power supply driver to provide support for caching of current limit and voltage. No values will be retained for the last current measurement made; we do not want to cache that type of information because current draw is subject to large changes and we do not want to feed old information back to a user. Other values that need to be retained in this class are the strings for the parent classes, GPIB Instrument, and IEEE 488.2 Instrument. The cluster definition is shown in Figure 10.21. The constructor for power supply will call the constructor for IEEE 488.2 Instrument. We need to retain a handle for this object, which will take the GPIB address information as an argument.

FIGURE 10.21

The GPIB addressing information does not need to be retained in the Typedef for power supply. The IEEE 488.2 Instrument will, in turn, call the constructor for GPIB Instrument, which takes the GPIB address information to its final destination, the VISA handle it is used to generate.

The Constructor VI for the power supply takes several inputs: the GPIB board number, GPIB primary address, GPIB secondary address, and the power supply model. The power supply constructor will then call the constructor for the IEEE 488.2 Instrument. As we have a higher base class, the IEEE 488.2 Instrument will call the constructor for the GPIB Instrument class. This class will initialize the VISA session and return a string burying this information from users and lower classes. This defensive programming restricts anyone from using this handle in a mechanism that is not directly supported by the base class. Constructor functions for each of the three classes in the chain are shown in Figure 10.22.

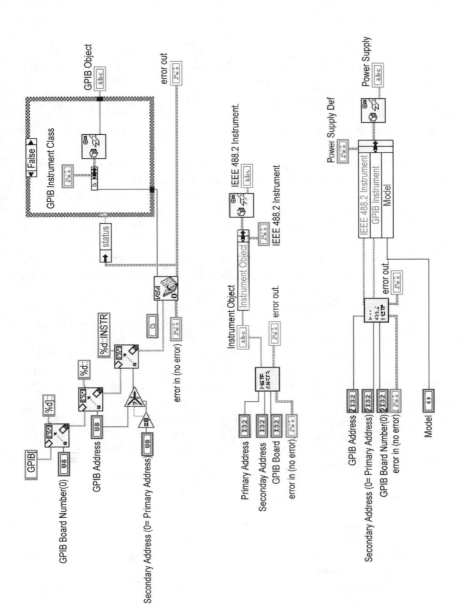

FIGURE 10.22

The next items that need to be developed are the Get and Set VIs for the properties' voltage and current. Get and Set methods will be supplied for voltage and current limits, but only a Get command needs to be supplied for the current draw property. Current draw is actually a measurement, but from an object design standpoint it is easily considered to be a property. Set Voltage is shown in Figure 10.23. The Get Current Draw method is shown in Figure 10.24.

We are not supplying Get or Set functions for the GPIB address or instrument model information. It is not likely that a user will switch out power supplies or change addressing information at run-time, so we will not support this type of operation. It is possible to supply a Set Address method that would close off the current VISA handle and create a new GPIB Instrument class to reflect the changes. This sounds like an interesting exercise for the reader, and it appears in the exercises at the end of this chapter.

One issue that we face is the lack of inheritance in our classes. The code to support inherited methods is somewhat bulky and limiting. The GPIB Write commands need to propagate from power supply to IEEE 488.2 Instrument to GPIB Instrument. This is a fairly large call chain for an act of writing a string to a GPIB bus. The wrapper VIs do not take a significant amount of time to write, but larger instruments with several hundred commands could require several hundred wrappers, which would take measurable engineering time to write and is a legitimate problem. In true object-oriented languages, the inherited functions are handled in a more elegant fashion.

The two interaction diagrams presented in Figures 10.25 and 10.26 show the sequence of VI calls needed to set the voltage of the power supply. The first diagram shows what happens when the Voltage Set method gives a value equal to the currently-cached voltage. The function simply returns without issuing the command. The second diagram shows that the Voltage Set method had an argument other than the currently-cached command and triggers the writing of a command to the GPIB instrument class.

In the next example, we consider a more complicated instrument. This example will introduce another concept in programming, "Friend classes." Friend classes are the scourge of true object-oriented purists, but they present shortcuts to solving some of our implementation problems.

10.7.1 COMPLEX INSTRUMENT DESIGNS

In this example we will implement the communications analyzer we performed the object analysis on in Section 10.3. Now that we understand how our object implementation works in LabVIEW, we can review some of the analysis decisions and move forward with a usable driver.

Some limitations arose in the power supply example we did previously, and we will need to address these limitations. First, power supplies are generally simple instruments on the order of 20 GPIB commands. Modern communications analyzers can have over 300 instrument commands and a dizzying array of screens to select before commands can be executed. The one driver model for a group of instruments will not be easily implemented; in fact, the IVI Foundation has not addressed

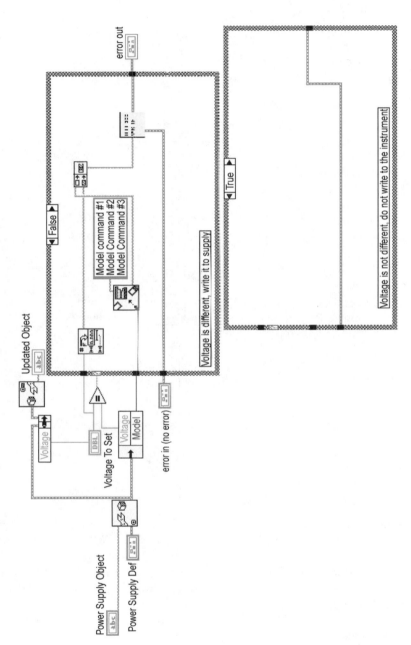

FIGURE 10.23

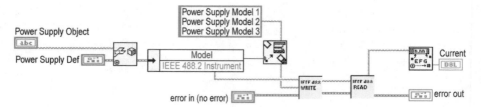

FIGURE 10.24

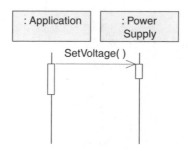

FIGURE 10.25

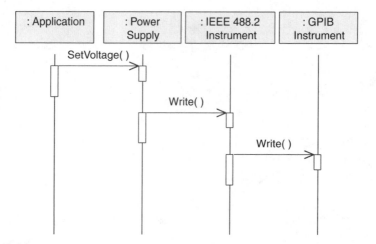

FIGURE 10.26

complex test instruments at this time. The matrix of command combinations to perform similar tasks across different manufacturer's instruments would be difficult to design, and an amazing amount of work to implement. In other words, standardizing complex instruments is a significant undertaking, and it is not one that the industry has completely addressed. It is also not a topic we should address.

The object analysis presented in Section 10.3 works very well in object-oriented languages. Our implementation does have some problems with inheritance, and we will have to simplify the design to make implementation easier. Abstract classes for the analyzers and generators have a single property and a handful of pure virtual methods. As virtual methods are not usable in our implementation, we will remove them. We will still make use of the IEEE 488.2 class and the GPIB Instrument class. The object hierarchy is greatly simplified. The communications analyzer descends from IEEE 488.2 Instrument, which is a natural choice. The generators, RF and AF, appear on the left side of the communications analyzer. This is an arbitrary choice as there is no special significance to this location. The analyzers appear on the right. The arrow connecting the communications analyzer to the component objects denotes aggregation. Recall that aggregated classes are properties of the owning class; they do not exist without the owning class. This is a natural extension of the object model: this model reflects the real configuration of the instrument. At this point we can claim a milestone in the development: the object analysis is now complete.

Hewlett Packard's HP-8920A communications analyzer will be the instrument that is implemented in this design. We are not endorsing any manufacturer's instruments in this book; its standard instrument driver appears on the LabVIEW CDs and allows us to easily compare the object-based driver to the standard driver.

The next difficulty that we need to address is the interaction between the aggregated components and the communications analyzer. This example will implement a relatively small subset of the HP-8920's entire command set. This command set is approximately 300 commands. Each component object has between 20 and 50 commands. A significant number of wrapper VIs need to be implemented as the aggregated components are private data members. This author is not about to write 300 VIs to prove an example and is not expecting the readers to do the same. We will need to find a better mechanism to expose the aggregate component's functionality to outside the object. The solution to this particular problem is actually simple. As the analyzers and generators are properties of the communications analyzer, we can use Get methods to give programmers direct access to the components. This does not violate any of our object-oriented principals, and eliminates the need for several hundred meaningless VIs. Figure 10.27 shows the list of the properties and methods the communications analyzer has. Properties and methods with a lock symbol next to them are private members and may only be accessed directly by the class itself.

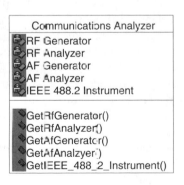

FIGURE 10.27

Now that we have a basic design down for the core object, it is time to examine a small matter of implementation. As we will give programmers access to the analyzers and generators, it stands to reason that the programmers will set properties and expect the components to send GPIB commands to the physical instrument. Reexamining the class hierarchy shows us that we do not have the ability to access the GPIB Instrument class directly from the generator and analyzers because they do not derive from GPIB Instrument. This presents a problem that needs to be resolved before we proceed with the design.

We require that the component objects be capable of accessing the GPIB Read and Write methods of the communications analyzer. It does not make sense to have each of the components derive from the communications analyzer — this does not satisfy the "is a" relationship requirement for subclasses. In other words, the phrase, "an RF generator 'is a' communications analyzer," is not true. Having the component objects derive from the communications analyzer is not a solution to this problem.

We can give the component objects a property for the communications analyzer. Then they can invoke its GPIB Write method. This solution does not realize the class hierarchy we developed in the object analysis phase, either. It is possible to go back and change the class hierarchy, but now we do not satisfy the "has a" requirement for aggregated components. An audio generator "has a" communications analyzer is not a true statement and, therefore, having a property of a communications analyzer does not make much sense.

As none of our solutions thus far make it in our given framework, we will need to expand it somewhat. An RF analyzer is a component in a communications analyzer and it does have access to the onboard controller (more or less). When performing an object analysis and design, it often helps to have the objects interact and behave as the "real" objects do. For the communications analyzer to give access to the GPIB Read and Write methods would violate the encapsulation rule, but it does make sense as it models reality. The keyword for this type of solution in C++ is called "friend." The friend keyword allows another class access to private methods and properties. We have not discussed it until this point because it is a technique that should be used sparingly. Encapsulation is an important concept and choosing to violate it should be justified. Our previous solutions to GPIB Read and Write access did not make much sense in the context of the problem, but this solution fits. In short, we are going to cheat and we have a legitimate reason for doing so. The moral to this example is to understand when violating programming rules makes sense. Keywords such as friend are included in languages like C++ because there are times when it is reasonable to deviate from design methodologies.

Now that we have the mechanism for access to the GPIB board well understood, we can begin implementation of the methods and properties of the component objects. We shall begin with the audio generator. Obvious properties for the audio generator are Frequency, Enabled, and Output Location. We will assume that users of this object will only be interested in generating sinusoids at baseband. Enabled indicates whether or not the generator will be active. Output Location will indicate where the generator will be directing its signal. Choices we will consider are the AM and FM modulators and Audio Out (front panel jacks to direct the signal to an external box). The last property that we need to add to the component is one for

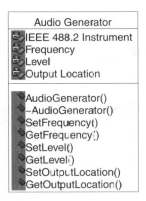

FIGURE 10.28

GPIB Instrument. As the communications analyzer derives from IEEE 488.2 Instrument, we do not have direct access to the GPIB Instrument base class. It will suffice to have the IEEE 488.2 Instrument as a property.

Methods for the audio generator will be primarily Get and Set methods for the Frequency, Enabled, and Output Location properties. Programmers have a need to change and obtain these properties. As this component borrowed the information regarding the GPIB bus, it is not appropriate to provide a Get method to IEEE 488.2 Instrument. A Set method for IEEE 488.2 Instrument does not make much sense either. The audio generator is a component of a communications analyzer and cannot be removed and transferred at will. The IEEE 488.2 Instrument property must be specified at creation as an argument for the constructor. The complete list of properties and methods for the audio generator appear in Figure 10.28.

The two last issues to consider for the audio generator are the abilities to cache and validate its property values. This is certainly possible and desirable. We will require the Set methods to examine the supplied value against that last value supplied. We also need to consider that the values supplied to amplitude are relative to where the generator is routing its output. A voltage is used for the audio out jacks, where FM deviation is used when the signal is applied to the FM modulator. When a user supplies an amplitude value, we need to validate that it is in a range the instrument is capable of supporting. This is not a mission-critical, show-stopping issue; the instrument will limit itself and generate an error, but this error will not be propagated back to the user's code. It makes more sense to have the object validate the inputs and generate an error when inputs are outside of acceptable ranges. RF generator will be subjected to similar requirements.

Last is the destructor for the function. Most of the properties are primitive, floating-point numbers and do not require a destructor. The IEEE 488.2 property has an embedded handle to a VISA session. It would make sense for the control to clean this up before it exits. In this case it is undesirable for the component objects to clean up the VISA handle. A total of five components have access to this property, and only one of them should be capable of destroying it. If we destroy the VISA handle in this object, it will be impossible to signal the other components that the GPIB interface is no longer available. As this component has "borrowed" access to

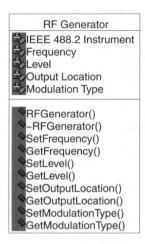

FIGURE 10.29

the GPIB object, it stands to reason that only the communications analyzer object should have the ability to close out the session. Therefore, this object does not need a destructor; everything including the GPIB object property can simply be released without a problem. We must be certain, however, that the destructor for the communications analyzer terminates the VISA session.

The audio and RF generators have a fair amount in common, so it is appropriate to design the RF generator next. The RF generator will need to have an IEEE 488.2 property that is assigned at creation, like the audio generator. In addition, the RF generator will require a Frequency, Amplitude, and Output Port property. Without rewriting the justification and discussion for the audio generator, we will present the object design for the RF generator in Figure 10.29.

Next on the list will be the audio analyzer. This component will not have any cached properties. The Get methods will be returning measurements gathered by the instrument. Caching these types of properties could yield very inaccurate results. The IEEE 488.2 Instrument will be a property that does not have a Get or Set method; it must be specified in the constructor. The audio analyzer will need a property that identifies where the measurement should be taken from. Measurements can be made from the AM and FM demodulator in addition to the audio input jacks. Measurements that are desirable are Distortion, SINAD, and Signal Level (Voltage). The Distortion and SINAD measurements will be made only from the demodulator, whereas the Signal Level measurement can only be made at the audio input jacks. These are requirements for this object.

The object should change the measurement location property when a user requests a measurement that is not appropriate for the current setting. For example, if a user requests the Signal Level property when the measurement location is pointing at the FM demodulator, the location property should be changed to Audio In before the measurement is performed. Measurement location can and should be a cached property. As measurement location is the only cached property, it is the only one of the measurement properties that should be included in this class. SINAD,

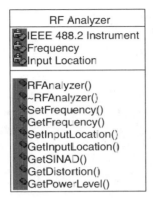

FIGURE 10.30

Distortion, and Signal Level will not be stored internally, and there is no reason why they need to appear in the cluster Typedef. The list of properties and methods for this class appears in Figure 10.30.

The analysis and design for the RF analyzer will be very similar to the audio analyzer. All measurement-related properties will not be cached, and the IEEE 488.2 method will not be available to the user through Get and Set methods. The properties that users will be interested in are Power Level, x, and y. Object Analysis is pretty much complete. The logic regarding the design of this component follows directly from the preceding three objects. The class diagram appears in Figure 10.31.

Now that we have a grasp of what properties and methods the objects have available, it is time to define their interactions. A short series of interaction diagrams will complete the design of the communications analyzer. First, let's consider construction of this object. The constructor will need to create the four component objects in addition to the IEEE 488.2 Instrument. Component constructors will be simple; they do not need to call other subVIs to execute. The IEEE 488.2 Instrument needs to call the constructor for GPIB Instrument. Sequentially, the communications

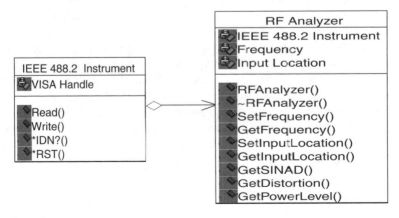

FIGURE 10.31

analyzer constructor will need to call the IEEE 488.2 constructor first; this property is handed to the component objects. The IEEE 488.2 constructor will call GPIB Instrument's constructor, and, lastly, the other four component constructors' can be called in any order. The sequence diagram for this chain of events is presented in Figure 10.32.

Figure 10.33 and 10.34 depict how the audio generator should behave when the audio generator's Set Frequency method is invoked. First, we will consider the call chain that should result when the new frequency value is equal to the old value. The method should determine that the property value should not be altered and the function should simply return. Figure 10.33 shows this interaction. Secondly, when the property has changed and we need to send a command to the physical instrument, a nontrivial sequence develops. First is the "outside" object placeholder. Set Frequency will then call IEEE 488.2's Write method to send the string. IEEE 488.2 will, in turn, call GPIB Instrument to send the string. This is a fairly straightforward call stack, but now that we have a description of the logic and sequence diagram there is absolutely no ambiguity for the programming. Figure 10.34 shows the second call sequence. Most of the call sequences for the other methods and properties follow similar procedures.

Now that each class, property, and method have been identified and specified in detail, it is time to complete this example by actually writing the code. Before we start writing the code, an e-mail should be sent to the project manager indicating that another milestone has been achieved. Writing code should be fun. Having a well-documented design will allow programmers to implement the design without having numerous design details creep up on them. Looking at the sequence diagrams, it would appear that we should start writing the constructors for the component objects first. Communication analyzer's constructor requires that these VIs be available, and we have already written and reused the IEEE 488.2 and GPIB Instrument objects. Audio generator's constructor diagram appears in Figure 10.35. The other constructors will be left to the exercises at the end of the chapter. Using skeleton VIs for these three constructors, we are ready to write the communication analyzer's constructor. The error cluster is used to force order of execution and realize the sequence diagram we presented in Figure 10.32. So far, this coding appears to be fairly easy; we are simply following the plan.

In implementing some of the functions for the analyzers and generators, we will encounter a problem. The Get and Set methods require that a particular screen be active before the instrument will accept the command. We left this detail out of the previous analysis and design to illustrate how to handle design issues in the programming phase. If you are collecting software metrics regarding defects, this would be a design defect. Coding should be stopped at this point and solutions should be proposed and considered. Not to worry, we will have two alternatives.

One possible solution is to send the command changing the instrument to the appropriate screen in advance of issuing the measurement command. This is plausible if the extra data on the GPIB bus will not slow the application down considerably. If GPIB bus throughput is not a limiting factor for application speed, this is the easiest fix to put in. Changes need to be made to the description of logic and that is about it. We can finish the coding with this solution.

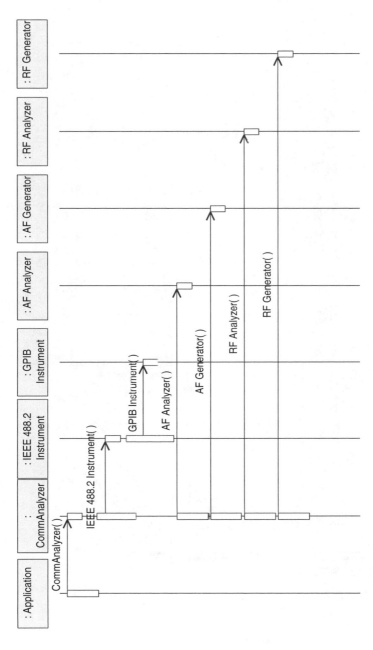

FIGURE 10.32

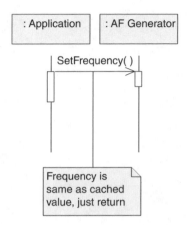

FIGURE 10.33

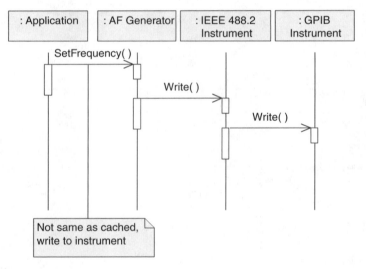

FIGURE 10.34

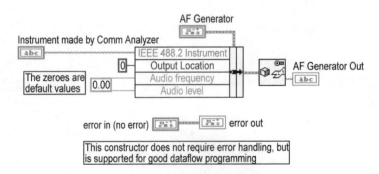

FIGURE 10.35

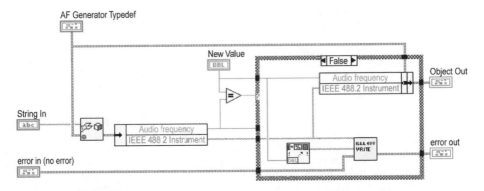

FIGURE 10.36

Another alternative is to include a screen property in the communications analyzer object. This property can cache information regarding which screen is active. If this is the case, programmers can Get/Set Screen before invoking the methods of the component objects. This will require us to go back and change all the sequence diagrams, the property list of communications analyzer, and document that users are responsible for the status of the screen.

The last alternative is to have communications analyzer define a global variable that all the component objects can access to check the status of the physical instrument's screen. This solution is going to be tossed out because it is a "flagrant technical foul" according to the object-oriented programming paradigm. Global data such as this can be accessed by anyone in any location of the program. Defensive programming is compromised and there is no way for the component objects to validate that this variable accurately reflects the state of the physical instrument. Having considered these three alternatives, we will go with the first and easiest solution. Measurement settling times will be orders of magnitude longer than the amount of time to send the screen change command. The transit time for the screen command is close to negligible for many applications. With this simple solution we present Figure 10.36, the code diagram for setting the audio generator's frequency. Remaining properties are implemented in a similar fashion.

10.8 OBJECT TEMPLATE

Many of the VIs that need to be implemented for an object design need to be custom written, but we can have a few standard templates to simplify the work we need to do. Additionally, objects we have written will begin to comprise a "trusted code base." Like other collections of VIs, such as instrument drivers, a library of VIs will allow us to gain some reduction in development time by reuse. The template VIs built in this section are available on the companion CD.

The template for constructors needs to be kept simple. As each constructor will take different inputs and use a different Typedef cluster, all we can do for the boilerplate is get the Flatten to String and Output terminals wired. Each constructor should return a string for the object and an error cluster. Simple objects such as our

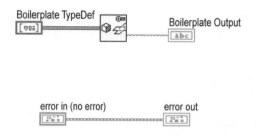

FIGURE 10.37

geometric shapes may not have any significant error information to return, but complex classes that encapsulate TCP conversations, Active X refnums, or VISA handles would want to use the error cluster information. Figure 10.37 shows the template constructor.

Get properties will be the next template we set up. We know that a string will be handed into the control. It will unflatten the string into the Typedef cluster and then access the property. This operation can be put into a VI template. It will not be executable, as we do not know what the Typedef cluster will be in advance, but a majority of the "grunt work" can be performed in advance. Because return types can be primitive or complex data types, we will build several templates for the more common types. Figure 10.38 shows the Get template for string values. This template will be useful for returning aggregated objects in addition to generic strings. Error clusters are used to allow us to validate the string reference to the object.

Set property templates will work very much like Get templates. As we do not have advance information about what the cluster type definition will be, these VIs will not be executable until we insert the correct type definition. Again, several

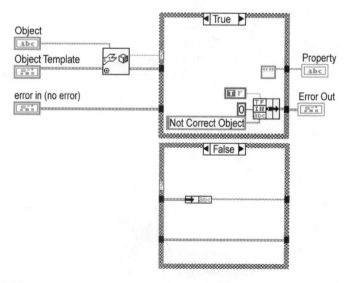

FIGURE 10.38

templates will be designed for common data types. Error clusters are used to validate that the string handed into the template is the correct flattened cluster.

10.9 EXERCISES

1. An object-based application for employee costs needs to be developed. Relevant cost items are salary, health insurance, dental insurance, and computer lease costs.
2. What properties should be included for a signal simulation object? The application will be used for a mathematical analysis of sinusoidal signals.
3. Construct an object-based instrument driver for a triple-output power supply based on the example power supply in Section 10.7.

BIBLIOGRAPHY

Bertand Meyer. *Object-Oriented Software Construction.* Prentice Hall, Englewood Cliffs, NJ, 1998.

Index

LIMITED WARRANTY

Taylor & Francis Group warrants the physical disk(s) enclosed herein to be free of defects in materials and workmanship for a period of thirty days from the date of purchase. If within the warranty period Taylor & Francis Group receives written notification of defects in materials or workmanship, and such notification is determined by Taylor & Francis Group to be correct, Taylor & Francis Group will replace the defective disk(s).

The entire and exclusive liability and remedy for breach of this Limited Warranty shall be limited to replacement of defective disk(s) and shall not include or extend to any claim for or right to cover any other damages, including but not limited to, loss of profit, data, or use of the software, or special, incidental, or consequential damages or other similar claims, even if Taylor & Francis Group has been specifically advised of the possibility of such damages. In no event will the liability of Taylor & Francis Group for any damages to you or any other person ever exceed the lower suggested list price or actual price paid for the software, regardless of any form of the claim.

Taylor & Francis Group specifically disclaims all other warranties, express or implied, including but not limited to, any implied warranty of merchantability or fitness for a particular purpose. Specifically, Taylor & Francis Group makes no representation or warranty that the software is fit for any particular purpose and any implied warranty of merchantability is limited to the thirty-day duration of the Limited Warranty covering the physical disk(s) only (and not the software) and is otherwise expressly and specifically disclaimed.

Since some states do not allow the exclusion of incidental or consequential damages, or the limitation on how long an implied warranty lasts, some of the above may not apply to you.

DISCLAIMER OF WARRANTY AND LIMITS OF LIABILITY

The author(s) of this book have used their best efforts in preparing this material. These efforts include the development, research, and testing of the theories and programs to determine their effectiveness. Neither the author(s) nor the publisher make warranties of any kind, express or implied, with regard to these programs or the documentation contained in this book, including without limitation warranties of merchantability or fitness for a particular purpose. No liability is accepted in any event for any damages, including incidental or consequential damages, lost profits, costs of lost data or program material, or otherwise in connection with or arising out of the furnishing, performance, or use of the programs in this book.